Compaction Quality Nondestructive Testing and Intelligent Control Technology for Infrastructure

Qinglong Zhang

Compaction Quality Nondestructive Testing and Intelligent Control Technology for Infrastructure

Springer

Qinglong Zhang
University of Science and Technology
Beijing
Beijing, China

ISBN 978-981-95-1263-8 ISBN 978-981-95-1264-5 (eBook)
https://doi.org/10.1007/978-981-95-1264-5

This Springer imprint is published by the registered company Springer Nature Singapore Pte Ltd.
The registered company address is: 152 Beach Road, #21-01/04 Gateway East, Singapore 189721, Singapore

Preface

In the context of rapid global economic development and accelerating urbanization, infrastructure construction as the cornerstone of social progress has reached unprecedented scales and quality standards. From crisscrossing transportation networks to towering hydraulic engineering projects, from densely clustered architectural complexes to extensive energy transmission pipelines, the successful implementation and long-term stable operation of every infrastructure project fundamentally rely on high-quality compaction operations as their solid guarantee. Compaction quality, being a critical factor influencing the durability, safety, and service life of infrastructure projects, has emerged as one of the core challenges demanding urgent resolution in modern engineering construction. The scientific detection and precise optimization control of compaction quality now stand as pivotal issues requiring comprehensive solutions in contemporary engineering practice.

Traditional methods for compaction quality detection predominantly rely on on-site sampling and laboratory testing. These approaches are not only time-consuming, labor-intensive, and costly but also struggle to achieve real-time, continuous monitoring of the compaction process. Moreover, they lack the capability to adjust compaction parameters promptly based on test results to optimize compaction effectiveness. With the continuous advancement of science and technology, non-destructive testing (NDT) technology has gradually emerged as a research focus and application trend in compaction quality assessment, owing to its significant advantages including non-invasiveness, high efficiency, and real-time capabilities. NDT technology enables rapid and accurate acquisition of physical and mechanical property parameters of filling materials while preserving their original structure, thereby providing reliable scientific foundations for the comprehensive assessment of compaction quality.

Meanwhile, the rapid advancement of intelligent control technology has provided robust support for the automation and intelligent control of compaction operations. By integrating advanced sensor technologies, data processing algorithms, and automated/intelligent control strategies, intelligent control systems enable real-time perception of parameter variations during compaction. Guided by predefined compaction quality standards, these systems autonomously adjust operational parameters of compaction machinery to achieve precise and optimized control. This "detection-decision-control" closed-loop management mode not only significantly enhances the efficiency and quality of compaction operations but also effectively reduces human interference on compaction outcomes. Such innovations establish a solid foundation for the high-quality construction of infrastructure projects.

The book *Compaction Quality Nondestructive Testing and Intelligent Control Technology for Infrastructure* emerges in response to these developments. It systematically synthesizes the latest research achievements and developmental trends in non-destructive testing for compaction quality while exploring the application principles and practical cases of intelligent control technologies in compaction operations. Designed as a professional reference, it integrates theoretical rigor, practical applicability, and forward-looking insights to serve scientific researchers, engineering professionals, and managerial personnel engaged in infrastructure development.

The book is organized into eight chapters: Chap. 1 outlines the technical requirements and development trends of non-destructive testing of compaction quality, compaction quality assessment and control; Chaps. 2–5 systematically discuss roller-integrated compaction monitoring technology based on maximum positive kinetic energy increment, roller-integrated compaction monitoring technology based on positive maximum velocity, non-contact-type roller-integrated acoustic wave detection technique, contact-type roller-integrated acoustic wave detection technology; Chaps. 6–7 introduce compaction quality interpolation assessment methods and models, and compaction quality intelligent assessment methods and models; Chap. 8 discusses compaction quality intelligent control technologies represented by unmanned rolling compaction and intelligent rolling compaction. The book comprehensively covers multiple aspects of non-destructive testing technologies for compaction quality, including their fundamental principles, detection methods, equipment selection and calibration, as well as intelligent control system architectures, control strategies, and optimization algorithms in compaction operations. Through extensive case studies and practical experience sharing, it not only demonstrates the remarkable effectiveness of these technologies in enhancing compaction quality but also explores future technological trajectories and application prospects. This work aims to contribute intellectual insights and technical empowerment to advance the high-quality development of infrastructure construction in China.

In the preparation of this book, we have strived to ensure accurate content, clear logical flow, and accessible language, enabling readers to effortlessly grasp the core principles and practical application techniques of non-destructive testing and intelligent control technologies for compaction quality. However, we are fully aware that technological progress is never-ending. The knowledge presented herein reflects current-stage research outcomes and practical summaries in the field. We eagerly anticipate collaborating with readers to explore emerging possibilities, collectively advance expertise, and contribute to the ongoing innovation and advancement of compaction quality testing and control technologies.

Finally, we extend our heartfelt gratitude to all experts, scholars, engineering professionals, and publishing colleagues whose support and contributions made this book possible. It is through your dedicated efforts and selfless commitment that this work has come to fruition. May this book serve as a trusted guide and companion for readers navigating the fields of infrastructure development and practical engineering. Together, let us forge a brighter future through collaborative innovation and shared aspirations in advancing compaction quality management technologies.

Beijing, China
May 2025

Qinglong Zhang

Competing Interests The author has no competing interests to declare that are relevant to the content of this manuscript.

Contents

Chapter 1
Introduction

1.1 Overview of Non-destructive Testing

1.1.1 The Meaning of Non-destructive Testing

As an emerging comprehensive application technology, non-destructive testing (NDT) refers to methods that assess various engineering materials, components, and structures without damaging or affecting their performance, shape, internal structure, or morphology. By utilizing the changes in responses to sound, heat, light, electricity, magnetism, and force caused by structural anomalies or defects within the materials, NDT effectively detects and evaluates the reliability, safety, continuity, integrity, and other physical properties of the objects under inspection.

As a branch of testing technology, non-destructive testing has a long-standing history, and with the advancement of social sciences and technology, it continues to evolve and progress, becoming an indispensable tool for industrial and societal development. The application range of modern non-destructive testing technology is exceedingly broad, particularly in fields such as mechanical manufacturing, healthcare and pharmaceuticals, aerospace, marine engineering, nuclear equipment, weapon systems, oil and chemicals, railways and shipping, metallurgy, armaments, energy, boilers, pressure vessels, and pipelines. Its usage is particularly prevalent in transportation, construction, water conservancy, and road infrastructure. It can be stated that NDT technology has been applied across virtually all sectors.

1.1.2 Characteristics of Non-destructive Testing

NDT technology possesses the following characteristics:

Q. Zhang, *Compaction Quality Nondestructive Testing and Intelligent Control Technology for Infrastructure*, https://doi.org/10.1007/978-981-95-1264-5_1

(1) Non-destructive. NDT technology operates under the premise of not damaging the object being inspected. It utilizes the phenomena arising from changes in the physical properties of materials due to internal structural anomalies or defects to ascertain whether there are any flaws present on the surface or within the object. This method does not inflict any harm on the materials, components, or structures being examined. Thus, non-destructiveness is a prominent characteristic of NDT, which is why it is often referred to as non-destructive inspection.
(2) Comprehensiveness. During the processing and utilization of any material, component, or structure, the occurrence of defects is unavoidable. Consequently, personnel must not only ascertain the presence of defects but also identify their location, size, and potential impact. Due to its non-destructive nature, NDT technology enables a comprehensive examination of the object being inspected, achieving 100% coverage during testing. In contrast, destructive testing typically relies on sampling methods, which do not permit thorough examination.
(3) Continuity. NDT technology can conduct comprehensive inspections throughout the entire process—from raw materials used in manufacturing, through intermediate stages, to the final products—without compromising the operational performance of the objects being inspected. It can also assess equipment in service. In contrast, destructive testing methods struggle to achieve continuous inspection throughout these stages.
(4) Universality. The application range of NDT technology is exceedingly broad. It is suitable for detecting defects in various equipment, pressure vessels, mechanical components, and more, regardless of whether the materials are metallic or non-metallic, forged, welded, cast, or in the form of plates, bars, or pipes. Appropriate NDT techniques can be employed for both internal and surface defect detection. Moreover, NDT technology permeates various fields, including military, scientific, industrial, and civilian applications, which underscores its extensive applicability.
(5) Diversity. The fundamental principles of NDT encompass nearly all branches of modern physics and information science, contributing to its diversity. There are over 70 distinct types of NDT technologies categorized by different principles and detection methods, including ultrasonic testing, X-ray testing, magnetic particle testing, thermal imaging, eddy current testing, magnetic testing, acoustic emission testing, infrared testing, laser testing, and electromagnetic testing, among others.
(6) Strong applicability. NDT is a highly practical technology that involves multiple disciplines and categories. In practice, it often requires inspectors to possess substantial engineering experience. Furthermore, with the continuous advancement and development of science and technology, automated and intelligent instruments are gradually emerging in the field of NDT. These automated and intelligent devices introduce more efficient, precise, and convenient detection methods, significantly enhancing production efficiency, reducing costs, and ensuring product quality and safety.

1.1.3 The Purpose and Significance of Non-destructive Testing

The primary objective of non-destructive testing technology is to conduct non-destructive inspections and evaluations of the object being examined, thereby identifying defects, damages, or other issues present on the surface or within the material. This ensures product quality, guarantees equipment safety, extends service life, and ultimately reduces production and maintenance costs. The significance of non-destructive testing technology is primarily reflected in the following aspects:

(1) Enhancing Product Quality

Through non-destructive testing technology, comprehensive and accurate inspections can be performed on products without compromising their integrity. This allows for the timely detection and rectification of defects, damages, or other issues on the surface or within the material, ensuring the completeness and reliability of the products, thereby enhancing their quality and competitiveness.

(2) Reducing Production Costs

While incorporating non-destructive testing into the production process may inevitably increase expenses and time, it ultimately proves beneficial when viewed from the perspective of the entire production cycle. Timely non-destructive inspections allow for the identification of defective products, helping to prevent waste in subsequent processes, thereby reducing scrap and rework rates, which significantly lowers overall production costs.

(3) Ensuring the Safety of In-Service Equipment

Although products and equipment may meet quality standards upon leaving the factory, various factors such as fatigue, corrosion, wear, and improper use can lead to the emergence of defects or damages that compromise the safe operation of in-service equipment. Such issues can result in frequent safety incidents. By employing non-destructive testing technology, regular inspections can be conducted on in-service equipment, facilitating the prompt identification and repair of internal defects, damages, or other problems. This proactive approach helps prevent equipment failures and accidents, ensures the safe operation of machinery, extends the service life of the equipment, reduces subsequent maintenance costs, and enhances both reliability and safety.

1.1.4 The Current Status and Development of Non-destructive Testing

NDT is a technique that employs physical or chemical methods to examine and assess materials, components, structures, or equipment without causing damage or affecting their usability. It finds extensive applications in various fields such as mechanical manufacturing, petrochemicals, aerospace, marine engineering, power systems, metallurgy, transportation, and nuclear energy. Particularly in sectors like transportation, construction, water conservancy, and road infrastructure, NDT plays a crucial role in ensuring product quality, maintaining the safe operation of equipment, mitigating accident risks, and extending service life. The evolution of non-destructive testing is closely intertwined with advancements in industrial technology, having undergone a long transformation from simple visual inspections to sophisticated instrument analyses.

(1) The Emergence Stage

The origins of non-destructive testing can be traced back to the late 19th and early twentieth centuries. Driven by the Industrial Revolution, the increasing demand for material performance and structural integrity prompted the exploration of more effective inspection methods. The earliest forms of non-destructive testing relied on simple observations, such as visual inspections. Skilled craftsmen would assess the surface of materials for defects like cracks and voids using the naked eye. Additionally, the hammering technique was commonly employed, where materials were struck, and the resulting sounds were analyzed to determine the presence of internal looseness or cracks. While these methods were straightforward, they relied heavily on subjective judgment, limiting their accuracy and reliability. In 1895, German physicist Wilhelm Röntgen [1] discovered X-rays while experimenting with a powered Crookes tube, laying the foundation for the advancement of non-destructive testing. X-rays possess the ability to penetrate substances, allowing for the visualization of internal structures, with early applications primarily focused on the medical field. Although early X-ray detection equipment was rudimentary and produced poor image quality, it opened a new avenue for detecting internal defects in materials. By the early twentieth century, some researchers [2–4] began experimenting with magnetic fields and magnetic particles to detect surface defects. They discovered that when a magnetized material had cracks on its surface, the magnetic field lines would distort, causing magnetic particles to be attracted to the cracks, thereby revealing the location of the defects.

(2) The Development Stage

With the continuous advancement of industrial technology, non-destructive testing techniques experienced significant development, leading to the refinement of various inspection methods and the expansion of their applications. Following the 1920s, improvements in the performance of X-ray tubes resulted in enhanced image quality. Consequently, X-ray testing technology became widely utilized for identifying welding defects, casting flaws, and other internal imperfections in metallic materials. During this phase, magnetic particle testing techniques were standardized, and inspection equipment and processes were gradually perfected, establishing it as a vital method for detecting surface defects, particularly in industries such as steel production and machinery manufacturing.

In the 1930s, Soviet scientist Sokolov [5] invented ultrasonic testing technology. By exploiting the principle of varying propagation speeds of ultrasonic waves in different media, he was able to assess the presence of internal defects in materials by measuring the travel time and reflection characteristics of the ultrasound. However, due to the limited electronic technology of the time, the early applications of ultrasonic testing were not widespread. Additionally, dye penetrant testing methods emerged during this period, utilizing penetrants to infiltrate surface-opening defects in materials, which were then revealed using a developer. This method was straightforward, cost-effective, and suitable for detecting surface-opening defects in various materials.

In 1943 and 1946, commercial A-type pulse reflection ultrasonic testing instruments were developed in the United States and the United Kingdom, respectively, and were gradually applied to the inspection of forged steel and thick steel plates. Furthermore, during World War II, the need to ensure the quality of weaponry and equipment spurred rapid advancements in non-destructive testing technology. Methods such as X-ray inspection, magnetic particle testing, and ultrasonic testing were extensively employed to evaluate components of military equipment, including aircraft, tanks, and ships, significantly enhancing the reliability of such weaponry.

(3) The Maturation Stage

The period from the 1960s to the 1980s marked the maturation phase of non-destructive testing technologies. With advancements in electronics, various non-destructive testing methods underwent further refinement, expanding their applications into high-demand fields such as aerospace and nuclear industries. The rapid progress in electronic technology led to significant improvements in ultrasonic testing. Enhancements in ultrasonic generators, receivers, and signal processing systems greatly increased the sensitivity and resolution of ultrasonic inspections. Consequently, ultrasonic testing emerged as a critical method for detecting internal defects, finding widespread use in the evaluation of metals, ceramics, and composite materials.

Eddy current testing began to be employed for assessing the conductivity, thickness, and defects in metallic materials. This method leverages the principle of electromagnetic induction, detecting the eddy currents generated within the material to ascertain its integrity. Eddy current testing is characterized by its non-contact, rapid, and highly sensitive nature, making it particularly suitable for identifying surface and near-surface defects in metals. To ensure the reliability and consistency of non-destructive testing results, various methods began to be standardized. Organizations such as the International Organization for Standardization (ISO) [6] and the American Society for Testing and Materials (ASTM) [7, 8] established a series of non-destructive testing standards, regulating aspects such as testing methods, equipment, and personnel qualifications.

With the advancement of computer technology, automated testing systems began to emerge. These systems are capable of automatically scanning materials or components, collecting data, and performing analysis and evaluation, significantly enhancing testing efficiency and reliability.

(4) The Modern Stage

Since the 1990s, with the rapid advancements in computer technology, sensor technology, and artificial intelligence, non-destructive testing has entered a new era characterized by intelligence, integration, and online monitoring. Technologies such as digital X-ray imaging and computed tomography (CT) have gained widespread application. Digital imaging technology allows for the acquisition of higher resolution and contrast images, which can be digitally processed to enhance detection accuracy and efficiency. Computed tomography (CT) provides three-dimensional imaging of objects, offering a clearer depiction of internal structures and defects.

Guided waves refer to ultrasonic waves that propagate within structures of specific geometries, primarily used for the inspection of long-distance pipelines and plates. This technique enables rapid assessment of extensive structures while maintaining high sensitivity to defects. Moreover, non-destructive testing has begun to be applied for online monitoring of structures, facilitating real-time evaluations of structural health [9]. By installing sensors on structures, parameters such as stress, strain, and temperature can be continuously monitored, while non-destructive testing techniques can be employed regularly or intermittently to detect potential defects promptly, thereby preventing accidents from occurring.

1.2 The Current Status and Development of Compaction Quality Detection

With the construction of highways, railways, airports, buildings, bridges, reservoirs, and dams around the globe, an increasing number of earthwork projects utilize various filling materials such as soil, asphalt mixtures, and soil-stone mixtures for subgrade and dam foundation construction. The compaction quality of these filling materials is a critical aspect of highway construction control, significantly influencing the strength and stability of the subgrade and pavement structures, which in turn affects the performance and service life of highways. Consequently, researchers both domestically and internationally have proposed numerous targeted methods for assessing the compaction quality of subgrades and dam foundations.

Currently, based on the different implementation approaches, the testing methods can be broadly categorized into three main types: The first category is direct methods, which include the sand replacement method, water replacement method, ring cutter method, as well as nuclear density gauges and non-nuclear density gauges. The second category comprises indirect methods, such as Rayleigh wave methods, on-board compaction detection techniques, resistivity testing, and impact response spectrum measurement methods. The third category encompasses simulation analysis methods, including linear regression prediction, nonlinear regression prediction, artificial neural network prediction, finite element method (FEM) simulations, and discrete element method (DEM) simulations.

1.2.1 Direct Methods

Direct methods involve the on-site measurement of the density and moisture content of fill materials to assess compaction quality. The sand replacement method is commonly used to measure the density and moisture content of the filling layer, comparing field measurements with the maximum dry density and optimal moisture content determined through standard Proctor tests conducted in the laboratory [10, 11]. This comparison allows for the calculation of compaction ratios and other performance indexes. Typically, the measured density in the field is either greater than or approximately equal to 98% of the maximum dry density, while the moisture content is usually within ±1% of the optimal moisture content [12].

This method is primarily employed for assessing the in-situ density of fine-grained soils, sandy soils, and gravelly soils, making it particularly suitable for determining the compaction quality of these materials. However, it is not applicable for assessing the compaction of materials like stone embankments, which may contain large voids or porous structures. Despite its utility, the sand replacement method has some limitations in practical engineering applications. Firstly, the procedure is somewhat cumbersome and requires experienced technicians to complete the test within approximately 30 min, making it time-consuming and labor-intensive. Additionally, construction activities around the testing site must be halted during the measurement, which can impact project timelines. Moreover, the accuracy of the results is significantly influenced by the precision of the technician's operations [13], making the measurements susceptible to human error.

The water replacement method shares a similar principle and operational steps with the sand replacement method, but it utilizes the incompressibility of water. This method is primarily suitable for assessing the compaction quality of coarser materials, such as coarse-grained soil, gravel, and rockfill. While the principle of the water replacement method is straightforward, the sampling process can be complex, presenting challenges in terms of execution and workload.

The ring cutter method employs a known mass and volume ring cutter to extract a soil sample, ensuring that the volume of the soil corresponds to that of the cutter. After weighing the sample and subtracting the mass of the cutter, the dry density of the soil can be calculated by dividing the mass by the volume. Although the ring cutter method involves relatively simple instrumentation and requires fewer samples, its applicability is limited. It cannot be used for stable soils containing coarse aggregates or loose materials [14]. Additionally, the accuracy of this method is directly affected by the operator's skill level, and the samples typically represent only the surface layer of the filling, resulting in measured compaction ratios that may slightly exceed the actual values.

To mitigate the influence of human factors on measurement results and to address the inefficiencies of traditional testing methods, researchers [15] have begun utilizing nuclear density gauges to determine the density and moisture content of filling materials on-site, allowing for the rapid acquisition of reliable test results. The nuclear density gauge method boasts significant advantages, including ease of operation, real-time detection, and rapid testing speeds, which effectively reduce the labor intensity for operators and accelerate construction progress.

However, this method is limited to random point measurements, capturing data only at discrete locations, which prevents a comprehensive assessment of compaction quality across the entire working surface. Furthermore, the use of nuclear density gauges involves radioactive materials, necessitating specific authorizations for transportation and operation, thereby increasing the overall costs associated with the equipment [16]. Additionally, the accuracy of the measurements relies on calibration using data obtained from the sand replacement method and laboratory drying techniques to enhance reliability [17].

As awareness of nuclear radiation and health protection has grown, non-nuclear density gauges have emerged as an alternative, employing non-radioactive principles to measure the density of filling materials, thus avoiding the risks of radioactive contamination and safety hazards associated with nuclear density gauges. Nevertheless, non-nuclear density gauge equipment is often prohibitively expensive, and the costs of maintenance and repair are significant, which considerably increases the overall testing expenditures.

As research continues to advance, scholars have discovered that in addition to utilizing the density of filling materials to characterize compaction quality, the strength or stiffness of the filling layer can also serve as an effective index. Currently, several techniques are available for the direct measurement of filling layer strength or stiffness, including the soil stiffness gauge method [18], the Clegg hammer test [19], the dynamic cone penetration test (DCP) [20], the lightweight deflectometer (LWD) [21], the lightweight falling weight deflectometer (LFWD) [22], and the portable falling weight deflectometer (PFWD) [23].

Among these, the DCP is particularly favored due to its straightforward operation and portability, making it suitable for on-site testing or the evaluation of bearing capacity in older subgrades. While it can serve as a supplementary method for assessing the compaction quality of subgrades, the relationship between penetration depth and compaction quality varies significantly across different fill materials, rendering this correlation not universally applicable.

In summary, by measuring data such as the density, moisture content, strength, and stiffness of fill materials, the compaction quality of materials like soil, aggregates, or hot mix asphalt (HMA) can be effectively evaluated. The primary direct methods for measuring density include the sand replacement method, water replacement method, ring cutter method, nuclear density gauge method, and non-nuclear density gauge method. For moisture content, the main direct measurement techniques consist of the sand replacement method, nuclear density gauge method, and field moisture content drying method. In terms of strength assessment, the DCP and Clegg hammer test are commonly utilized. Lastly, direct methods for measuring modulus include the plate load test (PLT), soil stiffness gauge method, LWD, and PFWD. To date, these methods have been extensively employed in the construction of roads, railways, airports, dams, and embankments. They exhibit a high degree of detection accuracy and have established themselves as the benchmark against which other methods are measured.

However, several significant issues remain to be addressed: (1) the compaction status of the entire working area cannot be accurately represented; (2) many methods are destructive, causing considerable disruption to the compaction area; (3) low efficiency and high costs severely impede construction timelines, diminish operational efficiency, and adversely affect project economics; (4) certain methods impose stringent requirements on operators and necessitate precise execution; (5) the compaction quality across the entire working area cannot be monitored in real time, resulting in a marked deficiency in data traceability.

1.2.2 Indirect Methods

To overcome the limitations of direct methods, achieving the assessment of compaction quality in fill materials without disturbing the filling body has become a focal point of research. With advancements in science and technology, indirect methods such as on-board compaction detection, ground-penetrating radar (GPR), Rayleigh wave methods, and resistivity methods have gradually been employed in earthworks projects, enhancing both the compaction quality of filling layers and the efficiency of construction operations.

The on-board compaction detection method involves mounting detection instruments on specific compaction machinery to indirectly assess the compactness in pavements or other filling bodies by measuring parameters such as compaction density, force/displacement, energy or machinery power output, number of passes, or temperature. This method is suitable for evaluating the compaction of fine-grained soils, coarse-grained soils, and large gravel.

Utilizing this approach allows for the rapid, continuous, and real-time assessment of compaction quality while significantly enhancing the efficiency and productivity of earthwork operations. It effectively improves the uniformity of compaction in pavements and other filling bodies, demonstrating excellent economic viability by reducing the life cycle costs associated with earthwork projects. In regions such as Europe and the Americas, this method has been successfully implemented in the construction of highways, airports, and railways, yielding substantial empirical data and insights from practical engineering applications.

The GPR method employs high-frequency electromagnetic waves to explore the propagation characteristics within subsurface media. It involves transmitting electromagnetic waves from a sending antenna into the ground and recording the reflected signals at a receiving antenna. Due to the varying reflective characteristics of fill layers with different compaction levels, analyzing the intensity, time, and waveform of the reflected waves enables an assessment of the compaction quality of the fill layer. This method is commonly used for evaluating the compaction of asphalt mixtures in pavement. Shuangguan et al. [24] proposed using the GPR method to assess the compaction of asphalt pavements, comparing the results with those obtained from nuclear density gauges. The findings validated the effectiveness of this method.

The GPR method is primarily employed to assess the moisture content and compaction of subgrade and other filling materials, evaluate the compaction of pavement, determine the thickness of structural layers, analyze pavement damage, and identify voids. This method is characterized by high detection efficiency and is classified as a non-destructive testing technique. However, its accuracy is insufficient, the data processing is intricate, and it often fails to meet the requirements of earthwork across various application scenarios. Given that the propagation speed of waves varies across different media, the compactness of soil can be estimated by establishing a relationship model between wave speed and soil density. The Shear Wave (SW) method is well-suited for detecting the compactness of fine-grained, coarse-grained, and large-grained soils. This approach is efficient, convenient, and does not disturb the surrounding soil. Nevertheless, due to the diverse and complex nature of soil, wave propagation differs among various soil types, and the underlying mechanisms remain inadequately explored. Moreover, the measurement results exhibit significant errors, necessitating multiple calibrations prior to use, which places high demands on operators and poses challenges for widespread implementation.

Moreover, the principal challenge encountered in the application of this method is the unknown impact of moisture sprayed on the surface during compaction on ground-penetrating radar signals. To address this issue, Shuangguan et al. [25] proposed a calibration algorithm based on a reference scanning technique, which extracts density information unaffected by surface moisture. By evaluating the performance of this algorithm through extensive empirical data, the results indicated a significant reduction in the average density prediction error, thereby greatly enhancing the accuracy of density data for asphalt pavements.

The Rayleigh wave method involves applying vibrations to the ground to generate Rayleigh waves, which are transverse waves that propagate along the surface of the medium. As the compaction level of the filling layer increases and its density rises, the propagation speed of the Rayleigh waves also increases. By measuring and analyzing the wave propagation speed in filling layers with varying degrees of compaction, a mathematical model can be constructed to correlate the density of the filling materials with wave speed, allowing for an indirect assessment of compaction quality.

This method is classified as a non-destructive testing technique; however, the complex and variable nature of filling materials results in distinct mathematical models for density and wave speed. Additionally, the process is intricate, requiring highly skilled personnel, and incurs substantial costs, leading to concerns regarding its economic feasibility.

The resistivity method is based on the correlation between the resistivity of filling materials and their physical properties, such as density and moisture content. During the compaction of filling materials, the application of compaction energy leads to closer contact between the particles, resulting in reduced porosity, increased conductivity, and consequently, decreased resistivity. In the detection process, electrodes are inserted into the filling material, and direct current is applied through the electrodes. The current and voltage passing through the filling material are measured to calculate the resistance value. Given the established functional relationship between the resistivity of filling materials and compaction density, empirical resistivity values can be converted to compaction levels using a previously developed standard curve (the resistivity-compaction relationship curve) from laboratory experiments. Islam et al. [26] employed the resistivity method to assess the compaction quality of soil, developing a relationship model between resistivity and maximum dry density. This method is classified as a destructive testing approach, typically used for non-mobile point measurements of filling layer compaction. The primary challenge lies in establishing a reliable mathematical model that correlates compaction density with resistivity.

1.2.3 Predictive and Simulation Analysis Methods

Given that the interaction between soil and roller displays intricate characteristics, including elastic-plasticity and non-linearity, alongside the highly uneven stress distribution within the soil profile, it becomes challenging to qualitatively or quantitatively characterize the compaction state of the soil.

In the realm of predictive methodologies, researchers have made notable advancements. It has been observed that when an average ground pressure is applied, greater wheel loads tend to result in tighter compaction at deeper soil layers [27–29]. Regarding soil compaction, several studies have highlighted the significant impact of tire parameters and wheel loads [30, 31]. Klos and Waszczyszyn [32] employed neural networks to forecast the compaction characteristics of coarse-grained soils. Patel and Mani [33] conducted an on-site investigation of sandy loam to evaluate subsoil compaction under varying wheel load ranges, multiple passes over the bulk density foundation, and various penetration resistance indexes. Through experimentation, Raper and Reeves [34] assessed the differences between soil bulk density and cone index (CI, a measure of soil hardness) under different conditions, including topsoil plowing, deep subsoil plowing, and fixed track tillage. Çarman [35] utilized the Mamdani fuzzy logic approach to examine the compaction of clay loam under varying conditions of pneumatic tire driving speed, wheel loads, and inflation pressures, suggesting that further research using artificial neural networks is essential for the system's advancement. Bayat et al. [36] compared three methodologies—artificial neural networks (ANN), linear regression, and non-linear regression—for predicting penetration resistance, demonstrating that ANN offers superior accuracy compared to multiple linear regression techniques.

Furthermore, the compaction of other materials presents challenges in both qualitative and quantitative descriptions of their compaction states, prompting numerous studies to address this issue. In the realm of asphalt material compaction, Xu et al. [37] proposed a sensor-information-statistics integrated model in conjunction with an intelligent compaction (IC) system for predicting the density of asphalt materials during highway construction. This model addresses the limitation of IC technology, which cannot directly measure material density to serve as an acceptance standard for owners. The integrated system is capable of real-time measurement of the mechanical responses of the roller and the material temperature, proposing a more reasonable model to estimate uncertainty-induced error terms. Additionally, they developed information management software for analyzing IC measurement data, with the statistical model being validated through data obtained from American highway engineering. The results demonstrated the model's rationality and precision.

Cosanti et al. [38] introduced an innovative approach to assess the compactness of river embankments. To tackle the unreliability of traditional quality inspection methods, which rely on random sampling points in the construction area and can lead to delays in construction operations and the timely improvement of compaction quality, Liu et al. [39] conducted a study on quality control in the compaction of earth-rock dams using real-time operational data. They proposed a quality inspection process capable of monitoring real-time compaction operations for earth-rock dams. Utilizing various independent variable data obtained from the real-time quality detection system—such as the number of compaction passes, driving speed, vibration state, and lift thickness, alongside the moisture content and gradation factors of the earth-rock mixture—they developed a multiple regression model to predict compaction quality. Finally, the Kriging interpolation method was employed to estimate the density at any point across the entire work area. This methodology was validated through the Nuoza River earth-rock dam project, demonstrating its effectiveness and feasibility, ensuring comprehensive coverage of compaction quality assessment over 100% of the work area, thereby reducing the uncertainty associated with random sampling and facilitating timely feedback for quality control, ultimately enhancing the construction of earth-rock dams. Awed et al. [40] constructed a model based on extensive laboratory measurements that predicts the laboratory compaction behavior of asphalt mixtures, incorporating parameters describing aggregate shape characteristics, aggregate gradation, and binder content.

Significant advancements have also been achieved in simulation analysis. Li and Schindler [41] employed the FEM to investigate soil compaction and tire fluid dynamics, developing two finite element tire models based on the actual geometry of Bridgestone tires, while analyzing the impact of axle load and tire pressure on soil compaction. Xia [42] established a FEM capable of simulating the soil compaction process and predicting spatial density. Simulation analyses derived from this model demonstrate that the proposed large-deformation FEM can adeptly predict the compaction density of the soil. Due to the limitations of the FEM as a continuum approach, which does not account for slip and the interlocking of aggregates during compaction, Chen et al. [43] employed the DEM to simulate the laboratory compaction of HMA. The results demonstrated a commendable consistency with the measurements obtained from laboratory compaction tests. Leveraging the viscoelastic properties of soil, Zolotarevskaya [44] derived a regression equation that accounts for density, moisture content, and linear compaction speed, subsequently performing mathematical simulations and calculations on soil compaction under dynamic loading conditions. Shoaib and Kari [45] utilized the DEM to conduct a nonlinear elastoplastic shock wave simulation for high-speed compaction. Ghanbari and Hamidi [46] performed a simulation analysis of the rapid impact compaction process involving loose sand. Additionally, Shangguan et al. [47] demonstrated the application of ANN-based pattern recognition to extract density information of asphalt pavement from simulated GPR signals.

Overall, the predictive methodologies primarily rely on ANN, linear regression, and non-linear regression to assess the compaction characteristics and conditions of soil. Some researchers have also employed CCC/RICM/IC technologies to forecast the compaction of filling materials and interpolate the compactness across the entire work area. Simulation analysis techniques typically utilize FEM and DEM to conduct mathematical simulations and analytical assessments of soil compaction characteristics under dynamic loading conditions. While these prediction and simulation analyses can offer a degree of qualitative or quantitative insight into soil-roller interactions and soil compaction, they lack time sensitivity and are unable to effectively monitor and manage compaction quality in real time.

1.3 Current Status and Development of Compaction Quality Assessment

1.3.1 Influencing Factor Analysis

Numerous factors influence the compaction effectiveness in earthwork construction. To ensure the robust stability and strength of the earthwork, it is imperative to identify and analyze the primary factors impacting compaction, followed by the implementation of targeted control measures to achieve optimal compaction outcomes. Zhang [48] examined several issues in the QC/QA of earth-rock dams, determining that soil type, moisture content, and compaction energy significantly affect the soil's compaction during construction. Guo [49] investigated the impact of various elements, including compaction machinery, driving speed, moisture content, the strength of the underlying layer, and rolling mode, on pavement compactness. Utilizing the surface vibration compaction method, Guo [50] conducted indoor compaction tests on natural gravel and identified the primary factors influencing its compaction effectiveness as rock content, gradation, and vibration parameters. By monitoring compaction parameters such as driving speed, the number of compaction passes, lift thickness, and the state of the exciting force, Liu et al. [51] assessed the factors affecting the compaction of earth-rock dams. For soil-rock mixtures, Tokiharu et al. [52] employed compaction testing to investigate the relationship between maximum particle size and maximum dry density, revealing a linear correlation between the two. Focusing on rockfill materials and soil-rock mixtures, Sitharam and Nimbkar [53] researched the effects of varying gradations and volume strains on the properties of granular materials. Liu and She [54] conducted compaction tests on different types of soil-rock mixtures, demonstrating that the compactness of these mixtures is closely linked to both compaction energy and compaction methodologies.

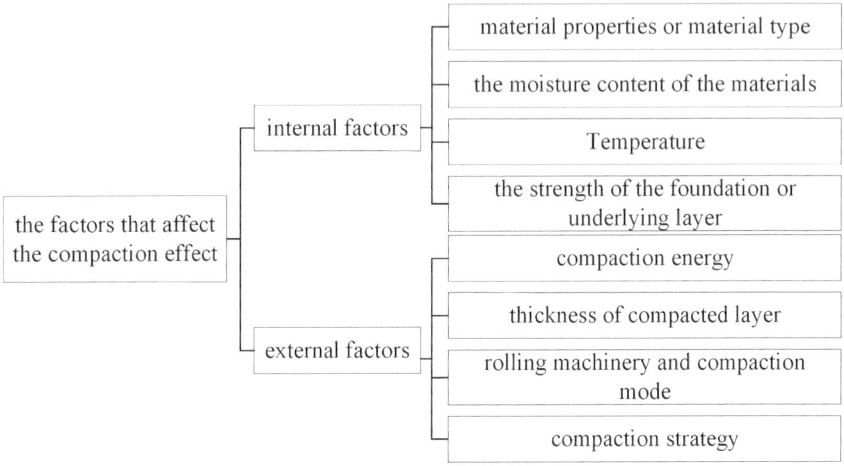

Fig. 1.1 The factors that influence compaction effectiveness

In summary of the aforementioned research, the factors influencing the compaction effectiveness of earthwork can be categorized into internal and external factors (refer to Fig. 1.1). Specifically, internal factors predominantly encompass material properties, material types, and the strength of the foundation or underlying layer. Conversely, external factors primarily include compaction energy, compaction parameters, rolling machinery, compaction methods, and compaction strategies. Within a consistent compaction energy framework, the properties or types of materials that dictate the compaction characteristics of filling materials can vary significantly. When other conditions are comparable, the moisture content of the materials emerges as a primary determinant of compaction effectiveness. Insufficient strength of the underlying layer or foundation will substantially diminish the effectiveness of the filling layer's compaction efforts. Compaction energy directly influences both the quality of compaction and the efficiency of construction, with appropriately calibrated compaction energy significantly enhancing construction efficiency and economic viability. Other influencing factors, whether internal or external, exert varying degrees of impact on the QC/QA of earthwork. Regardless of their classification, a thorough analysis of all influencing factors pertinent to QC/QA in earthwork is essential.

1.3.2 Correlation Between RICM Indexes and Factors Affecting Compaction Quality

Traditionally, the compaction quality of filling materials, such as soil, aggregates, or asphalt mixtures, is assessed through spot inspections of density, moisture content, strength, and modulus at select discrete points [55]. However, these methods exhibit notable drawbacks, including point sampling and destructive measurements, low efficiency, high costs, and heightened demands on operators. Moreover, such approaches fail to accurately represent the compaction status of the entire work area and are insufficient for meeting the requirements of mechanized rapid earthwork construction. To address the limitations of current detection methods, RICM technology is increasingly utilized for QC/QA in road construction [56–58]. The RICM value (RICMV) is derived from vibration data collected by sensors installed in a vibratory drum [59]; it serves to characterize the real-time compaction status of compacted materials, including fine-grained soils, granular soils, and HMA [56, 60–63]. Several RICM values are presently utilized, such as the compaction meter value (*CMV*) employed by Geodynamic [64], *kb* utilized by Ammann [60, 65], machine drive power (*MDP*) implemented by Caterpillar [61, 66, 67], and OMEGA and Bomag Variocontrol used by Bomag [68, 69]. Additionally, White et al. [66] suggested the use of E_{vib} as a vibration modulus for pavement layers during compaction, while Rinehart et al. [57, 70] reported that total harmonic distortion (*THD*) serves as a highly sensitive index for evaluating the compaction status of soil.

It can be asserted that the crux of RICM technology lies in characterizing the compaction quality of filling materials through RICM indexes. Therefore, in addition to controlling the material source prior to rolling and precisely managing construction parameters during the compaction process, it is essential to investigate the correlation between RICM indexes and factors influencing compaction quality. Identifying the primary factors affecting compaction effectiveness and utilizing them as reference data for characterizing compaction quality will facilitate the precise control of construction quality in filling layers.

White et al. [71] conducted a series of soil compaction measurement experiments using a sheep's foot roller, focusing on the *MDP* as a CCI. They performed box plot analyses to examine the relationship between net machine power and the number of passes across test strips with varying moisture contents, investigating how data dispersion in net machine power varied with the number of passes under different moisture conditions. Thompson and White [61] established a multiple regression model based on *MDP*, dry bulk density, and moisture content, using *MDP* values and moisture content to predict dry bulk density, DCP values, and EPLT values, subsequently conducting a comparative analysis with field point measurement results. Their findings indicated a high correlation in predicting field compaction measurements when moisture content was included in the regression model.

Addressing the issue of material heterogeneity, Vennapusa et al. [72] preliminarily attempted to combine RICM technology with geostatistical analysis methods to study the variations of the *CMV* index and its spatial statistics as a function of the number of passes. Liu et al. [73] conducted field rolling tests on earth-rock dams based on compaction values (*CV*), investigating the correlation between the *CV* index and the number of passes, which revealed a strong relationship between the two. Meehan et al. [74] executed compaction tests based on the *CMV* index and employed Kriging interpolation methods to analyze the correlations between the *CMV* index, random sampling measurements, and moisture content. White et al. [62] explored the accelerated application of intelligent compaction technology in subgrade soils, aggregate bases, and asphalt pavement materials, analyzing the correlations between CCI (CMV, MDP, E_{vib}) and material moisture content. Fan et al. [75] conducted an analysis of factors affecting the continuous compaction quality detection indicator CMV in high-speed railway subgrades, constructing a multiple regression model relating CMV to amplitude, vibration frequency, and driving speed, with regression analysis revealing that the CMV index is primarily influenced by factors such as vibration frequency, amplitude, drive direction, and speed.

Building upon the aforementioned research findings, current studies primarily focus on utilizing correlation analysis methods to investigate the relationships between CCI (CMV, MDP, E_{vib}, CCV, CV, k_s) and factors influencing compaction quality (the number of passes, material moisture content, particle gradation, vibration frequency/amplitude, driving speed, and lift thickness). Researchers aim to establish single-factor or multi-factor regression models correlating continuous compaction indexes with compaction quality, thereby facilitating a quantitative analysis of their interrelations and providing a basis for the assessment of compaction quality.

1.3.3 Compaction Quality Assessment

According to current standards, the compaction quality of fill materials during earthwork construction is primarily controlled through sampling methods that measure dry density or compactness, along with construction parameters during compaction (such as compaction trajectory and number of compaction passes) [76]. Given that different factors influencing compaction have varying impacts on construction quality, it is pertinent to conduct research on the establishment of a compaction quality assessment model. This model would incorporate CCI and construction parameters related to filling rolling, thereby allowing the effects of various factors on compaction quality to be reflected through the continuous compaction indexes.

Building upon the correlation analysis presented in Sect. 1.3.2, continuous compaction indicators for earthwork filling materials, alongside material characteristics (such as moisture content and P5) and other factors affecting compaction effectiveness, can be treated as variables to develop a regression model for the compaction quality of earthwork filling materials. Furthermore, by integrating geostatistical analysis methods (such as Kriging interpolation), it will be possible to assess compaction quality at any location within the rolling work area.

Currently, the prevalent compaction quality assessment models primarily encompass regression models [73, 77–80], neural network models [81], models founded on kernel methods [81–84], fuzzy control models [85, 86], and various other types of models [87–89]. Among these, regression models are categorized into simple linear regression (SLR), simple nonlinear regression (SNR), multiple linear regression (MLR), and multiple nonlinear regression (MNR). Neural network models chiefly comprise the radial basis function (RBF) neural network model, the ANN model, and a model based on bidirectional extreme learning machine (B-ELM).

Based on the CCI (*MDP*), Thompson and White [61] conducted research on the assessment of clay compaction, constructing a multiple linear regression model wherein the average *MDP* and moisture content served as independent variables, while dry bulk density, DCP indexes, ELWD, ESSG, and EPLT were treated as dependent variables. The results of the multiple regression analysis were utilized to evaluate field compaction conditions. Xu et al. [90] explored systematic methods for intelligent compaction data analysis and management, establishing a multiple regression model for compaction quality that included ICMVsb (the base compaction value of filling materials), T (HMA temperature), moisture content, Fhma (vibration frequency of the roller during HMA compaction), Fsb (vibration frequency of the roller during subgrade compaction), and D (density of the asphalt mixture) as independent variables to study the impact of each factor on the compaction quality of asphalt mixtures.

Liu et al. [39] conducted a control study on the compaction quality of earth-rock dams using real-time construction data, developing a multiple regression model with variables including the number of compaction passes, driving speed, vibration state, lift thickness, moisture content of the earth-rock mixture, and gradation. They also employed Kriging interpolation methods to evaluate compaction at any point within the work area. Furthermore, Liu et al. [73] utilized RICM technology to assess the compaction quality of filling materials in earth-rock dams, establishing a regression model for density based on the CCI (*CV*) and moisture content as independent variables for the core-wall gravel mixture. For filling materials, they developed a multiple regression model for dry density, where *CV* and P5 content were the independent variables. These regression models, combined with Kriging interpolation methods, enabled the assessment of compaction quality at any location within the rolling area of the earth-rock dam. Xu and Chang [63] investigated intelligent compaction assessments for asphalt materials, conducting a regression analysis between the CCI and the asphalt mixture base layer. Liu et al. [91] focused on real-time quality monitoring and control of highway compaction, establishing a multifactor linear regression model with CCI (*E* or *CMV*) as independent variables and dry density (ρ) as the dependent variable, facilitating real-time comprehensive evaluations of highway subgrade compaction quality.

Nie et al. [58] conducted research on anomaly detection in integrated compaction measurements using rollers, developing linear regression models for CCI (*CMV* and E_{vd}), both with and without anomaly data, to analyze quality control and inspection of highway subgrades. Meehan et al. [74] carried out compaction tests based on CCC technology, constructing corresponding linear regression models, and utilized Kriging interpolation methods to compare the differences between interpolation results and actual sampling measurements. Liu et al. [51] analyzed the correlation between real-time monitoring indexes of filling material compaction quality and compaction parameters, establishing a single-factor regression model with CCI (*CV*) as the dependent variable and the number of passes as the independent variable. Additionally, they developed a multifactor multiple regression model with *CV* as the dependent variable and compaction parameters (the number of compaction passes, lift thickness, and driving speed) as independent variables. In practical construction scenarios, the multitude of compaction parameters necessitates a more comprehensive assessment of compaction effectiveness, as single-factor regression models may not accurately reflect compaction outcomes; thus, multifactor multiple regression models better characterize compaction quality. Addressing the issue of compaction quality detection for high-speed railway subgrades, Dou [92] conducted field rolling test studies, constructing a multiple regression model based on field rolling data and numerical simulation CMV_k data to evaluate the compaction quality of filling layers and validate the efficacy of experimental data. Wang [93] researched rapid assessment methods for the compaction quality of earth-rock mixed subgrades, establishing single-factor regression models where the compactness of various soil types served as the dependent variable and settlement amount as the independent variable. They also formulated a multifactor regression model encompassing compactness, settlement amount, and moisture content, with regression prediction results indicating that the multifactor model provided superior accuracy in predicting the compactness of mixed earth-rock subgrades compared to single-factor models.

Utilizing support vector regression (SVR) in conjunction with a chaotic firefly algorithm (CFA) as the foundation, Liu et al. [82] and Wang et al. [83] developed a comprehensive evaluation method that integrates the improved analytic hierarchy process (i-AHP) and the enhanced geo-statistical analysis method (i-GAM), alongside an assessment method based on SVR with CFA. Building upon this framework, Wang et al. [84] introduced a compaction quality assessment method that combines a smart bacteria-foraging algorithm-based customized kernel support vector regression (SBFA-CKSVR) with an enhanced probabilistic neural network (EPNN). In addition to these assessment methods, a limited number of researchers have proposed several real-time compaction quality assessment techniques, such as a fuzzy evaluation method grounded in Dempster-Shafer (DS) evidence theory. A comparative analysis of the existing compaction quality assessment models and methods is presented in Table 1.1. The table indicates that current research on compaction quality assessment models and methods predominantly concentrates on dam construction. Furthermore, most existing compaction quality assessment models and methods generally lack real-time evaluation capabilities, rendering them non-time-sensitive.

Building upon the aforementioned research findings, and recognizing the close relationship between factors influencing compaction effectiveness and the compaction quality of earthwork filling, numerous scholars have employed correlation analysis to explore the relationships between CCI and factors affecting compaction effectiveness. This research has validated whether the compaction quality of earthwork filling materials can be represented by the factors influencing compaction effectiveness.

Among these factors, the number of passes, material moisture content, vibration frequency/amplitude, driving speed, and lift thickness are frequently utilized in correlation analyses with CCI such as CMV, MDP, E_{vib}, CCV, CV and k_s. Following the correlation analysis between CCI and the factors affecting compaction effectiveness, it is possible to establish single-factor or multi-factor regression models. These models can be employed to verify the representative relationship between the CCI

Table 1.1 Comparison of compaction quality assessment models and methods (√is involved)

Models	Indexes	Methods	Scenarios	Real-time	References
SLR/MLR	Compactness/ deflection/ *CMV*		Road		[94]
SLR	*CMV*	Combination of δ and R	Railway		[95]
SLR	Cone resistance				[44]
SLR/SNR	*CF*/*CMV*		Dam		[96]
SLR/SNR/MNR	Dry density/ Compactness	Geostatistics with *CV*	Dam		[89]
SLR/MLR/MNR	Dry density	Geostatistics with *SCV*	Dam		[97]
MNR	Compactness	Geostatistics-based	Dam		[98]
SVR with CFA	Compactness	Combination of *i*-AHP and *i*-GAM	Dam	√	[75]
SVR with CFA	*CMV*	SVR with CFA	Dam		[76]
SBFA-CKSVR		SBFA-CKSVR	Dam		[90]
Cloud-fuzzy		Cloud-fuzzy	Dam		[91]
SLR			Dam		[57]
MLR	*CV*		Dam		[99]
SLR/MLR	Compactness	Combination of CV_0 and E	Road		[58]
		Equivalent additional stress	Airport	√	[92]
SLR/MLR	Compactness/ E_{vd}	Geostatistics with *CMV* or *VCV*			[93]
CDD		CDD-based	Dam		[100]
SLR	*CMV*		Road		[101]
SLR/MLR	Compactness	*CV*-based	Dam		[102]
SLR/MLR/MNR	Dry density	Combination of E and *THD*	Dam		[103]
B-ELM	Compactness	B-ELM	Dam	√	[104]
SLR	Compactness	Geostatistics with *CV*	Road		[105]
Dual coupled	Dry density	Coupled with dry density and reliability	Dam		[106]
RBF	Relative density		Dam		[107]
MNR	Compactness		Dam	√	[108]

(continued)

Table 1.1 (continued)

Models	Indexes	Methods	Scenarios	Real-time	References
Fuzzy	*CV*	Fuzzy evaluation-based D-S	Dam		[109]
ANN	Compactness/ dry density	Based-ANN	Dam		[110]
KM + AC-BFA + FL	Compactness		Dam	√	[111]

values of earthwork filling materials and the factors considered in the regression model that influence compaction effectiveness.

In essence, a representational model can be developed to illustrate how factors impacting compaction effectiveness relate to CCI. This model can indirectly substantiate the correlation between CCI and the compaction quality of filling materials (such as gravel). Furthermore, it provides a scientific basis for subsequent research aimed at utilizing detection methods to derive CCI values that directly represent the compaction quality metrics of filling materials (such as dry density and compactness).

1.4 The Current Status and Development of Compaction Quality Control

Through an extensive review and analysis of related research, a classification, as depicted in Fig. 1.2, is presented to elucidate the evolutionary trajectory of QC/QA methods for earthwork compaction. This progression can be broadly delineated into four distinct eras: conventional compaction, digital rolling compaction, automatic rolling compaction, and intelligent control compaction. During the conventional compaction era, key characteristics included manual operation, manual monitoring, and point sampling detection. To facilitate real-time monitoring of compaction parameters and quality, researchers introduced digital rolling compaction technology. The QC/QA methods have evolved from the 1.0 era to the 2.0 era, characterized by manual driving, digital monitoring and management, point sampling detection, and spatial positioning. With an emphasis on precision control of compaction parameters, scholars proposed automatic rolling compaction technology, which partially achieves active control over the compaction process, transitioning from manual to automatic driving modes. As technology advances, it is anticipated that QC/QA challenges associated with existing digital and automatic rolling compaction will be addressed. Grounded in a framework of material-machine-information-machine intelligent decision-making, intelligent control compaction methods usher in the 4.0 era of QC/QA. In this new epoch, the overarching goal is to comprehensively mitigate the influence of human factors on compaction quality control and verification.

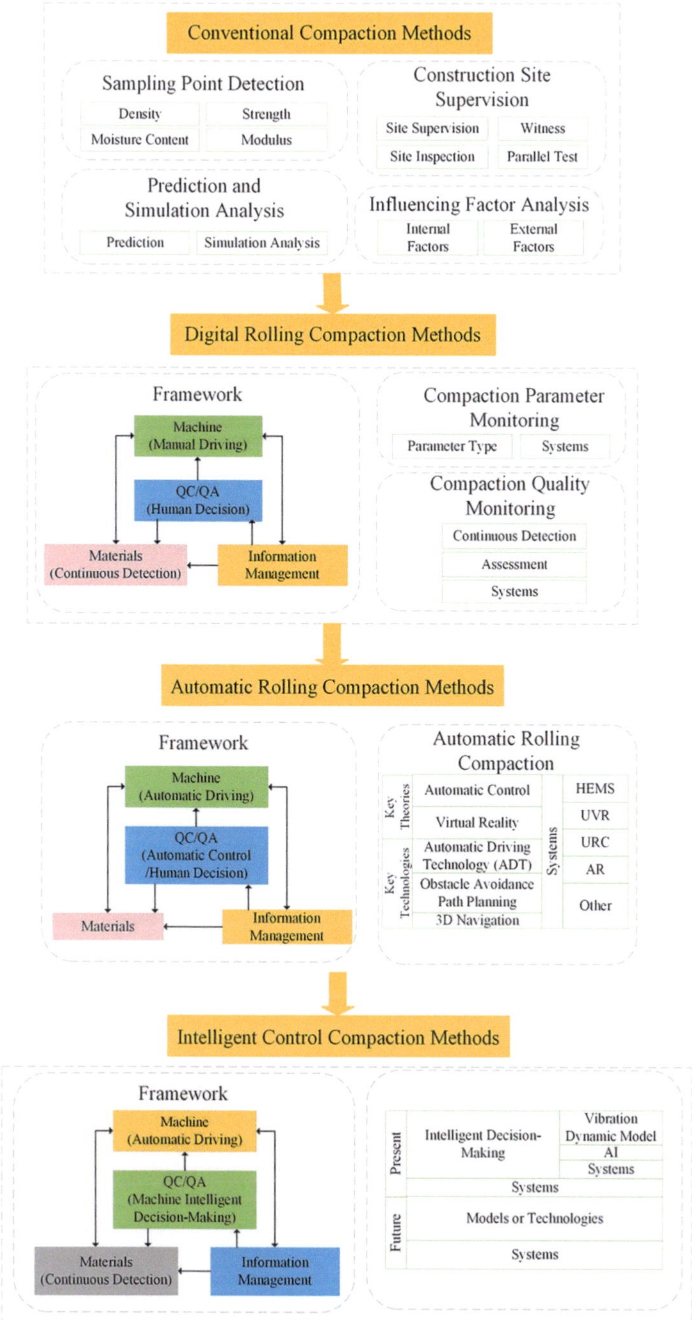

Fig. 1.2 Classification of compaction quality control and assurance methods for earthwork

1.4.1 Conventional Compaction Method

As an empirical control method, conventional compaction is characterized by manual operation, on-site supervision, manual documentation, and point sampling detection [73, 112]. Presently, conventional compaction methods can be categorized into four types: point sampling detection, predictive and simulation analysis, construction site supervision, and analysis of influencing factors [112–117]. In practical engineering applications, conventional compaction methods predominantly depend on the manual regulation of compaction parameters—such as the number of compaction passes, compaction trajectory, vibration frequency, lift thickness, and driving speed—during construction, as well as point sampling detection (e.g., compactness or dry density) at specified locations following construction [55, 113, 118] to ensure the compaction quality of earthwork. The methods of point sampling detection are varied and widely utilized, with many having been codified into specifications and standards [113–117]. Predictive and simulation analysis methods offer qualitative or quantitative insights into soil-roller interactions and soil compaction [33, 43]. During specific phases of a construction project, supervisors may dedicate all or part of their time to on-site monitoring and supervision of rolling construction activities, which constitutes construction site supervision. Regarding the analysis of influencing factors, addressing this aspect is fundamental to developing a robust compaction quality assessment model.

1.4.2 Digital Rolling Compaction Method

According to existing specifications, the QC/QA of earthwork primarily rely on the management of compaction parameters—such as the number of compaction passes, compaction trajectory, vibration frequency, lift thickness, and driving speed—during construction, along with random spot tests conducted post-construction [113–117]. To facilitate real-time monitoring of compaction parameters and overall quality, researchers have introduced various digital rolling compaction methods and technologies, including CCC [119, 120], RICM [56, 121], and IC [90, 122, 123]. The digital rolling compaction approach is executed within the framework of material-machine-information-manual decision-making. Given that QC/QA of earthwork are intricately linked to compaction parameters and quality, research on digital rolling compaction methods predominantly centers on the monitoring of these compaction parameters and the corresponding quality of compaction (refer to Fig. 1.2).

(1) Compaction parameter monitoring

Compaction parameter monitoring begins with the determination of compaction parameters through on-site compaction testing, subsequently enabling real-time monitoring of these parameters via an RTK base station, satellite, master control center, and GPS positioning apparatus installed on the roller. In 1989, Thurner and Sandstrom [124] developed a compaction documentation system tailored for unbound aggregates. Pampagnin et al. [125] proposed a GPS-based guiding system architecture and advanced a computer-integrated road construction approach for compaction (CIRCOM). To investigate the correlation between vibration characteristics and underlying soil properties, Mooney and Rinehart [57] conducted a field monitoring study on roller vibrations during the compaction of subgrade soil. Following this, they established a comprehensive instrumentation system (CIS) to monitor the three-dimensional vibrations of roller compactor components and the phase lag of the drum response relative to the eccentric force [70].

In addition to CIRCOM, other researchers have engaged in compaction parameter monitoring studies for the QC/QA of road construction, leading to the development of various systems, including the compactor tracking system (CTS), pavement compaction process monitoring system (PCPMS), remote intelligent monitoring system (RIMS), and intelligent compaction monitoring system (ICMS). With regard to earth-rock dams, much of the research has concentrated on the monitoring of mechanical and construction parameters during compaction. Several systems [126–131] have been developed to enhance the compaction quality and construction efficiency of earth-rock dams, such as the compaction quality monitoring system based on positioning compensation technology (CQMS), the Georobot/WLAN-based intelligent monitoring system (GWIMS), the real-time supervisory system (RTSS), the continuous compaction monitoring system (CCMS), and the compaction quality monitoring system for rockfill dams (RDCQMS). To ensure the compaction efficiency of earth-rock dam construction, Liu et al. [126] developed an automatic control and real-time monitoring system (ACRM) aimed at the automatic regulation of moisture content in earth-rock dam materials. Liu et al. [127] created a cyber-physical system for the collaborative control of multiple rollers (MRCM). Furthermore, several researchers have conducted compaction parameter monitoring studies on embankments [132], high-filled channels [133], and earthworks[70], resulting in the development of compaction equipment equipped with global positioning system technology (GEGPS), a monitoring and control system for roller compaction quality (CQMCS), and a comprehensive instrumentation system (CIS), respectively.

By synthesizing the existing specifications pertinent to the QC/QA of earthwork [113] alongside current research focused on compaction parameter monitoring (refer to Table 1.2), the monitored compaction parameters can be broadly categorized into three types: material properties, mechanical parameters, and construction parameters. Material properties primarily encompass moisture content and particle gradation; mechanical parameters include excitation force, vibration frequency, and amplitude; while construction parameters consist of lift thickness, the number of compaction passes, compaction trajectory, driving speed, moisture content control standards, and the volume of water added to the filling materials.

In pursuit of enhanced monitoring of various compaction parameters, researchers have developed an array of compaction parameter monitoring systems leveraging computer technology, automatic control technology, communication technology, three-dimensional modeling technology, and GPS/BDS positioning technology, all with the objective of improving compaction quality and construction efficiency.

Table. 1.2 Related work contributions and comparison (√is involved)

System	Parameter type			Contribution	References
	Material properties	Mechanical parameters	Construction parameters		
CDS	–	√	√	Compaction documentation system for unbound aggregates	[124]
CIRCOM	–	–	√	Computer integrated road construction of compaction	[125]
MSEEF	–	√	–	Monitoring system for the eccentric excitation force	[57]
CIS	–	√	–	Monitor the three-dimensional vibration	[70]
ACRM	√	–	√	Automatic control and monitoring for truck watering	[126]
MRCM	–	√	√	Cyber-physical monitoring for multi-roller compaction	[127]
CQMS	–	√	√	Theory and mathematical model of CQMS based on PCT	[128]
GWIMS	–	–	√	Georobot/WLAN-based intelligent monitoring	[129]
RCQSS	–	–	√	Monitoring the number of compaction passes	[130]
CCMS	–	√	√	Continuous compaction monitoring system based BDS	[131]
CEGPS	–	–	√	Monitoring field lift thickness	[132]
CQMCS	–	√	√	Monitoring and control of compaction quality	[133]
RDCQMS	–	√	√	GPS-based monitoring for construction quality	[134]

Nevertheless, the existing compaction parameter monitoring methods and systems continue to rely on the operator to manage the roller's operations, resulting in the compaction quality of earthwork being significantly influenced by human factors. While real-time monitoring of compaction parameters is feasible, achieving active and precise control over these parameters remains elusive, leading to issues such as rolling omissions, cross-rolling, and repeated rolling at the interfaces of adjacent layers.

(2) Compaction quality monitoring

Compaction quality monitoring offers an innovative solution to some of the challenges linked to conventional compaction methods [91]. Research in this domain has primarily concentrated on continuous detection methods, compaction quality assessment methodologies, and compaction quality monitoring systems (see Fig. 1.2). Presently, the most widely employed continuous detection methods for assessing compaction quality include the acceleration method, GPR method, SW method, force method, deformation method, energy method, fiber Bragg grating (FBG) method, acoustic wave method, and electrical resistivity (ER) method, among others. At the heart of continuous detection lies the CCI, which is utilized to characterize the compaction quality of the filling materials [27].

Building on the CCI, researchers have developed various types of compaction quality assessment models and subsequently proposed corresponding assessment methods [73]. To achieve comprehensive, real-time, online, and automatic monitoring of compaction quality, compaction quality monitoring systems that integrate CCI or compaction quality assessment methods have been introduced. Currently, a diverse array of compaction quality monitoring systems is extensively employed in earthwork, particularly in the quality control of road construction [56, 60, 121–123, 135–139]. With the widespread adoption and implementation of compaction quality monitoring technology in earthwork, several issues associated with conventional compaction methods have been partially mitigated.

1.4.3 Automatic Rolling Compaction Method

To address the challenges posed by manual operation, some researchers have sought to harness automatic control technology and other essential technologies, such as path planning, for rolling operations, subsequently proposing various automatic rolling compaction methods [140–150] aimed at mitigating the impact of human factors on QC/QA. In the 1980s, Japan became the first country to implement road rolling machinery equipped with automatic control devices and autonomous rolling (AR) methods in asphalt pavement construction [140].

As theoretical frameworks and technologies have advanced, diverse automatic rolling compaction methods and URC systems have been employed across various infrastructures, including roads, railways, dams, and airports. Huang et al. [146] designed and executed a programmable logic controller (PLC)-based autonomous construction system utilizing unmanned vibratory rollers (UVR) for road construction. Zou et al. [148] validated an obstacle detection method grounded in Dempster-Shafer evidence theory. To alleviate driver fatigue, a trajectory tracking method that is resistant to disturbances was discussed [141]. In pursuit of high-quality road construction, the efficacy of autonomous tandem rollers for optimizing asphalt compaction was evaluated [144].

In the realm of airport engineering, a high-embankment monitoring system (HEMS) focused on the compaction of high embankments was developed, which incorporates optimal path algorithms and unmanned vehicle control technology [118, 142, 143]. Additionally, other researchers concentrated on path tracking control and the automatic operation of unmanned vibratory rollers for earth-rock dam construction [145, 147, 151, 153]. In the context of collaborative operations involving multiple unmanned rollers, Shi et al. [154] undertook preliminary exploratory and application research. Regarding roller compacted concrete (RCC) dams, Shi [155] conducted an exploratory study on the implementation and application of unmanned roller compacted dam construction technology.

A variety of automatic rolling compaction methods and URC systems have emerged for earthwork applications in roads, railways, airports, dams, and embankment construction (see Table 1.3). These methods and systems are executed within the framework of material-machine-information-manual decision-making. The fundamental theories underpinning the realization of automatic rolling compaction encompass automatic control, virtual reality, navigation and positioning, and path planning.

Key technologies integral to the URC system include automatic driving technology, obstacle avoidance and path planning, three-dimensional (3D) navigation, human–machine interface (HMI), and GPS/BDS positioning technology. Some of these systems incorporate functionalities such as path planning (PP), obstacle avoidance (OA), and collaborative rolling (CR).

Despite the advancements in automatic rolling compaction technology, which have partially addressed the limitations of traditional compaction methods, the technical capabilities remain confined to the realm of automatic feedback control. Engineering applications reveal several challenges inherent to the automatic rolling compaction method: construction processes continue to adhere to established quality standards and parameters; the construction process is characterized by sequential procedures; QC/QA are still reliant on manual offline decision-making assessments; varying construction areas are subjected to fixed compaction parameters; and compaction parameters cannot be independently optimized.

1.4.4 Intelligent Control Compaction Method

Current research indicates that intelligent control compaction methods closely resemble digital rolling compaction methods, which represent another focal point in the QC/QA methodologies for earthwork. Drawing upon the concepts and principles of intelligent control within control theory, scholars have proposed various intelligent control compaction (ICC) methods structured around the material-machine-information-machine intelligent decision-making interaction system. In these methods, a controller (or system) is designed to incorporate functions such as learning, abstraction, reasoning, and decision-making, enabling it to adaptively respond to changes in environmental information, thereby facilitating the autonomous execution of rolling operations without human intervention.

ICC methods exhibit three prominent features: intelligent perception of compaction information, intelligent decision-making regarding working parameters, and unmanned rolling compaction. In comparison to digital rolling compaction and automated rolling compaction, ICC effectively addresses compaction control challenges characterized by uncertain mathematical models and complex, highly non-linear task requirements. A pivotal aspect of intelligent decision-making lies in

Table. 1.3 Automatic rolling compaction methods and unmanned rolling compaction systems (√is involved)

Author	Contribution	PP	OA	CR
Sun [140]	Automatic control devices and rolling driving methods	√	–	–
Yao et al. [118]	HEMS, mainly including optimal path algorithm and unmanned vehicle control	√	–	–
Yao et al. [141]	Accurate trajectory tracking for self-driving vibratory roller	√	–	–
Song and Zhang [142]	A simulation model build based on the pure pursuit algorithm	√	–	–
Zhang et al. [143]	Optimal path planning of impact roller	√	–	–
Husemann et al. [144]	The evaluation of the impact of different road compaction strategie	√	√	√
Song and Xie [145]	A composite disturbance rejection for the path-following control of rollers	√	–	–
Huang et al. [146]	Autonomous construction system for an unmanned vibratory roller	√	–	–
Bian et al. [147]	Path following a control method based on fuzzy algorithm	√	–	–
Zou et al. [148]	A method of obstacle detection based on D-S evidence theory	√	√	–
Fang et al. [151]	A path following control model for an unmanned vibratory roller	√	–	–
Yang et al. [155]	A novel and effective path tracking control of articulated road roller	√	–	–
Chen et al. [152]	An improved technology for unmanned driving	√	√	–
Shi et al. [153]	Unmanned roller group collaborative complete coverage path planning	√	√	√
Shi [155]	Unmanned rolling dam construction technology of high arch dams	√	–	–
Zhang et al. [156]	Unmanned rolling compaction system, including an unmanned roller, RTK-GPS system, wireless	√	√	√

establishing control criteria for the compaction parameters. Through a comprehensive review of the existing literature, ICC methods can be broadly categorized into three types based on the different approaches to formulating control criteria [149, 150, 157].

(1) Vibration Compaction Model

One category is model-based physical methods, which utilize a vibration compaction model to simulate the vibration compaction process and optimize the working parameters of the roller [158–160]. Through these simulations, the optimization of working parameters can be achieved. The fundamental premise of the vibration compaction model is to accurately replicate the vibration compaction process, calculate the dynamic responses of the soil-roller interactions, and ascertain the elastic–plastic deformation of the filling materials, as well as the effects of various parameters on the compaction outcome, among other factors.

Extensive vibration compaction tests have led some scholars to discover a correlation between the compaction effect and the working parameters of the roller [158–160]. To deepen the understanding of the mechanisms underlying vibration compaction, relevant researchers have investigated the vibratory roller-soil interaction system and established vibration compaction models that analyze how different compaction parameters influence the dynamic characteristics and compaction effects of the interactive system, thereby providing a theoretical foundation for the optimization of working parameters [158, 161]. Currently, vibration compaction models can be classified into viscoelastic models and viscoelastic plastic models. The former encompasses both linear elastic models and nonlinear elastic models, while the latter includes asymmetric hysteretic models and viscoelastic models that incorporate plastic components.

In the realm of viscoelastic modeling, Yoo and Selig [162] introduced a classic two-degree-of-freedom (2-DOF) vibration compaction model in the 1970s. This model is grounded in linear elastic vibration theory and employs a mass-spring-damping system to represent the vibratory roller-soil interaction. Its straightforward structure, minimal parameters, and clear interpretability have led to widespread application in analyzing the interaction between the roller and the soil, providing valuable guidance for monitoring soil compaction status, optimizing compaction process parameters, and designing roller structures.

Building upon several assumptions, Fang et al. [151] established a 2-DOF dynamic model and subsequently developed a dynamic model of the vibratory roller influenced by the exciting force. Mooney and Rinehart [57] found that a 2-DOF lumped parameter vibration theory utilizing a viscoelastic soil model served as a reliable predictor of roller-soil responses, despite slightly underpredicting phase lag. However, the model remains oversimplified, with certain assumptions—such as the complete elasticity of soil and constant contact between the vibratory wheel and the soil—being inconsistent with actual compaction scenarios.

In the 1980s, a novel compaction technology emerged that achieved compaction through the application of horizontal rubbing forces to the soil, leading to the development of the oscillating roller. Thurner [163] proposed a single-degree-of-freedom (1-DOF) oscillating compaction model, which employs horizontal stiffness and damping to simulate the interaction between the oscillating wheel and the soil. Paulmichl et al. [164] advanced this field by establishing a lumped parameter model with a three-degree-of-freedom (3-DOF) representation of the interacting oscillation roller-subsoil system.

As compaction theory and machinery have evolved, both vibratory and oscillating rollers have emerged, necessitating more sophisticated models than the simple vibration or oscillation models. Consequently, an overall dynamic model has been proposed [165, 166]. This model typically employs a linear spring-damper system to characterize the soil properties and articulate the interaction between the vibratory wheel and the soil. By considering both vertical and horizontal vibrations, the model offers a more comprehensive evaluation of the overall performance of vibratory and oscillating rollers. Nonetheless, its complexity hinders effective improvement in simulation accuracy concerning roller-soil interactions and soil compaction effects.

In instances where soil stiffness is considerable, the vibratory roller may detach from the soil during the rolling process, resulting in a phenomenon termed "jumping vibration" [167, 168]. Upon separation, the vibratory wheel ceases to interact with the soil, necessitating corrections to the linear elastic model. The rolling state is thus delineated into two segments: contact and separation, leading to the establishment of a piecewise linear model [169]. Furthermore, experiments reveal that the shear modulus of soil diminishes as shear strain increases, indicating that soil stiffness is related to both damping and shear modulus. As such, the stiffness and damping of soil are not linear but contingent upon strain. Wersäll et al. [170] formulated a vibration compaction model that accounts for the characteristics of soil strain softening, based on the 2-DOF model, and proposed an iterative procedure for calculating strain-related stiffness and damping. The results of this model exhibit greater alignment with the test results obtained from a small indoor vertical vibration compactor, compared to those derived from the linear elastic model.

Assuming the soil is entirely elastic and does not experience plastic deformation, the viscoelastic model is applicable primarily in scenarios where the soil is fully compacted or nearly so. During the compaction process, soil transitions from a loose to a compacted state, undergoing plastic deformation, which renders the viscoelastic model unsuitable for uncompacted soil and incapable of analyzing the compaction deformation process. To more accurately depict dynamic characteristics and the compaction deformation process, certain scholars have proposed vibration compaction models that incorporate plastic deformation.

These models are primarily categorized based on varying approaches to account for plastic deformation: one approach involves integrating plastic deformation elements into the vibration compaction model to create a viscoelastic-plastic model; the other employs an asymmetric hysteretic model to characterize the differing loading and unloading stiffness of the soil resulting from plastic deformation. Given that soil experiences elastoplastic deformation upon loading, with elastic rebound occurring solely during unloading, the force–deformation relationship exhibits nonlinear hysteresis characteristics. To address this asymmetry, some researchers have proposed using an asymmetric hysteretic curve to describe the force–deformation relationship of the soil and have established an asymmetric hysteretic vibration compaction model. Grade introduced a triangular hysteretic model, while Shen et al. [171] employed the Bouc-Wen hysteresis model to describe the hysteresis characteristics of soil.

Compared to the linear elastic model, the asymmetric hysteretic model better accounts for the distinct mechanical properties of soil during loading and unloading, aligning more closely with the dynamic characteristics observed in actual compaction processes. However, this model necessitates the specification of displacements at multiple split points (or yield points), as the displacements corresponding to different compactness vary, complicating the determination of model parameters.

In the 1990s, Pietzsch et al. [172] proposed a four-degree-of-freedom (4-DOF) viscoelastic-plastic vibration compaction model, which comprises a roller dynamics analysis sub-module and a soil property proton module. This 4-DOF viscoelastic-plastic model effectively describes the dynamic and elastoplastic deformation characteristics of soil during compaction, providing a more accurate representation than the viscoelastic model. To optimize the compacting operation of intelligent vibratory rollers across different compaction stages, Gong et al. [173] developed a 4-DOF dynamic model of the vibratory wheel-soil system under coupled operating conditions. However, due to the multitude of parameters involved—particularly the mass, stiffness, and damping of the vibrational soil and the additional soil—this model is seldom utilized in practice.

To enhance the practicality of the vibration compaction model, some researchers have introduced a three-degree-of-freedom (3-DOF) viscoelastic-plastic model [167, 174, 175], which serves as a simplification of the 4-DOF viscoelastic-plastic model. Beainy et al. [150, 157, 176] proposed a viscoelastic-plastic compaction model that employs the Burgers model to characterize the deformation features of asphalt materials. They contended that the deformation of asphalt materials under compressive strength can be segmented into three components: viscoelastic deformation, instantaneous elastic deformation, and plastic deformation. The viscoelastic deformation is represented by a parallel spring and damper, the instantaneous elastic deformation by an elastic spring, and the plastic deformation by a viscous damper. The model parameters are derived from laboratory tests.

Imran [177] advanced a viscoelastic-plastic block model based on the viscoelastic-plastic vibration compaction model, thereby rendering the simulation process more reflective of the actual vibration compaction process. Depending on the roller's driving direction, the model divides the contact area between the vibratory wheel and the asphalt material into three segments: a rear block, a center block, and a front block, each composed of Burgers material. It is assumed that the vibratory wheel maintains contact with all three blocks, with the interaction force between the vibratory wheel and the soil being the sum of the forces exerted by these blocks. Building on this, Imran [177] proposed a vibration compaction model that considers the movement of the roller, segmenting the pavement into multiple blocks, each with a width of $d/3$. As the roller advances, each strip sequentially traverses the front, directly beneath, and behind the contact area between the vibratory wheel and the soil. The duration of each stage is determined by the vehicle's speed and the wheel's contact width. The block deformation is computed through the integration of strain over time across these three stages, with the initial condition for each stage's integration being the result from the previous stage. This approach allows the model to account for the effects of the roller on the vibratory wheel-soil interaction, as well as the impact of vehicle speed on the compaction outcome. Comparative analyses with field tests demonstrate that the model effectively simulates the interaction between the roller and the asphalt pavement, as well as the deformation characteristics of the asphalt.

Building upon the aforementioned analysis, it is evident that significant advancements have been achieved in the vibration compaction models employed for QC/QA in earthwork. Nevertheless, both research and practical applications have demonstrated that the current vibration compaction models fall short in accounting for the elastoplastic deformation of filling materials. Additionally, they lack the capability to adaptively modify model parameters and intelligently adjust to optimal compaction parameters in response to varying compaction conditions.

(2) Prediction model based on AI

Another approach involves mathematical models, wherein predictive models based on artificial intelligence (AI) algorithms have endeavored to adaptively adjust working parameters to enhance the compaction process. Following the identification of key influencing factors on compaction parameters, Li [178] proposed a neural network model tailored for intelligent rollers. This model comprises a prediction network, which determines the maximum compactness for the subsequent pass based on system state variables, and a control network that outputs the frequency, amplitude, and roller speed.

Zhang and Goh [179, 180] utilized backpropagation neural networks and multivariate adaptive regression splines to evaluate pile drivability in relation to predicting maximum compressive stresses, maximum tensile stresses, and blows per foot, thereby facilitating the optimization of dynamic compaction for high-performance (HP) piles. Zhang et al. [181] introduced a fuzzy control strategy aimed at optimizing compaction parameters, wherein fuzzy control rules were established from on-site data and subsequently refined through a genetic algorithm (GA).

Wang et al. [84] focused on managing compaction quality assessment and control for earth-rock dams by developing a three-dimensional, real-time monitoring system grounded in a smart bacteria-foraging algorithm, customized kernel-support vector regression, and an enhanced probabilistic neural network. Isik and Ozden [182], along with Günaydin [183], presented ANN prediction models for estimating soil compaction parameters. Ranasinghe et al. [184, 185] introduced a novel predictive tool developed through ANNs to assess the effectiveness of rolling compaction and enhance ground compaction capabilities. Taghavifar et al. [186] employed a hybrid ANN and imperialist competitive algorithm to forecast soil compaction. Furthermore, Xu and Chang [187] devised an innovative integrated material-machine-information and human-decision system, capable of providing feedback to construction personnel to optimize compaction efforts and facilitate adaptive control of pavement material quality.

Despite the advancements in AI-based prediction models for optimizing quality control and assurance in earthwork, several challenges remain unresolved. Current research predominantly centers on road construction, indicating a deficiency in AI algorithms tailored for optimizing the compaction processes associated with railways, airports, dams, and embankments. There has yet to be a comprehensive study on the effective integration and fusion for the continuous monitoring of compaction quality and precise control of compaction parameters. Additionally, AI-based prediction models often struggle to accurately depict the compaction process as influenced by varying filling materials and the interactions between the roller and the soil.

(3) Compound model

By merging the distinct advantages of the vibration compaction model and AI-based prediction models, several scholars have embarked on research surrounding hybrid models. To enhance the intelligence of subgrade and pavement rolling compaction technology, Lin and Wang [112] established an intelligent rolling compaction model that integrates a viscoelastic-plastic vibration compaction model with an intelligent prediction model for compaction parameters. Prokopev et al. [188] introduced the concept of a "cyber-physical road construction system" designed for the continuous control of asphalt mixture compaction. Within the hierarchical "study" layer, measurement and forecast results are generated and made accessible to relevant specialists, serving as a basis for theoretical analysis through a mathematical model of the object, thereby facilitating the adjustment of optimal operating modes.

Building on the vibration compaction process model and a comprehensive evaluation method for compaction quality, An et al. [189] proposed a dynamic optimization method for the compaction process of rockfill materials, achieving dynamic optimization of compaction parameters by considering an overall optimal solution. Zhang et al. [190] developed an intelligent compaction control theory and an intelligent rolling compaction method aimed at ensuring compaction quality for earth-rock dams, demonstrating significant application potential in the construction of highways, railways, and airports.

Preliminary explorations of compound models in QC/QA for earthwork suggest that they can effectively leverage the strengths of both the vibration compaction model and AI-based prediction models. Such hybrid models possess the capability to intelligently adjust optimal compaction parameters in response to varying compaction conditions, showcasing considerable promise in advancing the intelligence of QC/QA processes for earthwork. However, their application remains limited in the contexts of road, railway, airport, dam, and embankment construction.

1.5 The Main Content

With the acceleration of global urbanization and the continuous expansion of infrastructure, geotechnical structures represented by roads, dams, railways, and airports have become an important cornerstone of modern social development. As the core link of earthwork engineering in infrastructure construction, the quality of compaction directly affects the safety, durability and life cycle cost of engineering structures. Traditional compaction quality control relies on manual experience judgment and sampling destructive testing (such as sand replacement method and cutting ring method). There are problems such as low detection efficiency, insufficient representativeness, and poor real-time performance, which makes it difficult to meet the needs of modern construction with high precision and high reliability. Simultaneously, the rapid development of intelligent construction technology has provided a new breakthrough for compaction quality control. By integrating the Internet of Things, artificial intelligence and advanced sensing technology, automatic and intelligent control technology, the non-destructive, digital and intelligent transformation of the compaction process has become an inevitable trend of technological innovation in the industry.

This book focuses on nondestructive testing and intelligent control technology of infrastructure compaction quality, and its significance is reflected in the following three aspects:

(1) technological innovation: breaking through the limitations of traditional detection methods, a dynamic perception technology of compaction status based on multimodal sensing (kinetic energy, velocity, and sound wave) is proposed to construct an evaluation model of "physical mechanism driven + data intelligent fusion", forming a complete technical chain from real-time monitoring to intelligent decision-making.
(2) engineering practice value: through the research and development of integrated detection devices and intelligent control systems for roller, the controllability of construction quality and operating efficiency can be significantly improved, the rework rate and resource waste can be reduced, and the infrastructure construction can be upgraded towards green and intelligent directions.
(3) interdisciplinary contributions: deeply integrate mechanical dynamics, signal processing, artificial intelligence and geotechnical engineering, expand the theoretical boundaries and methodological system of compaction quality control research, and provide a reference example for interdisciplinary integration and paradigm innovation in the field of intelligent construction.

The results of this book not only provide a systematic solution for compaction quality control, but also respond to the urgent needs for high-quality development of global infrastructure from the dimensions of technology, economy and sustainability. It has important strategic significance for promoting technological innovation in the industry and helping achieve the "dual carbon" goals.

This book focuses on the core theories and engineering practices of infrastructure compaction quality non-destructive testing and intelligent control technology. It is divided into eight chapters with progressive content, covering technological innovation, method research and practical application. The main contents of each chapter are summarized as follows:

This chapter explains the importance of compaction quality detection and control in infrastructure construction, sorts out the development status and challenges of non-destructive testing and intelligent control technology, and clarifies the research objectives and structural framework of this book.

Chapter 2 proposes a theoretical model of energy change with the dynamic response of rolling machinery as the core, analyzes the relationship between the maximum positive kinetic energy increment of the vibratory wheel and the compaction parameter and compaction effect, analyzes the correlation mechanism between the maximum positive kinetic energy increment and the compaction state, constructs a compaction quality characterization index based on the maximum positive kinetic energy increment and a roller-integrated compaction monitoring technology, and verifies its applicability and effectiveness through engineering.

Chapter 3 proposes a speed change theoretical model with the dynamic response of the rolling machinery as the core, analyzes the relationship between the maximum forward speed of the vibratory wheel and the compaction parameter and compaction effect, analyzes the correlation mechanism between the maximum forward speed and the compaction state, constructs a compaction quality characterization index based on the maximum forward velocity and a roller-integrated compaction monitoring technology, and verifies its applicability and effectiveness through engineering.

Chapter 4 proposes a theoretical model of non contact-type acoustic wave changes with the dynamic response of rolling machinery as the core, analyzes the relationship between the characteristic physical quantities of acoustic wave and the compaction parameter and compaction effect, resolves the correlation mechanism between acoustic wave and compaction state, constructs compaction quality characterization index based on the characteristic physical quantities of acoustic wave and non-contact roller integrated acoustic wave detection technology, and verifies its applicability and effectiveness through engineering.

Chapter 5 proposes a theoretical model of contact-type acoustic wave changes with the dynamic response of rolling machinery as the core, analyzes the relationship between the characteristic physical quantities of acoustic wave and the compaction parameter and compaction effect, resolves the correlation mechanism between acoustic wave and compaction state, constructs compaction quality characterization index based on the characteristic physical quantities of acoustic wave and roller-integrated acoustic wave detection technology, and verifies its applicability and effectiveness through engineering.

Chapters 6 and 7 propose a compaction quality assessment method based on geostatistical analysis and a compaction quality assessment method based on Green spline spatial interpolation, providing efficient computing tools for global quality assessment, and verifying its applicability and effectiveness through engineering; integrating machine learning and deep learning technologies to construct a multi-source data-driven compaction quality intelligent assessment framework and method, and further propose a compaction quality intelligent assessment framework and method that considers small sample uncertainties, and verifying its applicability and effectiveness through engineering.

Chapter 8, relying on high-precision positioning systems (such as Beidou/GPS), intelligent sensors, automated control system and remote monitoring platform, realize autonomous navigation, path planning and precise operation of rolling equipment, build unmanned rolling compaction technology and system, and verify its applicability and effectiveness through engineering; propose adaptive control strategies for rolling processes, integrate feedback optimization and decision-making algorithm, form a "detection-decision-control" closed-loop system, build intelligent rolling technology and system, verify its applicability and effectiveness through engineering, and promote the intelligent upgrade of the construction process.

The book is based on theoretical innovation and guided by engineering implementation. It systematically presents the technical chain from basic nondestructive testing and monitoring technology to intelligent decision-making and control method. It aims to provide a forward-looking and practical reference guide for scientific researchers, engineers and technicians in the infrastructure field.

References

1. Röntgen WC (1896) On a new kind of rays. Science 3(59):227–231
2. Förster F (1986) New findings in the field of non-destructive magnetic leakage field inspection. NDT Int 19(1):3–14
3. BakunovAS KurozaevVP (2004) Increasing the reliability of magnetic-particle testing by means of a MJIK-10 automated unit for magnetic fluorescent-penetrant inspection of pipe end faces. Russ J Nondestr Test 87(3):54–59
4. Muzhitskii VF, Bakunov AS (2004) A set of portable equipment for fluorescent magnetic particle testing. Russ J Nondestr Test 34(8):90–96
5. Shcherbinskii V (2005) How ultrasonic nondestructive testing of welded joints gained its independence. Russ J Nondestr Test 41(6)
6. ISO 17635(2016) 2016, Non-destructive testing of welds-General rules for metallic materials, Switzerland
7. ASTM (2017) Standard guide for acoustic emission examination of concrete structure. E3100-22, West Conshohocken, PA
8. ATTM (2021) Standard practice for measuring ultrasonic velocity in materials by comparative pulse-echo method. E494-20, West Conshohocken, PA
9. Yu R (2018) Research on large-scale steel structure non-destructive cloud detection technology based on Hadoop. South China University of Technology
10. ASTM (2000) Standard test methods for laboratory compaction characteristics of soils using standard effort. D698, West Conshohocken, PA
11. ASTM (2009) Standard test methods for laboratory compaction characteristics of soil using modified effort. D1557, West Conshohocken, PA
12. Elias PEV, Christopher BR, Berg RR et al (2001) Mechanically stabilized earth walls and reinforced slopes design and construction guidelines. FHWA technical report, U.S. DOT, Washington, DC
13. Krebs RD, Waller ED (1971) Highway materials. McGraw-Hill, New York
14. Juan GZ (2011) Relativity analysis on the detection methods of compaction degree for expressway subgrade. Adv Mater Res 413(413–413):252–257
15. ASTM (2007) Standard test method for in-place density and water content of soil and soil-aggregate by nuclear methods. D6938, West Conshohocken, PA
16. Shangguan P, Al-Qadi IL (2015) Calibration of FDTD simulation of GPR signal for asphalt pavement compaction monitoring. IEEE Trans Geosci Remote Sen 53(03):1538–1548
17. Berney ES, Kyzar JD (2012) Evaluation of non-nuclear soil moisture and density devices for field quality control. Transportation research record 2310, Transportation Research Board, Washington, DC, pp 18–26
18. ASTM (2007) Standard test methods for measuring stiffness and apparent modulus of soil and soil-aggregate in-place by an electro-mechanical method. D6758-02 West Conshohocken, PA
19. ASTM (2007) Standard test method for determination of the impact value (IV) of a soil. D5874, American standard Testing Methods, West Conshohocken, PA, pp 1–9
20. ASTM (2007) Standard test methods for use of the dynamic cone penetrometer in shallow pavement applications. D6951M-09, West Conshohocken, PA

21. Kongkitkul W, Saisawang T, Thitithavoranan P et al (2014) Correlations between the surface stiffness evaluated by light-weight deflectometer and degree of compaction. Tunn Undergr Constr Shanghai
22. Zabielska-Adamska K, Sulewska MJ (2014) Dynamic CBR test to assess the soil compaction. J Test Eval 43(05):1028–1036
23. Junhui Z, Le D, Ling Z et al (2020) Using portable falling weight deflectometer to determine treatment depth of subgrades in highway reconstruction of Southern China. Transp Saf Environ 2(1):18–28
24. Shangguan P, A1-Qadi I, Leng Z et al (2013) Innovative approach for asphalt pavement compaction monitoring with ground-penetrating radar. Transp Res Rec J Transp Res Board 79–87
25. Shuangguan P, A1-Qadi I, Coenen A et al (2016) Algorithm development for the application of ground-penetrating radar on asphalt pavement compaction monitoring. Int J Pavement Eng 17(03):189–200
26. Islam T, Chink Z, Mustafa MM et al (2012) Modeling of electrical resistivity and maximum dry density in soil compaction measurement. Environ Earth Sci 67(5):1299–1305
27. Arvidsson J, Keller T (2007) Sill stress as affected by wheel load and tyre inflation pressure. Soil Till Res 96:284–291
28. Lamandé M, Schjønning P, Tøgersen FA (2007) Mechanical behaviour of an undisturbed soil subjected to loadings: effects of load and contact area. Soil Tillage Res 97:91–106
29. Lamandé M, Schjønning P (2011) Transmission of vertical stress in a real soil profile. Part II: effect of tyre size, inflation pressure and wheel load. Soil Tillage Res 114:71–77
30. Ansorge D, Godwin R (2007) The effect of tyres and a rubber track at high axle loads on soil compaction, Part 1: single axle-studies. Biosyst Eng 98:115–126
31. Ansorge D, Godwin R (2008) The effect of tyres and a rubber track at high axle loads on soil compaction—Part 2: Multi-axle machine studies. Biosyst Eng 99:338–347
32. Klos M, Waszczyszyn Z (2010) Prediction of compaction characteristics of granular soils by neural networks. In: Artificial neural networks-ICANN 2010, lecture notes in computer science. Springer, Berlin, Heidelberg, pp 42–45
33. Patel S, Mani I (2011) Effect of multiple passes of tractor with varying normal load on subsoil compaction. J Terramech 48:277–284
34. Raper RL, Reeves DW (2007) In-row subsoiling and controlled traffic effects on coastal plain soils. Trans ASABE 50:1109–1115
35. Arman K (2008) Prediction of soil compaction under pneumatic tires a using fuzzy logic approach. J Terramech 45:103–108
36. Bayat H, Neyshabouri MR, Hajabbasi MA, Mahboubi AA, Mosaddeghi MR (2008) Comparing neural networks, linear and nonlinear regression techniques to model penetration resistance. Turk J Agric For 32:425–433
37. Xu QW, Chang GK, Gallivan VL (2015) A sensing-information-statistics integrated model to predict asphalt material density with intelligent compaction system. IEEE/ASME Trans Mechatron 20(06):3204–3211
38. Cosanti B, Presti DCFL, Squeglia N (2015) An innovative method to evaluate degree of compaction of river embankments. Eng Geol Soc Territ 2:853–856
39. Liu DH, Sun J, Zhong DH et al (2012) Compaction quality control of earth-rock dam construction using real-time field operation data. J Constr Eng Manag 138(09):1085–1094
40. Awed A, Kassem E, Masad E et al (2015) Method for predicting the laboratory compaction behavior of asphalt mixtures. J Mater Civ Eng 27(11):04015016
41. Li H, Schindler C (2013) Analysis of soil compaction and tire mobility with finite element method. Proc Inst Mech Eng Part K J Multi Body Dyn, 227:275–291
42. Xia K (2014) Numerical prediction of soil compaction in geotechnical engineering. Comptes Rendus Mécanique 342:208–219
43. Chen J, Huang B, Shu X et al (2014) DEM simulation of laboratory compaction of asphalt mixtures using an open source code. J Mater Civ Eng 27(03):04014130

44. Zolotarevskaya DI (2011) Mathematical simulation and calculation of the soil compaction under dynamic loads. Eurasian Soil Sci 44:407–416
45. Shoaib M, Kari L (2012) Non-linear elasto-plastic shock wave simulation in high-velocity compaction by discrete element method. AIP Conf Proc 1474:171–174
46. Ghanbari E, Hamidi A (2014) Numerical modeling of rapid impact compaction in loose sands. Géoméch. Eng. 6:487–502
47. Shangguan P, Al-Qadi I, Lahouar S (2014) Pattern recognition algorithms for density estimation of asphalt pavement during compaction: a simulation study. J Appl Geophys 107:8–15
48. Zhang YQ (2007) Discussion on several problems in compaction quality control of roller compacted earth-rock dam. China Rural Water Hydropower 6:122–124
49. Guo LX (2006) Some factors influencing the compaction effect of highway's roadbed. Sci Tech Inf Dev Econ 16:268–269
50. Guo XL (2014) Research of compaction quality control and rapid detection technology of the natural gravel subgrade. Master's thesis, Chang'an University, Xi'An, China
51. Liu DH, Li ZL, Wang AG (2013) Correlation analysis of rolling parameters and real-time monitoring index for rockfill dam compaction quality evaluation. J Tianjin Univ Sci Technol 46:361–366
52. Tokiharu N, Akitoshi M, Osamu S (1994) Evaluation of density from compaction tests on coarse grained soils. Doboku Gakkai Ronbunshu 28:177–185
53. Sitharam TG, Nimbkar MS (2000) Micromechanical modelling of granular materials: effect of particle size and gradation. Geotech Geol Eng 18:91–117
54. Liu LP, She XS (2006) Study on compaction property of earth-rock mixture. Chin J Rock Mech Eng 25:206–210
55. Barman M, Imran S A, Nazari M et al (2015) Intelligent compaction of stabilized subgrade of flexible pavement. IFCEE 2015 2554–2566
56. White DJ, Vennapusa PKR, Gieselman HH (2011) Field assessment and specification review for roller-integrated compaction monitoring technologies. Adv Civ Eng 2011(1):783836
57. Mooney MA, Rinehart RV (2007) Field monitoring of roller vibration during compaction of subgrade soil. J Geotech Geoenviron Eng 133(3):257–265
58. Nie Z, Wang X, Jiao T (2016) Anomalous data detection for roller-integrated compaction measurement. Int J Geomech 16(1):B4015004
59. Facas NMM (2010) Position reporting error of intelligent compaction and continuous compaction control roller-measured soil properties. J Test Eval 38(1):1
60. White D, Vennapusa P, Thompson M (2007) Field validation of intelligent compaction monitoring technology for unbound materials
61. Thompson JM, White JD (2008) Estimating compaction of cohesive soils from machine drive power. J Geotech Geoenviron Eng 134(12):1771–1777
62. White DJ, Vennapusa P, Gieselman H et al (2008) Accelerated implementation of intelligent compaction technology for embankment subgrade soils, aggregate base, and asphalt pavement materials. Transp Pooled Fund (TPF) No, 2008, 954
63. Xu Q, Chang GK (2013) Evaluation of intelligent compaction for asphalt materials. Autom Constr 30:104–112
64. White JD, Thompson JM (2008) Relationships between in situ and roller-integrated compaction measurements for granular soils. J Geotech Geoenviron Eng 134(12):1763–1770
65. Anderegg R, von Felten DA, Kaufmann K (2006) Compaction monitoring using intelligent soil compactors. In: Geocongress 2006: geotechnical engineering in the information technology age, pp 1–6
66. White DJ, Jaselskis EJ, Schaefer VR et al (2005) Real-time compaction monitoring in cohesive soils from machine response. Transp Res Rec 1936(1):172–180
67. Komandi G (1999) An evaluation of the concept of rolling resistance. J Terramech 36(3):159–166
68. Brandl H, Adam D (1999) Sophisticated continuous compaction control of soils and granular materials. In: International conference on soil mechanics and foundation engineering, pp 31–36

69. Kloubert HJ (2002) Asphalt manager with high efficient compaction system for better roads. In: 3rd international conference bituminous mixtures and pavements, Thessaloniki, Greece
70. Rinehart RV, Mooney MA (2008) Instrumentation of a roller compactor to monitor vibration behavior during earthwork compaction. Autom Constr 17(2):144–150
71. White DJ, Morris M, Thompson M (2006) Power-based compaction monitoring using vibratory padfoot roller. In: GeoCongress 2006: geotechnical engineering in the information technology age, pp 1–6
72. Vennapusa RKP, White JD, Morris DM (2010) Geostatistical analysis for spatially referenced roller-integrated compaction measurements. J Geotech Geoenviron Eng 136(6):813–822
73. Liu D, Li Z, Lian Z (2014) Compaction quality assessment of earth-rock dam materials using roller-integrated compaction monitoring technology. Autom Constr 44:234–246
74. Meehan LC, Cacciola VD, Tehrani SF et al (2017) Assessing soil compaction using continuous compaction control and location-specific in situ tests. Autom Constru 73:31–44
75. Fan J, Song XD, Tian LF et al (2015) Study on factors of continuous compaction measured value in subgrade of high-speed railway. J Railw Sci Eng 12(03):463–467
76. National Energy Administration of the People's Republic of China (2009) DL/T 5128-2009 code for construction of concrete faced rockfill dams. China Electric Power Press, Beijing
77. Liu DH, Li ZL, Wang AG (2014) Roller working-based real-time monitoring and rapid assessment of rock-fill dam compaction quality. J Hydraul Eng 45:1223–1230
78. Zhu XY, Bai SJ, Xue GP, Yang J, Cai YS, Hu W, Jia XY, Huang BS (2018) Assessment of compaction quality of multi-layer pavement structure based on intelligent compaction technology. Constr Build Mater 161:316–329
79. Liu DH, Gao L, Lin M, Li ZL (2018) Real-time monitoring and assessment of compaction quality for highway asphalt pavements. J Hohai Univ Nat Sci 46:307–313
80. Wu LL, Jiang HH, Tang JW, Gao MX, Fan SF, Yan XX (2020) Continuous compaction monitoring technology based on multiple regression analysis. Rock Soil Mech 41:2081–2090
81. An ZZ, Liu TY, Huangfu ZH, Zhang QL, Li CS, Li QB (2020) Neural network model for evaluating compaction quality of rockfill materials by compaction meter value. J Hydroelectr Eng 39:110–120
82. Liu M, Wang X, Wang J, Cui B, Deng B, Shi M (2019) Comprehensive evaluation for real-time compaction quality using i-AHP and i-GAM: case study of earth-rock dam. Appl Sci 9:1543
83. Wang J, Zhong D, Wu B, Shi M (2018) Evaluation of compaction quality based on SVR with CFA: case study on compaction quality of earth-rock dam. J Comput Civ Eng 32:05018001
84. Wang J, Zhong D, Adeli H, Wang D, Liu M (2018) Smart bacteria-foraging algorithm-based customized kernel support vector regression and enhanced probabilistic neural network for compaction quality assessment and control of earth-rock dam. Expert Syst 35:e12357
85. Liu DH, Wu Y (2019) Fuzzy comprehensive evaluation of earth-rock dam compaction quality based on real-time monitoring. J Hydroelectr Eng 38:142–153
86. Wang F, Zhong D-H, Yan Y-L, Ren B-Y, Wu B-P (2018) Rockfill dam compaction quality evaluation based on cloud-fuzzy model. J Zhejiang Univ A 19:289–303
87. Yao YP, Liu BY (2016) An evaluation algorithm of compaction quality for high embankments airports. Ind Constr 46:42–45
88. Wang JJ, Zhong DH, Wu BP, Liu MH, Zhang ZL (2019) Method of updating compaction quality evaluation model of earth-rock dam using concept drift detection. J Tianjin Univ Sci Technol 52:492–500
89. Wang XL, Zhou L, Ren BY, Cui B, Zhang NM (2015) Dual evaluation on rolling compaction quality of rock-fill dam based on real-time monitoring. J Hydroelectr Eng 34:164–170
90. Xu QW, Chang GK, Gallivan VL (2012) Development of a systematic method for intelligent compaction data analysis and management. Constr Build Mater 37:470–480
91. Liu DH, Lin M, Li S (2016) Real-time quality monitoring and control of highway compaction. Autom Constru 62:114–123
92. Dou P (2014) Correlation verification experimental investigation of high-railway subgrade compaction quality test value. Central South University

93. Wang J (2007) A method study of rapid determination on compaction quality of soil-rock embankment. Shanghai Jiao Tong University
94. Yoon S, Hastak M, Lee J (2015) Intelligent compaction of asphalt pavement implementation. In: Joint transportation research program Publication No. FHWA/IN/JTRP-2015/05
95. Thurner H, Sandström A (1980) A new device for instant compaction control. In: Proceedings of the international conference on compaction, Paris, France, vol 2, pp 611–614
96. Hua T, Yang X, Yao Q, Li H (2018) Assessment of real-time compaction quality test indexes for rockfill material based on roller vibratory acceleration analysis. Adv Mater Sci Eng 2018:1–15
97. Kröber W (1988) Untersuchung der dynamischen Vorgänge bei der Vibrationseverdichtung von Böden. PhD dissertation, Schriftenreihe, Helf 11, Lehrstuhl und Prüfamt für Grundbau, Bodenmechanik und Felsmechanic der Technischen Universität München (in German)
98. Du G, Chen B, Yang H (2012) Application of GPS roller compaction monitoring system in the construction quality management of earth-rock dams. Des Hydroelectr Power Stat 28:1–6
99. Floss R, Gruber N, Obermayer J (1983) A dynamical test method for continuous compaction control. In: Rathmayer HG, Saari KHO (eds) Proceedings of 8th European conference on soil mechanics and foundation engineering, Helsinki, pp 25–30
100. Duan LQ (2009) Study on continuous compaction quality control of rock-fill embankment. Southwest Jiaotong University
101. Do W, Oloufa A, Thomas HR (1999) Evaluation of global positioning system devices for a quality control system for compaction operations. Transp Res Rec 1675:67–74
102. Shangguan P, Al-Qadi I, Coenen A et al (2016) Algorithm development for the application of ground-penetrating radar on asphalt pavement compaction monitoring. Int J Pavement Eng 17(03):189–200
103. Lefebvre G, Karray M (1998) New developments in in-situ characterization using Rayleigh waves. In: Proceedings of the 51st Canadian geotechnical conference, Edmonton, Alta., BiTech Publishers Ltd., Richmond, B.C, pp 821–828
104. Oloufa A (2002) Qualtiy control of asphalt compaction using GPS-based system architecture. IEEE Robot Autom Mag 9:29–35
105. Karray M (1999) Utilisation de l'analyse modale des ondes de Rayleigh comme outil d'investigation géotechinique in-situ. Thèse de doctorat en génie civil (PhD thesis), Département de génie, Université de Sherbrooke, Sherbrooke, Que
106. Oloufa A, Do W, Thomas HR (1997) Autmated monitoring of compaction using GPS. In: Proceedings of the 1997 5th ASCE construction congress: managing engineered construction in expanding global markets, pp 1004–1011
107. Huang SX, Liu JN, Wu XM (2005) GPS real-time supervisory system and its preliminary application in the construction of face rockfill dam. Geomat Inf Sci Wuhan Univ 30(09):813–816
108. Bharath KK, Hui-Ping T, Robert LS et al (1998) AutoPave: towards an automated paving system for asphalt pavement compaction operation. Autom Constr 8:165–180
109. Chiang CH, Lai J, Wu S (2011) Evaluating the compaction quality of backfills by stress wave velocities. J Test Eval 39(05):1–7
110. Facas NW, van Susante PJ, Mooney MA (2010) Influence of rocking motion on vibratory roller-based measurement of soil stiffness. J Eng Mech 136(07):898–905
111. Zhong DH, Liu DH, Cui B (2011) Real-time compaction quality monitoring of high core rockfill dam. Sci China Technol Sci 54(07):1906–1913
112. Lin SM, Wang GY (2018) An intelligent rolling compaction system for subgrade and pavement. CN Patent CN109356003A
113. China Electricity Council (2013) D/T5129-2013-specifications for rolled earth-rockfill dam construction. China Electric Power Press, Beijing, China
114. Research Institute of Highway Ministry of Transport (2015) JTG/T F20–2015-technical guidelines for construction of highway roadbases. China Communications Press, Beijing, China
115. CCCC Third Highway Engineering Co. Ltd (2015) JTG/T 3610-2019-Technical specifications for construction of highway subgrades. China Communications Press, Beijing, China

116. FHWA Generic Soils IC Specification (2014) Intelligent compaction technology for soil application; federal highway administration. US, Department of Transportation, Washington, DC, USA
117. Southwest Jiaotong University (2015) Q/CR 9210–2015-technical specification for continuous compaction control of fill engineering of railway earth structure. China Railway Publishing House, Beijing, China
118. Yao Y-P, Ruan Y-Z, Chen J, Geng Y, Zhang X, Liu B-Y, Zong X-P, Yu G-Z (2018) Research on a real-time monitoring platform for compaction of high embankment in airport engineering. J Constr Eng Manag 144:04017096
119. Adam, D (1997) Continuous Compaction Control (CCC) with vibratory rollers. Environmental geotechnics. In: Balkema AA (ed) Proceedings of the 1st Australia-New Zealand conference on environmental geotechnics-geoenvironment 97, Melbourne, Victoria, Australia, 26–28 November 1997, Rotterdam, The Netherlands, pp 245–250
120. Pistrol J, Villwock S, Völkel W, Kopf F, Adam D (2016) Continuous Compaction Control (CCC) with oscillating rollers. Procedia Eng 143:514–521
121. White DJ, Vennapusa PK (2010) A review of roller-integrated compaction monitoring technologies for earthworks. Final report ER10-04. Iowa State University, Ames, IA, USA, pp 1–36
122. Briaud J, Seo J (2003) Intelligent compaction: overview and research needs. Texas A&M University, College Station, TX, USA, pp 1–84
123. Anderegg R, Kaufmann K (2004) Intelligent compaction with vibratory rollers: feedback control systems in automatic compaction and compaction control. Transport Res Rec 1868:124–134
124. Thurner HF, Sandström A (1989) Compaction meter and compaction documentation system. Unbound Aggreg Roads 46–51
125. Pampagnin LH, Peyret F, Garcia G (1988) Architecture of a GPS-based guiding system for road compaction. In: Proceedings of the 1988 IEEE international conference on robotics and automation, Philadelphia, PA, USA, 24–29 April 1988, pp 2422–2427
126. Liu DH, Cui B, Liu YG, Zhong DH (2013) Automatic control and real-time monitoring system for earth-rock dam material truck watering. Autom Constr 30:70–80
127. Liu D, Chen J, Li S (2018) Collaborative operation and real-time control of roller fleet for asphalt pavement compaction. Autom Constr 98:16–29
128. Zhong DH, Li XC, Cui B, Wu BP, Liu YX (2018) Technology and application of real-time compaction quality monitoring for earth-rockfill dam construction in deep narrow valley. Autom Constr 90:23–38
129. Zeng HE, Liu JP (2018) Preliminary experiment study on Georobot/WLAN-based intelligent monitoring system for roller compaction. Water Resour Hydropower Eng 49:141–147
130. Huang S, Zhang W (2018) A fast calculation method of rolling times in the GNSS real-time compaction quality supervisory system. Adv Eng Softw 128:20–33
131. Xu J, Liu FP, Wang JX (2020) Research and application of continuous compaction monitoring system based on BDS. Agric Eng 10:59–63
132. Meehan CL, Khosravi M, Cacciola D (2013) Monitoring field lift thickness using compaction equipment instrumented with global positioning system (GPS) technology. Geotech Test J 36:755–767
133. Yu Y, Cui B, Zhou L (2013) Application of real-time monitoring and control system of roller compaction quality to construction of South Xinzheng section of mid-route of south-to-north water transfer project. Water Resour Hydro Power Eng 44:73–76
134. Ma HQ (2012) Key technologies on construction quality control of rockfill dam with core wall filled by clay soil mixed with gravel in Nuozhadu hydropower station project. Water Power 38:12–15
135. White DJ, Vennapusa PKR (2009) Report of the workshop on intelligent construction for earthworks; EERC Publication ER09–02; Sheraton West Des Moines Hotel: West Des Moines, IA, USA, pp 1–156

136. Mooney M, Adam D (2017) Vibratory roller integrated measurement of earthwork compaction: an overview. In: Proceedings of the FMGM 2007: seventh international symposium on field measurement in geomechanics, Boston, MA, USA, 27 September 2017, pp 1–12
137. Hu W, Shu X, Jia XY, Huang BS (2018) Geostatistical analysis of intelligent compaction measurements for asphalt pavement compaction. Autom Constr 89:162–169
138. Wang SQ, Zhao S, Al-Qadi IL (2018) Continuous real-time monitoring of flexible pavement layer density and thickness using ground penetrating radar. NDT E INT. 100:48–54
139. Barman M, Nazari M, Imran SA, Commuri S, Zaman M (2016) Quality improvement of subgrade through intelligent compaction. Transp Res Rec J Transp Res Board 2579, 59–69
140. Sun M (1985) Automatic control device and rolling driving method of Japanese road rolling machinery. Road Mach Constr Mech 4:34–37
141. Yao D, Xie H, Qiang W, Liu Y, Xiong S (2018) Accurate trajectory tracking with disturbance-resistant and heading estimation method for self-driving vibratory roller. IFAC PapersOnLine 51:754–758
142. Song E-B, Zhang X (2020) Research on the path optimization of unmanned rolling impaction for high embankment of airport. Jpn Geotech Soc Spéc Publ 8:142–148
143. Zhang X, Luo T, Song EB, Geng Y (2020) Algorithm for optimal path planning of impact roller in high-embankment air-port. Jpn Geotech Soc Spec Publ 8:159–163
144. Husemann J, Wolf P, Vierling A, Berns K, Decker P (2020) Towards high-quality road construction: using autonomous tandem rollers for asphalt compaction optimization. In: Proceedings of the 37th international symposium on automation and robotics in construction, Kitakyshu, Japan, 27 October 2020
145. Song K, Xie H (2020) Path-following control for unmanned rollers: a composite disturbance rejection-based framework. In: Proceedings of the 2020 IEEE 9th data driven control and learning systems conference, Liuzhou, China, 20–22 November 2020
146. Huang WZ, Wang D, Xiao ZD, Yao Q, Du DJ (2020) Design and implementation of PLC-based autonomous construction system of unmanned vibratory roller. Comput Mater Continua 65:897–912
147. Bian YM, Fang XJ, Yang M, Zhao Z (2017) Automatic rolling control for unmanned vibratory roller based on fuzzy algorithm. J Tongji Univ Nat Sci 45:1830–1838
148. Zou B, Yu SL, Wang KW (2019) Obstacle detection for automatic driving roller compacting vehicle. Digit Manuf Sci 17:245–250
149. Teng YN, Li XP, Ma H, Wen BC (2009) Study on vibration friction mechanism and vibration response analysis based on vibration compaction system. Appl Mech Mater 16–19:84–87
150. Beainy F, Commuri S, Zaman M, Syed I (2013) Viscoelastic-plastic model of asphalt-roller interaction. Int J Géoméch 13:581–594
151. Fang X, Bian Y, Yang M, Liu G (2018) Development of a path following control model for an unmanned vibratory roller in vibration compaction. Adv Mech Eng 10:1–16
152. Chen ZY, Zhao YF, Zhou B, Jiang L, Zhao HM (2019) Study and application of unmanned driving technology or filling and rolling construction of earth-rockfill dam. Water Resour Hydropower Eng 50:1–7
153. Shi MN, Cui B, Wang JJ, Guan T, Tong DW, Ren BY (2020) Research on cooperative complete coverage path planning for unmanned roller group under complex construction conditions. J Hydraul Eng 51:1544–1557
154. Shi GS (2020) Implementation and application of unmanned rolling dam construction technology for high arch dam. Build Tech Dev 47:31–33
155. Yang M, Bian Y, Liu G, Zhang H (2020) Path tracking control of an articulated road roller with sideslip compensation. IEEE Access 8:127981–127992
156. Zhang Q, An Z, Liu T, Zhang Z, Huangfu Z, Li Q, Yang Q, Liu J (2020) Intelligent rolling compaction system for earth-rock dams. Autom Constr 116:103246
157. Beainy F, Commuri S, Zaman M (2014) Dynamical response of vibratory rollers during the compaction of asphalt pavements. J Eng Mech 140:04014039

158. Zhang QZ (2010) Study on the Model and Parameters of the Vibration Compaction System in the Soil Subgrade. PhD thesis, Chang'an University, Xi'An, China
159. Wersäll C, Nordfelt I, Larsson S (2017) Soil compaction by vibratory roller with variable frequency. Géotechnique 67:272–278
160. Wersäll C, Nordfelt I, Larsson S (2018) Resonant roller compaction of gravel in full-scale tests. Transp Geotech 14:93–97
161. Li QB, Zhang ZS, Liu TY, Huangfu ZH, Zhang QL, Ma WM, Li CS, An ZZ (2020) Review of vibration compaction model of soil and rock materials. Yellow River. 42:105–111
162. Yoo T-S, Selig ET (1979) Dynamics of vibratory-roller compaction. J Geotech Eng Div 105:1211–1231
163. Gong T (2013) Dynamic analysis and vibration comfort research for vibratory roller. PhD thesis, Xiangtan University, Xiangtan, China
164. Paulmichl I, Adam D, Adam C, Völkel W (2020) Assessment of a compaction indictor for oscillation rollers with a lumped parameter model. Geotech Eng 173:302–318
165. Zheng SH, Lin SW (2011) Dynamical property modeling and analysis on intelligent vibratory rollers. Chin J Constr Mach 9:398–403
166. Shen PH, Lin SW (2008) Coupling analysis for integrated dynamics characteristics of intelligent compaction system. China Mech Eng 19:2395–2399
167. Xin LL, Liang JH, Wen BC (2008) Dynamic property analysis on pulsating vibratory roller. Chin J Constr Mach 6:161–166
168. He C, Zhang ZF (1992) Numerical simulation of the condition of jump of the roller and choice of compactor's mechanical and vibrational parameters. J Wuhan Inst Technol 14:7–19
169. Rose KD, Kumar K, Rana RS, Sahni A, Smith T (1978) Vibratory compaction of bituminous mixes in France. Int J Rock Mech Min Sci Geomech Abstr 15, A118
170. Wersäll C, Larsson S, Bodare A (2014) Dynamic response of vertically oscillating foundations at large strain, computer methods and recent advances in geomechanics. In: Proceedings of the 14th international conference of international association for computer methods and recent advances in geomechanics, London, UK, 13 December 2014, pp 643–647
171. Liu H, Xue R (2005) On legal affairs concerning side-standing supervision. J Shanxi Univ Philos Soc Sci 28:50–53
172. Pietzsch D, Poppy W (1992) Simulation of soil compaction with vibratory rollers. J Terramech 29:585–597
173. Gong CX, Liu BX, Gong T (2014) Dynamic modeling and parameter optimization of intelligent vibratory roller. Mech Sci Technol Aerosp Eng 33:27–31
174. Thurner H (2001) Quality assurance and self control in road construction, advanced measurement technology. In Proceedings of the 4th international symposium infrastructure construction systems and technologies, Bauma, Munich, Germany, pp 73–81
175. Adam D, Kopf F (2000) Theoretical analysis of dynamically loaded soil. In: European workshop compaction of soils and granular materials; Presses Ponts et Chaussies, Paris, France, pp 207–220
176. Imran SA, Commuri S, Zaman M (2015) A 2 dimensional dynamic model of asphalt-roller interaction during vibratory compaction. In: Proceedings of the 12th international conference on informatics in control, automation and robotics (ICINCO), Colmar, France, 10 December 2015, pp 533–540
177. Imran SA (2016) Modeling and analysis of the interaction between roller drum and pavement material during compaction. PhD thesis. University of Oklahoma, Oklahoma, Norman
178. Li J (2012) Study on control techniques of the vibratory roller based on soil compactness characteristic. PhD thesis, Xi'an University of Technology, Xi'an, China
179. Zhang WG, Goh ATC (2016) Multivariate adaptive regression splines and neural network models for prediction of pile drivability. Geosci Front 7:45–52
180. Zhang W, Goh A (2012) Multivariate adaptive regression splines for analysis of geotechnical engineering systems. Comput Geotech 48:82–95
181. Zhang Y, Zhang J, Shu XZ, Guo L, Shi Y, Liu XB (2009) Optimization of intelligent compactness control rule of vibratory roller based on genetic algorithm method. In: Proceedings of the

2009 fifth international joint conference on INC, IMS and IDC, Seoul, Korea, 25–27 August 2009, pp 1943–1947
182. Isik F, Ozden G (2013) Estimating compaction parameters of fine- and coarse-grained soils by means of artificial neural net-works. Environ Earth Sci 69:2287–2297
183. Günaydin O (2009) Estimation of soil compaction parameters by using statistical analyses and artificial neural networks. Environ Geol 57:203–215
184. Ranasinghe RATM, Jaksa MB, Nejad FP, Kuo YL (2019) Prediction of the effectiveness of rolling dynamic compaction using artificial neural networks and cone penetration test data. Chin J Rock Mech Eng 38:153–170
185. Ranasinghe R, Jaksa M, Kuo Y, Nejad FP (2017) Application of artificial neural networks for predicting the impact of rolling dynamic compaction using dynamic cone penetrometer test results. J Rock Mech Geotech Eng. 9:340–349
186. Taghavifar H, Mardani A, Taghavifar L (2013) A hybridized artificial neural network and imperialist competitive algorithm optimization approach for prediction of soil compaction in soil bin facility. Measurement 46:2288–2299
187. Xu Q, Chang KG (2016) Adaptive quality control and acceptance of pavement material density for intelligent road construction. Autom Constr 6278–6288
188. Prokopev AP, Nabizhanov IZ, Emelyanov RT, Ivanchura IV (2020) Concept of a new method for continuous non-destructive control of asphalt road pavements compaction. J Phys Conf Ser 1515:1–6
189. An Z, Liu T, Zhang Z, Zhang Q, Huangfu Z, Li Q (2019) Dynamic optimization of compaction process for rockfill materials. Autom Constr 110:103038
190. Liu TY, Zhang QL, Li YD, Zhang ZS, Huangfu ZH (2017) An intelligent rolling compaction robot. CN Patent CN105002810B

Chapter 2
Roller-Integrated Compaction Monitoring Technology Based on Maximum Positive Kinetic Energy Increment

Currently, the primary materials used for subgrade and pavement construction include soil, asphalt mixtures, and soil-stone mixtures. The control of compaction quality for these filling materials is a critical aspect that directly influences the performance and service life of the filling body. Conventional methods for detecting compaction quality are typically destructive, resulting in irreversible structural damage to the filling body during implementation. These methods are also time-consuming and labor-intensive, often yield limited sample collection, and are significantly impacted by human factors, leading to considerable disturbances. Consequently, they fail to provide a real-time reflection of the compaction status across the entire working area, which does not meet the increasing demands for high efficiency and precision in construction.

Therefore, it is imperative to propose a monitoring method for the compaction quality of filling materials characterized by continuity, real-time capabilities, and high accuracy, and to subsequently develop an appropriate system based on this method. Through the study of the energy conversion patterns between the vibratory roller and the filling materials within a single cycle, it has been discovered that the trend in the change of the roller's positive kinetic energy can reflect variations in the stiffness of the filling body during the compaction process. Based on this finding, this chapter introduces an integrated compaction monitoring technology for rollers that is founded on the principle of maximizing positive kinetic energy increments.

2.1 Theoretical Analysis

The roller-integrated compaction monitoring (RICM) technology is increasingly being applied in road construction [1, 2]. RICM refers to the utilization of sensors installed on rollers to monitor vibration information, which is then processed into

Q. Zhang, *Compaction Quality Nondestructive Testing and Intelligent Control Technology for Infrastructure*, https://doi.org/10.1007/978-981-95-1264-5_2

RICM index values to represent the real-time compaction state of filling materials, such as soil, aggregates, or asphalt mixtures [3].

To date, various RICM indicators have been proposed to characterize the compaction quality of different types of filling materials, including unit volume compaction work E [1], compaction meter value (CMV) [4], material stiffness (k_s) [5], machine drive power (MDP) [6], vibration modulus (E_{vib}) [7], total harmonic distortion (THD) [8], compaction control value (CCV) [9], sound compaction value (SCV) [10] and longitudinal wave velocity (V_p) [11, 12], among others. Research has shown that these indexes exhibit a certain degree of correlation with the stiffness or modulus of filling materials. However, due to the complex interactions between the vibratory roller and the subgrade/pavement structure, these various indexes must be established on specific theoretical assumptions, which leads to certain limitations in their practical engineering applications [1].

Moreover, there is a notable lack of research regarding the applicability of RICM indicators for low liquid limit silt, characterized by low clay content, low plasticity index, and low strength. Thus, this section introduces a new compaction monitoring indicator based on maximum positive kinetic energy increment—Kinetic Energy Compaction Value ($KECV$).

(1) Unit Volume Compaction Work (E)

According to vibration theory [13], the force acting in the vertical direction on the vibratory roller during the vibration process can be expressed as follows:

$$P = W + F\sin \omega t \tag{2.1}$$

In the equation, W represents the radial load of the vibratory roller; F denotes the excitation force during vibration; ω signifies the angular velocity of the vibratory roller; and t indicates the working time of the vibratory roller.

Based on the two-degree-of-freedom vibration compaction model [14, 15], during the vibration process, a phase difference φ exists between deformation and load due to the inherent damping effect of the soil [1]. Therefore, the compaction work performed by the vibratory roller during a single vibration cycle can be expressed as follows:

$$E_0 = WA(\sin\varphi + 1) + \frac{FA}{2}\left[\left(\varphi + \frac{\pi}{2}\right)\sin\varphi + \cos\varphi\right] \tag{2.2}$$

In the equation, A represents the working amplitude of the vibratory roller. Previous studies [16] have indicated that when the vibration frequency of the roller approaches the natural frequency of the "vibratory roller-soil" system, the phase difference φ can be approximated as $2/\pi$. Substituting $\varphi = 2/\pi$ into the previous equation, the compaction work performed by the vibratory roller during a single vibration cycle can be simplified as follows:

$$E_0 = 2A\left(W + \frac{\pi F}{4}\right) \tag{2.3}$$

After n passes of compaction, the unit volume compaction work of the compacted filling material is given by:

$$E = 2A\left(W + \frac{\pi F}{4}\right)\frac{fn}{Bhv} \tag{2.4}$$

In the equation, B denotes the width of the vibratory roller; f represents the vibration frequency; h indicates the thickness of the filling layer; and v signifies the driving speed.

(2) Kinetic Energy Compaction Value ($KECV$)

Research [2] has demonstrated that the interaction process between the vibratory roller and the soil is a dynamic energy transformation process. Within a single vibration cycle, the vibratory roller exhibits three distinct downward movement states: starting from its initial position (Position 0, as illustrated in Fig. 2.1), assuming that the downward direction is positive, the first movement state of the roller is determined by the excitation force and radial load, resulting in an increase in the roller's positive velocity and kinetic energy.

Upon reaching the second movement state, the roller makes contact with the soil (as shown in Position 1 of Fig. 2.1) and continues its downward motion. During this interaction, the soil is compressed, while the net vertical force acting on the roller begins to decrease, leading to a deceleration in the rate of increase of the roller's kinetic energy until it no longer increases and reaches its maximum.

In the third movement state, the roller continues its downward motion, generating a net vertical force directed upward, which causes a reduction in the roller's kinetic energy, ultimately bringing it to a stop at zero, followed by an observable upward rebound of the roller. Throughout the contact process between the vibratory roller and the soil, the soil is compressed, absorbing energy and thereby enhancing its own stiffness.

During the initial compaction phase, the soil exhibits a relatively loose state (Fig. 2.1a), and the acceleration signal shows a relatively smooth curve (Fig. 2.1d). At this stage, the soil's stiffness is low, resulting in a significant absorption of energy during compression, which leads to a smaller increment in the roller's positive kinetic energy (Fig. 2.1g).

However, as the number of passes increases, the compaction level of the filling material gradually rises (Fig. 2.1b), accompanied by an emergence of higher harmonic components that steadily increase in number (Fig. 2.1e), challenging the compression of the soil. Despite the increase in both radial load and excitation force on the roller with the rising amplitude, the proportion of energy absorbed by the soil during compression decreases, while the proportion of the roller's kinetic energy increases (Fig. 2.1h).

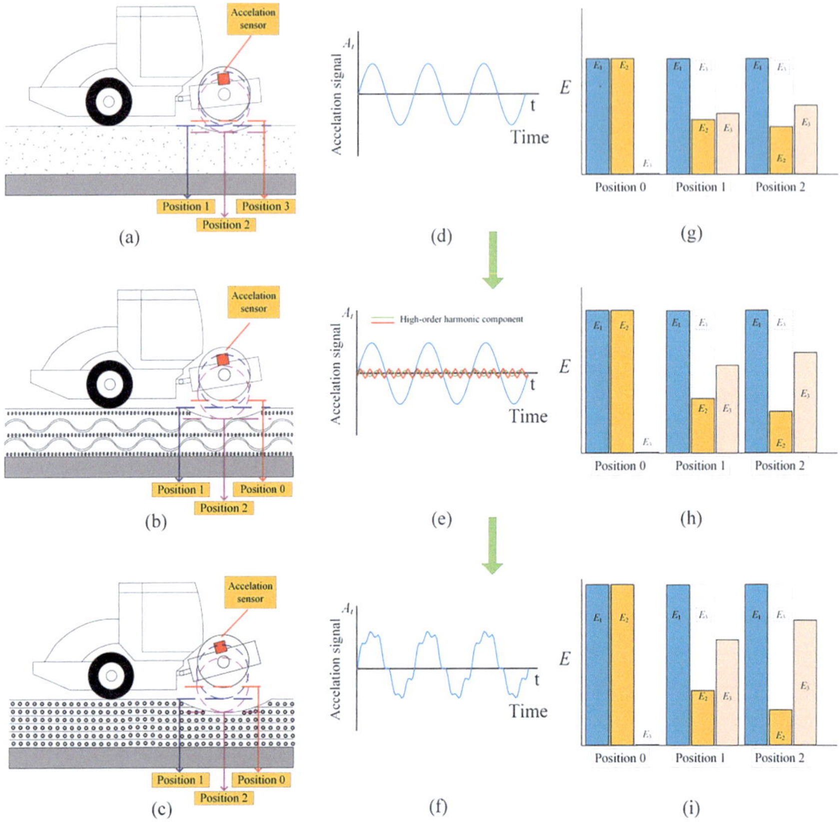

Fig. 2.1 Variation of work done by the vibratory roller with changes in soil stiffness

As the number of passes reaches a certain threshold, the filling material becomes denser and gradually stabilizes (Fig. 2.1c), with the higher harmonic components also tending toward stability (Fig. 2.1f). The balance between the energy expended for soil compression and the fluctuations in the vibratory roller's kinetic energy becomes evident, resulting in the stabilization of the roller's kinetic energy (Fig. 2.1i).

① Compaction Work E_1

During a single vibration cycle, the compaction work performed by the vertical force as the vibratory roller moves downward from its highest position (Position 0, as shown in Fig. 2.1) to Position 2 (as depicted in Fig. 2.1) is denoted as E_1. When the vibratory roller falls from its highest point at time t_1, transitions to motion state two at time t_2, and reaches Position 2 at time t_3, the vertical net force acting on the roller is zero, and the positive kinetic energy is at its maximum. Therefore, based on Eqs. (2.1) and (2.2), E_1 can be expressed as:

$$E_1 = -\int_{t_1}^{t_2} (W + F\sin\omega t)A\omega\cos(\omega t - \varphi)dt + \int_{t_2}^{t_3} (W + F\sin\omega t)A\omega\cos(\omega t - \varphi)dt$$
$$= WA(2\sin\varphi - 1) + \frac{FA}{4}(4\varphi\sin\varphi - \pi\sin\varphi + 2\cos\varphi) \tag{2.5}$$

To eliminate the influence of the roller's initial startup, from ($t_1 = (3\pi/2)/\omega$) to ($t_2 = 2\pi/\omega$), and from ($t_2 = 2\pi/\omega$) to ($t_3 = (2\pi + \varphi)/\omega$), the vibratory roller moves downward from its highest position, with its velocity continually increasing until it reaches its maximum positive velocity. After (t_3), as the roller continues its downward motion, the velocity will decrease to zero as it reaches the vertical lowest point.

Substituting ($\varphi = 2/\pi$) into Eq. (2.5), the compaction work performed by the vibratory roller during the attainment of maximum positive kinetic energy can be simplified as follows:

$$E_1 = A\left(W + \frac{\pi F}{4}\right) \tag{2.6}$$

Let the discrete acceleration signal of the vibratory roller be denoted as $\{a(n)\}$ (where ($n = 0, 1, 2, \ldots, N - 1$)), with ($N$) representing the sample size and (f_s) the sampling frequency. Through frequency domain integration, the acceleration signal of the vibratory roller can be transformed into the vertical displacement signal of the roller, which corresponds to the amplitude [17]:

$$A(n) = \sum_{k=0}^{N-1} -\frac{1}{(2\pi k\Delta f)^2} H(k)X(k)e^{j2\pi kn/N} \tag{2.7}$$

where:

$$H(k) = \{ \begin{matrix} 1 & (f_d \le k\Delta f \le f_u) \\ 0 & (其他) \end{matrix} \tag{2.8}$$

In the equation, (f_d) and (f_u) represent the lower and upper cutoff frequencies, respectively; ($X(k)$) denotes the Fourier transform of ($a(n)$); and ($\Delta f = f_s/N$) signifies the frequency resolution. Furthermore, research has indicated that as the soil stiffness increases, both the vibratory roller's acceleration and displacement amplitudes gradually increase, meaning that with each additional pass of compaction, both (A) and (W) progressively rise. The magnitude of the excitation force can be calculated using the excitation force frequency and Eq. (2.9) [1].

$$F = \frac{Me4\pi^2 f^2}{Me4\pi^2 f_R^2} F_R = \frac{f^2}{f_R^2} F_R \tag{2.9}$$

In the equation, (f_R) and (F_R) correspond to the rated excitation force frequency and rated excitation force of the compactor in its output state, with specific values determined by the model of the compactor. During the compaction process, the actual working frequency of the vibratory roller remains relatively constant, and the calculated actual excitation force (F) also remains largely unchanged, resulting in minimal impact on the variation of the compaction work (E_1). Through the aforementioned analysis, it is evident that as the number of compaction passes increases, (E_1) gradually rises.

② The energy absorbed by the compressed soil E_2

When the soil is subjected to the stress ($\sigma(t) = \sigma_s \cos(\omega_s t)$), the energy absorbed during compression can be simplified to a calculation of viscous energy absorption based on the Kelvin model [18], which is expressed as follows:

$$E_2 = \frac{\sigma_s^2}{2} \frac{\eta \omega_s^2}{E_s^2 + \eta^2 \omega_s^2} V_s \tag{2.10}$$

In the equation, (σ_s) represents the stress amplitude; (ω_s) denotes the angular velocity of the filling material's vibration; (E_s) is the composite elastic modulus of the soil; (η) signifies the composite viscosity coefficient of the soil; and (V_s) refers to the volume of the soil involved in the compression [19].

According to existing research [20, 21], the influence of compactness and moisture content on the viscosity coefficient (η) of the soil is minimal, allowing (η) to be treated as a constant throughout the compaction process. Previous studies [18] have indicated that the stress amplitude (σ_s) is related to the output energy of the compactor during the compaction process. Generally, if the output energy of the compactor remains constant during compaction, then the stress amplitude (σ_s) applied to the soil also remains unchanged. As the compaction progresses, the stiffness of the soil increases, resulting in a continuous rise in its natural frequency $\sqrt{k/m}$ (where (k) and (m) represent the stiffness and mass of the soil, respectively) [15]. Furthermore, with an increasing number of compaction passes, the elastic modulus (E_s) gradually increases [1], leading to a gradual reduction in the volume of the same layer of filling material.

From the above analysis, it can be concluded that as the number of compaction passes increases, based on Eq. (2.11), (E_s^2/ω_s^2) gradually increases while (E_2) gradually decreases.

$$E_2 = \frac{\sigma_s^2}{2} \frac{\eta}{\frac{E_s^2}{\omega_s^2} + \eta^2} V_s \tag{2.11}$$

③ The increment of maximum positive kinetic energy of the vibratory roller E_3

During a single vibration cycle, when the vertical net force on the vibratory roller is zero, the increment of the roller's positive kinetic energy reaches its maximum, expressed as follows:

$$E_3 = \Delta W \tag{2.12}$$

$$\Delta W = E_{k1} - E_{k0} \tag{2.13}$$

In the equation, (ΔW) represents the increment of maximum positive kinetic energy of the vibratory roller; (E_{k0}) is the kinetic energy of the roller at its highest point; and (E_{k1}) denotes the maximum kinetic energy of the roller in the vertical direction.

④ Kinetic Energy Compaction Value

According to existing research [22], the energy losses in the form of sound, light, and heat produced by native geological materials such as soil and rock under dynamic loading impacts can be considered negligible. Therefore, during the compaction process, the compaction work (E_1), the energy absorbed by the soil during compression (E_2), and the kinetic energy of the vibratory roller (E_3) are interconvertible, with their total quantity being conserved. Based on the law of conservation of energy, the relationship among the three is expressed in Eq. (2.14).

$$E_1 = E_2 + E_3 \tag{2.14}$$

From the above analysis, it is evident that as the number of compaction passes increases, (E_1) gradually increases while (E_2) gradually decreases. According to the law of conservation of energy, it can be concluded that (E_3) will gradually increase, as illustrated in Fig. 2.1g–i. Therefore, the variation in the kinetic energy of the vibratory roller can reflect changes in the soil stiffness and the compaction quality.

Based on the aforementioned principles, the integrated compaction monitoring index for the compactor, derived from the maximum positive kinetic energy increment, is termed the Kinetic Energy Compaction Value ($KECV$). The specific calculation formula is as follows:

$$KECV = \overline{E_{k_{\max}}} \tag{2.15}$$

In the equation, the term $\overline{E_{k_{\max}}}$ represents the average increment of maximum positive kinetic energy of the vibratory roller over (M) cycles, with the calculation formula as follows:

$$\overline{E_{k_{\max}}} = \frac{E_{3_1} + E_{3_2} + \cdots + E_{3_i} + \cdots + E_{3_M}}{M} \tag{2.16}$$

In the equation, the term E_{3_i} represents the increment of maximum positive kinetic energy for the (i)-th cycle, where ($i = 0, 1, 2, \ldots, M$).

2.2 Roller-Integrated Compaction Monitoring System

The roller-integrated compaction monitoring system utilizes collection devices installed on the roller to monitor the information of the vibratory roller, determining the corresponding RICM indexes. By combining this data with the real-time location information of the vibratory roller recorded via satellite navigation technology, the system employs visualization techniques (such as color-coding) to display compaction parameters, roller positions, and RICM indexes in real time. On-site construction personnel can utilize this real-time cloud map to assess the next steps in the construction process. Additionally, satellite navigation technology facilitates the long-distance transmission of cloud map data and other information, ensuring continuous communication and timely feedback with remote terminals.

2.2.1 System Configuration

The roller-integrated compaction monitoring system is primarily composed of three components: the positioning system, the collection system, and the analysis system, as illustrated in Fig. 2.2. Among these, the flow station on the roller is equipped with two BDS receivers, which are responsible for determining the location and angular orientation.

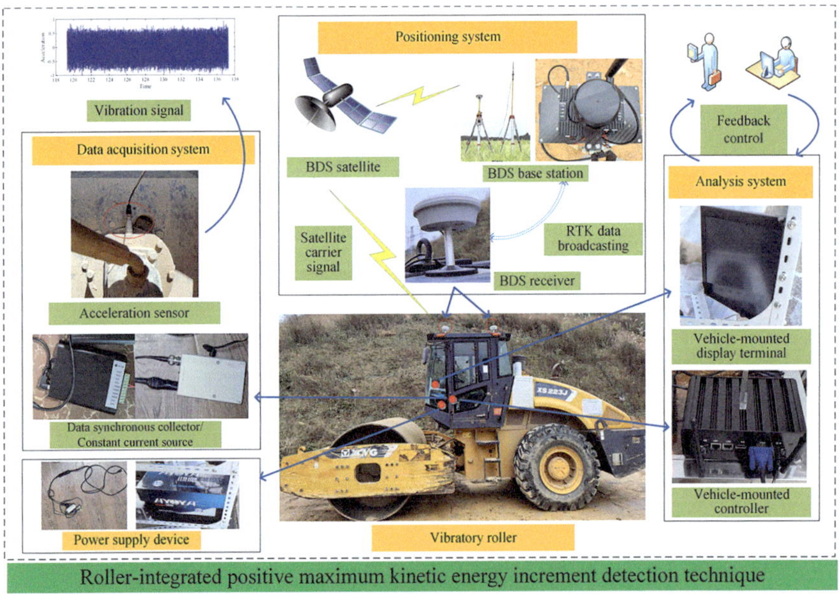

Fig. 2.2 Roller-integrated compaction monitoring system

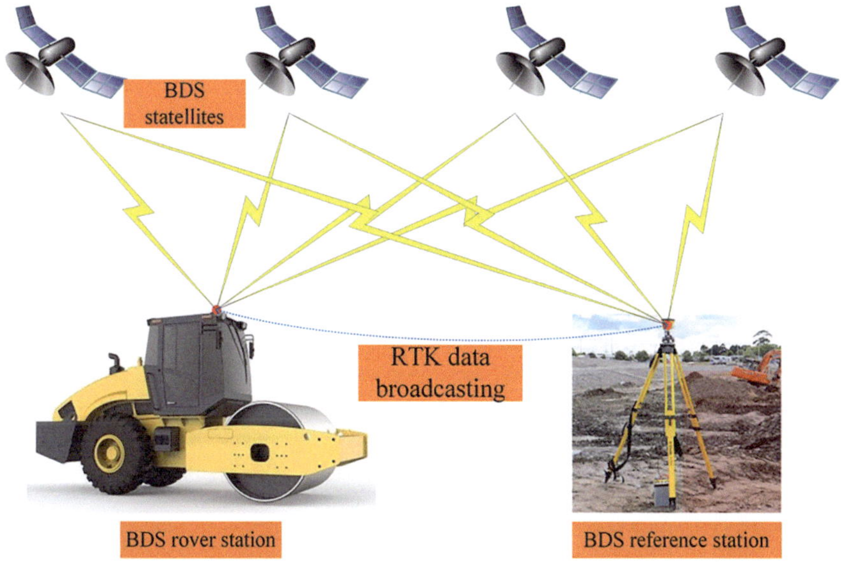

Fig. 2.3 Principle of BDS-RTK technology implementation

(1) Positioning System

The positioning system primarily consists of Beidou satellites, a Beidou reference station, and two Beidou mobile station receivers. By utilizing Real-Time Kinematic (RTK) differential positioning technology, the system achieves centimeter-level high-precision differential positioning, as illustrated in Fig. 2.3. Its main function is to receive signals from Beidou satellites via data broadcasting, employing differential algorithms to accurately calculate the real-time position of the compactor. This information is then used to obtain compaction data such as the number of passes and driving speed, allowing for long-distance transmission through data broadcasting.

① Beidou Satellite Navigation System (BDS)

The Beidou Satellite Navigation System (BDS) commenced development of its experimental system in 2000, achieving coverage of the Asia–Pacific region by 2012 and beginning to provide services such as positioning and navigation. Ultimately, in 2020, it achieved global coverage, signifying the successful establishment of a fully operational global satellite navigation system independently developed by China. The abundance of navigation satellites and the variety of satellite orbits endow the Beidou system with excellent obstruction resistance and multi-frequency navigation signals, enabling higher precision services for users, particularly those in low-latitude regions. Furthermore, the innovative integration of navigation and communication has made real-time navigation, rapid and precise positioning, location reporting, and precise timing a reality. This advancement allows BDS to accurately determine the position, trajectory, and speed of vehicles such as compactors, facilitating upgrades and transformations in traditional construction and transportation methods.

The positioning mechanism of the BDS system relies on the ability of the receiver to establish communication with at least four satellites at any given time and location. By utilizing the pre-established positions of the satellites and the time difference between the satellites and the receiver, the geographical location of the receiver—including longitude, latitude, and altitude—can be determined by solving four equations. Specifically, satellites scattered across the sky continuously transmit carrier signals containing position and time information. The advanced navigational information transmitters and computers onboard the satellites ensure the high precision of the orbital position and system time information carried by these signals. After a brief time interval, the ground-based receiver captures the carrier signals from the satellites; this time interval is referred to as the time delay. By determining the magnitude of this time delay and combining it with the speed of light, the distance between the satellite and the receiver can be obtained. The calculation formula is as follows:

$$R = \frac{1}{S}\sum_{i=1}^{S}\hat{y}^2 - \left(\frac{1}{S}\sum_{i=1}^{S}\hat{y}\right)^2 + \frac{1}{S}\sum_{i=1}^{S}\sigma_i^2 \tag{2.17}$$

In the equation, (r) represents the distance between the satellite and the ground receiver; (x_j, y_j, z_j) denotes the position coordinates of a particular satellite in the Earth coordinate system; (x, y, z) indicates the position coordinates of the ground receiver in the Earth coordinate system; (c) is the speed of light; (Δt) is the time delay; (t_1) is the time at which the receiver captures the carrier signal; and (t_2) is the time at which the satellite transmits the carrier signal. By applying the distance formula in three-dimensional space, three equations can be constructed based on the distances r between the receiver and three satellites. Considering the timing discrepancies t_0 between the satellite system clock and the receiver clock, the time delay Δt must be corrected, as follows:

$$\Delta t = t_1 + \Delta t_0 - t_2 \tag{2.18}$$

Therefore, it is necessary to utilize the distance between another satellite and the receiver to form four equations to solve for the four unknowns: (x), (y), (z), and (t_0). In practice, the ground receiver can establish connections with more than four satellites. In this scenario, depending on the satellite constellation distribution, the connected satellites are grouped into several sets, each containing four satellites. Each group can yield a set of receiver position coordinates. By employing an algorithm to select the group with the smallest error, the output receiver position coordinates can be determined, thereby enhancing positioning accuracy.

Based on the analysis above, it is evident that the entire positioning process relies on the distance calculation formula. Aside from the speed of light, which remains essentially constant, the factors influencing the positioning results are primarily those affecting the accuracy of time and the satellite's position. The position coordinates of the satellites in the Earth coordinate system and the frequency of rubidium atoms on the system's atomic clock are calibrated at least once a day to ensure the precision of the information and timing accuracy of the satellite's carrier signal output. The transmission of satellite carrier signals through the Earth's atmosphere is subject to the effects of the troposphere and ionosphere. Additionally, the time at which the receiver captures the signals may vary due to changes in the receiver's position and the inherent precision of the equipment. To minimize errors as much as possible, differential positioning technology can be employed to achieve a finer level of accuracy.

② Differential Positioning Technology

Differential positioning technology is a novel technique that enhances positioning accuracy by processing the location data from a reference receiver (base station) and a ground receiver through real-time or post-processing differential methods. This approach effectively eliminates common errors, such as satellite orbital inaccuracies, atomic clock discrepancies, and the effects of the troposphere and ionosphere. Based on the correction information provided by the differential reference station, differential positioning technology can be categorized into three main types: position differential, pseudorange differential, and carrier phase differential, ordered from lower to higher precision.

Position differential technology is the simplest and most adaptable differential method. Once the receiver installed at the reference station connects with four satellites, it can determine the position coordinates of the reference station within the Earth coordinate system. By comparing these coordinates with the actual known coordinates, the correction parameters can be established and subsequently transmitted to the receiver (mobile station) for the adjustment of its position coordinates, thereby achieving enhanced precision in positioning. Position differential technology effectively eliminates common errors between the reference station and the mobile station; however, it requires the condition that both stations simultaneously connect to the same group of satellites. Therefore, the distance between the reference station and the mobile station should be maintained within 10 km.

Pseudorange differential technology (DGPS, also known as RTD) is widely utilized in commercial differential receivers and serves a broad range of applications. Although it is influenced by the distance between the reference station and the mobile station, this method can achieve positioning accuracy at the sub-meter level. The underlying principle builds upon position differential technology, determining the actual distances r between the receiver at the reference station and all connected satellites. By comparing the measured values (i.e., pseudoranges) calculated based on the speed of light and time with the actual distances, correction values and their rates of change for the pseudoranges can be obtained. The receiver at the mobile station, by receiving correction information from the reference station, can compute the adjusted pseudoranges. By establishing the corresponding set of equations, the specific position of the mobile station receiver in the geographic coordinate system can be determined. Furthermore, more commercially advanced reference station receivers utilize additional $\alpha - \beta$ filters to refine the differential values and ascertain the associated errors, which are then transmitted to the mobile station receiver for correction. In comparison to position differential technology, the mathematical model for pseudorange differential correction calculations is more complex, involving a greater volume of data, and does not necessitate that the reference station and mobile station receivers connect to the same group of satellites.

Carrier phase differential technology is a method for real-time dynamic positioning that utilizes carrier phase measurements, commonly referred to as RTK (Real-Time Kinematic) technology. This widely used satellite positioning technique achieves centimeter-level precision in real-time localization. The underlying principle builds upon pseudorange differential technology, wherein the reference station receiver acquires carrier phase observations from satellites and transmits this data along with its own coordinate information in real time via data broadcasting. The mobile station receiver not only receives carrier phase observations from the satellites but also the data broadcast signal from the reference station. Utilizing built-in RTK software, it performs a series of real-time differential operations, including resolving double-differenced ambiguities, calculating baseline vectors, and executing coordinate transformations, resulting in centimeter-level precision for the mobile station's three-dimensional coordinate positioning.

In practice, carrier phase differential technology can be categorized into two main types: the correction method and the differential method. The correction method resembles pseudorange differential technology, whereby the reference station calculates the error correction for the satellite's carrier phase observations, and the mobile station applies this correction to its carrier phase measurements, classifying it as quasi-RTK technology. The differential method involves transmitting the reference station's carrier phase observations to the mobile station, which then processes this signal alongside its own received satellite signals to derive the final three-dimensional precise positioning coordinates, representing true RTK technology.

As scientific and technological advancements continue, RTK technology has evolved from local differential and multi-reference station local differential methods to wide-area differential techniques, significantly enhancing the measurement range and data transmission capabilities of RTK. Furthermore, the mobile station's operation is unaffected by its motion state, allowing for initialization in a stationary state and enabling the resolution of ambiguities and real-time processing in a dynamic state upon startup.

By integrating the Beidou Satellite Navigation System and the carrier phase differential module as part of the positioning system for the roller-integrated monitoring system, real-time centimeter-level positioning of the roller can be achieved. This setup provides crucial construction parameters such as the number of passes and driving speed during the compaction, thereby offering effective technical support for the visualization of compaction quality assessment.

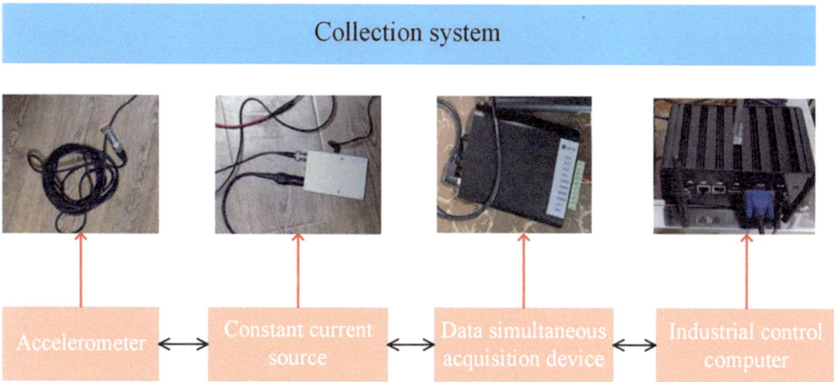

Fig. 2.4 Acquisition system

(2) Acquisition System

The acquisition system of the roller-integrated monitoring system primarily consists of accelerometers and a data synchronization acquisition device. Externally, it interfaces with the vibratory drum, while internally, it connects to the industrial control computer, as illustrated in Fig. 2.4. The working principle involves the sensor converting a specific motion parameter of the vibratory drum (such as displacement, velocity, or acceleration) into a voltage signal. The data synchronization acquisition device collects and stores these voltage signals temporally. Consequently, the acquisition system is designed based on the characterization indexes for the quality of compaction of the filling material, with these indexes determining the types of sensors used, which in turn influences the accuracy of the data synchronization acquisition device.

① Accelerometer

When selecting the RICM index based on the acceleration signals from the vibratory drum to characterize compaction quality, the accelerometer is chosen as the sensor. The specifications of the sensor are determined according to the vibration frequency of the drum, and the location of the sensor is established based on the vibration principles of the compactor. Typically, the vibration frequency of compactors used at construction sites ranges from 25 to 35 Hz; therefore, this system selects a piezoelectric accelerometer, which utilizes the direct piezoelectric effect for measurement. It features an integrated IEPE isolated two-wire low-resistance voltage output. As direct connection to direct current is prohibited, a constant current source is employed.

The excitation force generated by the rotation of the eccentric mass on the vibratory drum facilitates the vertical oscillation of the steel drum, thereby compacting the filling material. Consequently, the sensor is optimally installed on the internal frame directly connected to the compactor's vibratory drum to monitor acceleration signals with minimal error, as depicted in Fig. 2.5.

Fig. 2.5 Installation of the accelerometer

② Data synchronization acquisition device

One end of the data synchronization acquisition device is connected to the accelerometer via a constant current source, while the other end interfaces with the industrial control computer, all integrated within a compact black box for ease of installation and use, as illustrated in Fig. 2.6a. The specifications of the data synchronization acquisition device can be determined by two parameters: sampling frequency and ADC resolution. According to the sampling theorem, the sampling frequency must typically be four times or more than the nominal excitation frequency of the vibratory drum to obtain a high-precision acceleration sine curve. To meet the stringent requirements for signal processing accuracy, a 16-bit ADC resolution is selected instead of the conventional 12-bit resolution.

Based on this, the acquisition device in this system has a measurement range of ±5 and ±10 V, with dimensions of 126 mm × 106 mm × 26 mm. It contains a

(a)

(b)

	A	B	C	D	E
1	date	accelation signal	2	3	4
2	8:36:43.252	1.4099997	3.222112	3.215465	9.999456
3	8:36:43.252	0.7107555	1.656789	1.653125	6.010201
4	8:36:43.252	0.7105554	1.656789	1.683831	6.019563
5	8:36:43.252	0.7113336	1.721595	1.718564	6.020304
6	8:36:43.252	0.7105554	1.741564	1.744315	6.023366
7	8:36:43.252	0.7105554	1.736665	1.766632	6.024555
8	8:36:43.252	0.7105554	1.732223	1.800256	6.025567
9	8:36:43.252	0.7105554	1.745865	1.814231	6.026579
10	8:36:43.252	0.7105554	1.803335	1.823654	6.027591
11	8:36:43.252	0.7105554	1.813325	1.833302	6.028603
12	8:36:43.252	0.7105554	1.823456	1.844236	6.029615
13	8:36:43.252	0.7112655	1.823145	1.855234	6.030627
14	8:36:43.252	0.7116667	1.832564	1.855354	6.031639
15	8:36:43.252	0.7113335	1.847223	1.856752	6.032651
16	8:36:43.252	0.7115552	1.853336	1.857354	6.033663
	13.252	0.7113334	1.864445	1.858423	6.034675
	13.252	0.7113332	1.865646	1.865234	6.035687

Fig. 2.6 Data synchronization acquisition device and acceleration signal data

16-bit A/D conversion circuit capable of synchronously collecting data from four single-ended ports at a sampling rate of 200 kSa/s. The primary function of the data synchronization acquisition device is to transmit the monitored acceleration signals to the industrial control computer and to synchronously relay this data along with the spatial location information of the roller provided by the positioning system to the analysis module in real time.

Another operational thread saves the signals according to the system's naming convention as "capture_card_data_YYYY-MM-DD_HH_MM_SS.dat". Upon opening the file in an Excel spreadsheet, the first column represents the timestamps of the real-time collected data, the second column contains the acceleration data from the vibrating drum, and the subsequent three columns hold invalid data from the remaining three collection ports of the acquisition card, as shown in Fig. 2.6b. To collect additional data, one simply needs to connect the sensor to any of the acquisition card's input ports to gather and save the data seamlessly.

(3) Analysis system

The analysis system primarily consists of an on-board controller and an on-board display terminal. The main function of the on-board controller is to utilize developed software to process, analyze, and compute the transmitted acceleration signals and position information of roller. The on-board display terminal is responsible for visualizing the compaction quality information (such as the *KECV* index) for the current compaction work area, as well as the compaction parameters of the roller (such as the number of passes, driving speed, and compaction trajectory). Additionally, it integrates a feedback module within the on-board controller to provide real-time updates on the compaction operations to both the roller operators and the onsite supervisory personnel.

① On-board controller

The on-board controller, serving as the core of the embedded processing system, significantly influences the speed and accuracy of information processing. This system utilizes an Intel Core i5-6200U industrial computer paired with a Linux operating system as its on-board controller, as illustrated in Fig. 2.5. The 8 GB of RAM supports the operation of the developed software system, while the 256 GB of storage memory can retain the acceleration data from the vibratory drum generated during a total operational time of 8 h daily over a week.

As a compact computer, this industrial unit is equipped with four USB 3.0 ports; aside from the connections for its own antenna and the Beidou receiver antenna, the remaining ports are designated for copying acceleration signal data and the compaction quality cloud map of the work area. It also features one VGA and one HDMI port for connecting to the on-board display terminal. The size of the industrial computer is intentionally kept compact to facilitate integration with the data synchronization acquisition device, constant current source, BDS receiver module, and other components in a single location, thereby optimizing spatial resources. The chosen dimensions for this system's industrial computer are 180 mm × 135 mm × 54 mm.

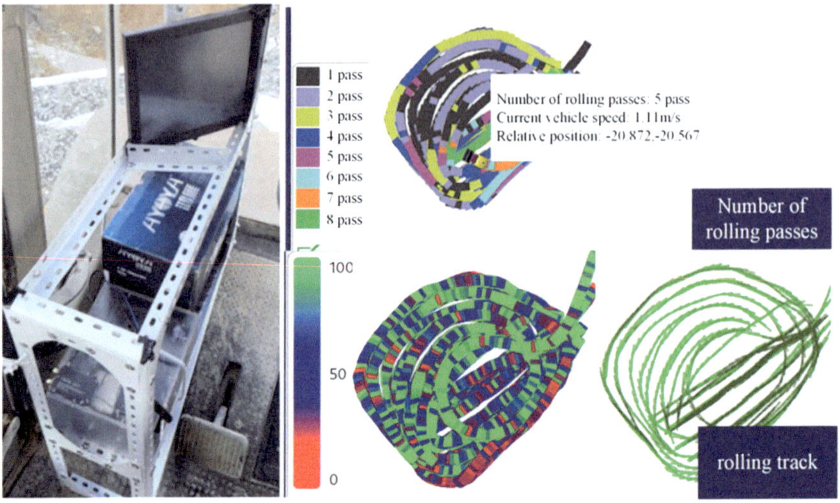

Fig. 2.7 On-board display terminal and partial program interfaces

Furthermore, a software system titled "Digitalized Compaction Intelligent Control System for Subgrade and Pavement" has been developed, which can be installed as a software package on any computer equipped with a Linux operating system. This software system comprises seven sections: the main interface, number of passes, compaction trajectory, *KECV* index, elevation, raw waveform, and parameter settings.

② On-board display terminal

The on-board display terminal is presented as an independent monitor, as shown in Fig. 2.7. Serving as the visual interface for the industrial computer's content, it facilitates information exchange with the industrial computer and features four USB 3.0 ports, one VGA port, and one HDMI port. To accommodate the system's operating voltage, a monitor with a 12 V power requirement has been selected. Additionally, to facilitate subsequent experiments and operational tasks, a monitor with a narrow aspect ratio and touchscreen functionality has been chosen. The display dimensions of this system's on-board display terminal are 271.5 mm × 219 mm, with a main frame size of 211.5 mm × 159 mm and a resolution of 1024 × 768.

2.2.2 System Functions

The roller-integrated compaction monitoring system is primarily divided into three components: the positioning system, the acquisition system, and the analysis system. The positioning system comprises the Beidou satellite, Beidou reference station, and Beidou receiver, which collectively capture the high-precision dynamic coordinates of the roller in real time. The acquisition system consists of the accelerometer and the data synchronization acquisition device, which continuously capture the acceleration signals from the vibratory drum and synchronize the signal data with the spatial location information of the compactor for real-time transmission to the analysis system.

The analysis system is made up of the on-board controller and the on-board display terminal, responsible for processing, analyzing, and computing the transmitted information regarding the compaction process. It visualizes and displays the current compaction quality information (such as the *KECV* index) and the compaction parameters of the roller (such as the number of passes and compaction trajectory), thereby facilitating information feedback. Through the roller-integrated compaction monitoring system, real-time monitoring and feedback control of the compaction parameters can be achieved, enhancing compaction quality. It allows for timely re-compaction of areas that do not meet quality standards, ensuring project quality while reducing overall costs.

2.3 Field Testing

In light of the evident lack of research regarding the applicability of the current roller-integrated compaction monitoring (RICM) metrics for low liquid limit silt with low clay content, low plasticity index, and low strength characteristics, the previous section proposed a novel index based on the maximum increment of forward kinetic energy—*KECV* (Kinetic Energy Compaction Value)—as a metric for assessing compaction quality in gravelly low liquid limit clay. This section relies on the construction of the fifth section of the Huan-Yong Expressway, focusing on the subgrade and water-stabilized layers of the highway project to validate the feasibility of using the *KECV* index as a representation of compaction quality for cement-stabilized gravel and gravelly low liquid limit clay.

2.3.1 Testing Site and Materials

The Hengyong Expressway, located in Hunan Province, China, spans a total length of 106.227 km from Yang to Yongzhou, with a subgrade width of 26.5 m, meeting the standards of a first-class highway. It serves as a vital component of Hunan's "six verticals and seven horizontals" expressway network, connecting four existing highways and enhancing the structure of the highway network. This development further strengthens the key transportation routes between East and Central China, significantly promoting resource development and coordinated economic and social growth in the cities of Hengyang, Yongzhou, and the broader Hunan region.

The field testing for this project is conducted along the fifth section of the Hengyong Expressway, specifically between K50 + 000 and K59 + 860, covering a total length of 9.86 km. The section of the highway that traverses the towns of Lingguan and Shuangqiao is designed as a dual-lane, four-carriageway, with a width of 26.5 m, allowing for a maximum vehicle speed of up to 120 km/h. The water-stabilized layer of this expressway project comprises both the base and sub-base layers, with the base layer being notably thicker and constructed in two compacted layers. The filling material for the water-stabilized layer consists of cement-stabilized gravel. Following the principle of proximity, the filling material for the sub-base layer has been selected as low liquid limit silt.

2.3.2 *Analysis of the Physical and Mechanical Properties of Materials*

(1) Sieve analysis test

① Base layer

To minimize shrinkage cracks, the cement-stabilized gravel base layer used in this project must comply with the specifications outlined in the "Technical Guidelines for Construction of Highway Roadbases" (JTG/T F20-2015) [23]. Specifically, the maximum particle size must not exceed 31.5 mm, the mass proportion of particles larger than 5 mm must be no less than 60%, and the content of particles smaller than 0.075 mm in the aggregate must be controlled to below 5%. The cement-to-gravel ratio is set at 5:100.

Through the design of a synthesized gradation and the results of unconfined compressive strength tests at varying cement contents, while considering the stability, safety, and economic costs of the filling materials, a cement dosage of 5.0% was selected as the optimal amount. The specific aggregate mix is detailed in Table 2.1. The particle gradation curve of the base layer filling materials obtained from the sieve analysis test is illustrated in Fig. 2.8. From the figure, it can be observed that the maximum particle size of the cement-stabilized gravel base layer is approximately 28 mm, with the proportion of particles smaller than 0.075 mm accounting for about 4.7–5%.

② Sub-base layer

To mitigate shrinkage cracks, the cement-stabilized gravel sub-base layer used in this project must adhere to the requirements set forth in the "Technical Guidelines

Table 2.1 Mix proportions for cement-stabilized gravel base layer

Specification/mm	19–26.5	9.5–19	4.75–9.5	0–4.75
Proportion/%	22	33	13	32

Fig. 2.8 Gradation curve of cement-stabilized gravel base layer

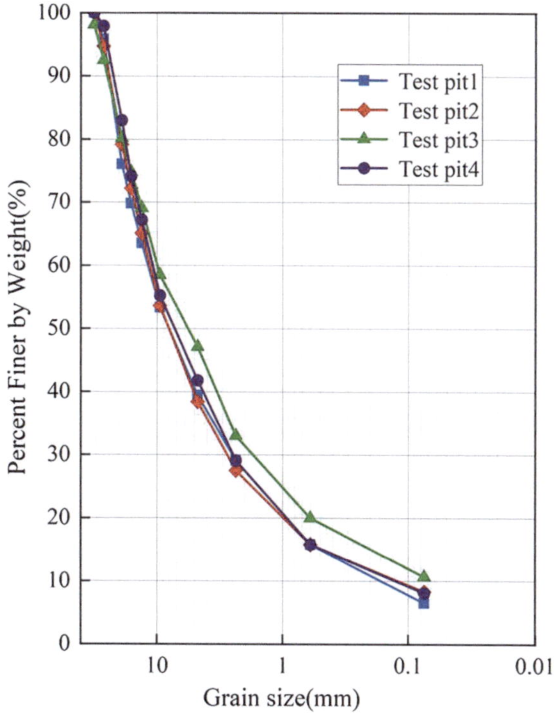

for Construction of Highway Roadbases" (JTG/T F20-2015) [23]. Specifically, the maximum particle size must not exceed 31.5 mm, the mass proportion of particles larger than 5 mm must be no less than 50%, and the content of particles smaller than 0.075 mm in the aggregate must be kept below 5%. The cement-to-gravel ratio is established at 4:100.

Through the design of a synthesized gradation and the results of unconfined compressive strength tests at various cement contents, while considering the stability, safety, and economic viability of the filling materials, a cement dosage of 4.0% was determined to be optimal. The specific aggregate mix is presented in Table 2.2. The particle gradation curve of the sub-base layer filling materials obtained from the sieve analysis test is illustrated in Fig. 2.9. From the figure, it can be observed that the maximum particle size of the cement-stabilized gravel sub-base layer is approximately 28 mm, with the proportion of particles smaller than 0.075 mm ranging from about 6.5 to 10.7%.

③ Subgrade

By selecting four samples of gravelly low liquid limit silt of equal mass from random locations within the on-site compaction area (with an aim to maintain a weight variance within 1 g), each of the four soil samples was subjected to a sieve analysis using a 2 mm aperture sieve. The material retained on the sieve was analyzed through

Table 2.2 Mix proportions for cement-stabilized gravel sub-base layer

Specification/mm	19–26.5	9.5–19	4.75–9.5	0–4.75
Proportion/%	26	33	14	27

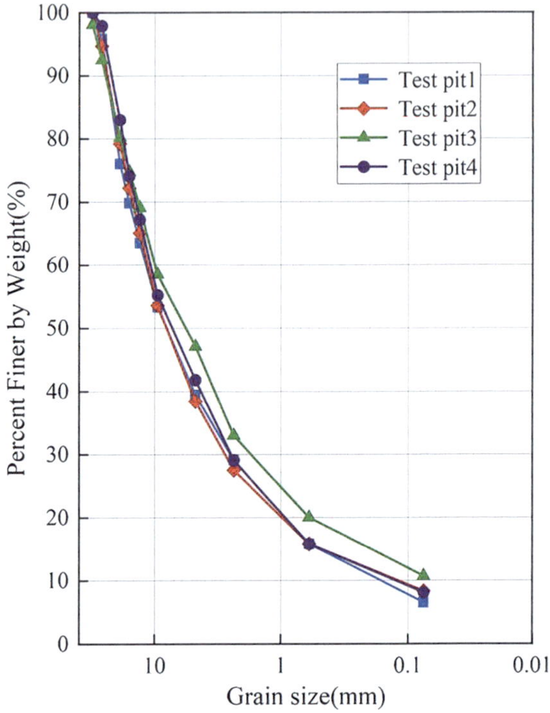

Fig. 2.9 Gradation curve of cement-stabilized gravel sub-base layer

coarse sieve analysis, while the material passing through was subjected to fine sieve analysis. Upon completion of the sieving process, the weight of the soil samples on each sieve was recorded according to the size of the sieve apertures, thereby determining the particle composition of the gravelly low liquid limit silt. The gradation curve is illustrated in Fig. 2.10. From the figure, it can be observed that the maximum particle size of the gravelly low liquid limit silt is approximately 30 mm, with the proportion of particles smaller than 0.075 mm ranging from about 61.2–66.6%.

(2) Limit moisture content test

In accordance with the "Test Methods of Soils for Highway Engineering" (JTG 3430-2020) [24], the liquid and plastic limits of soil samples retrieved from the field were determined using a digital liquid-plastic limit apparatus and other equipment. The measured penetration depth of the cone probe into the soil and the corresponding moisture content results are presented in Table 2.3. From the table, it can be noted

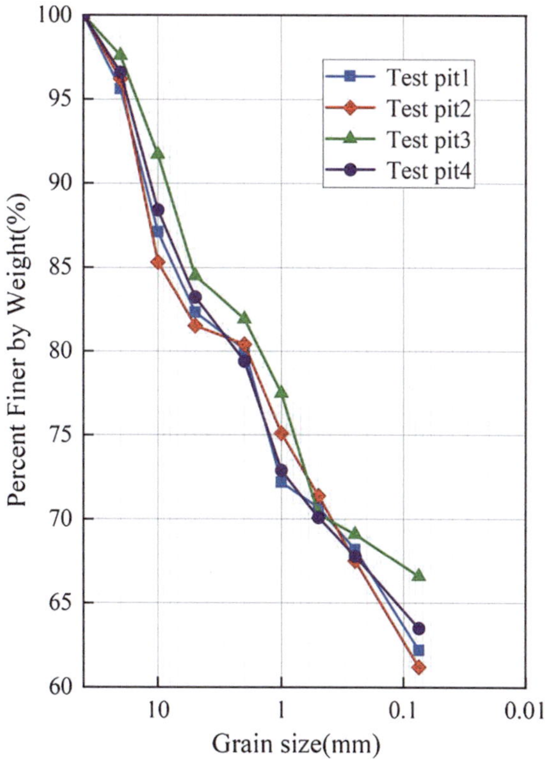

Fig. 2.10 Gradation curve of gravelly low liquid limit silt

Table 2.3 Results of limit moisture content test

Test item	Number of tests		
	1	2	3
Penetration depth $1/2(h_1 + h_2)$/(mm)	4.4	10.0	19.9
Average moisture content/(%)	22.8	27.2	31.8

that the liquid limit of the soil sample collected from the site is 31.8%, the plastic limit is 21.6%, and the plasticity index is 10.2.

(3) Proctor compaction test

Moisture content is a critical physical parameter of filling materials, significantly influencing their properties and representing their engineering characteristics. Consequently, when the moisture content of the filling materials during actual construction closely aligns with the optimum compaction moisture content, the compactness achieved can approach the maximum possible level, resulting in optimal compaction outcomes. Therefore, determining the optimum moisture content and maximum dry density of filling materials is essential for the success of construction projects.

The compaction test simulates the compaction effect of a roller on filling materials using a compaction apparatus in a laboratory setting, thereby establishing the relationship between moisture content and dry density. This allows for the determination of the maximum dry density achievable by the filling materials and the corresponding optimum moisture content, providing vital data support for subsequent on-site vibratory compaction (including the design of compaction parameters and selection of RICM indexes).

Compaction tests are categorized into two types: lightweight and heavyweight compaction tests. Lightweight compaction tests are employed for materials with smaller particle size ranges, while heavyweight compaction tests are utilized for materials with larger particle size distributions.

① Base layer

Based on the results of the sieve analysis, the heavyweight compaction test method was employed to determine the optimum moisture content and maximum dry density of the cement-stabilized gravel base layer. The instruments used for the test included an electric heavyweight compaction apparatus, an electric drying oven, and an electronic balance, with specific operational procedures referenced in the "Test Methods of Soils for Highway Engineering" (JTG 3430-2020) [24].

The compaction curve for the cement-stabilized gravel base layer obtained from the compaction test is illustrated in Fig. 2.11. From the figure, it can be observed that the optimum moisture content and maximum dry density for Soil Sample 1 are 4.9% and 2.412 g/cm^3, respectively, while for Sample 2, these values are 4.6 and 2.411 g/cm^3. Consequently, the optimum moisture content and maximum dry density for the cement-stabilized gravel base layer in the field area are determined to be 4.8% and 2.411 g/cm^3, respectively.

② Sub-base layer

Based on the results of the sieve analysis, the heavyweight compaction test method was utilized to determine the optimum moisture content and maximum dry density of the cement-stabilized gravel sub-base layer. Following the compaction test procedures for the base layer, the compaction curve for the cement-stabilized gravel sub-base layer is illustrated in Fig. 2.12. From the figure, it is evident that the optimum moisture content and maximum dry density for Soil Sample 3 are 4.7% and 2.391 g/cm^3, respectively, while for Soil Sample 4, these values are 4.4% and 2.390 g/cm^3. Consequently, the optimum moisture content and maximum dry density for the cement-stabilized gravel sub-base layer in the field area are determined to be 4.6% and 2.391 g/cm^3, respectively.

③ Subgrade

Based on the results of the sieve analysis, the heavyweight compaction test method was employed to determine the optimum moisture content and maximum dry density

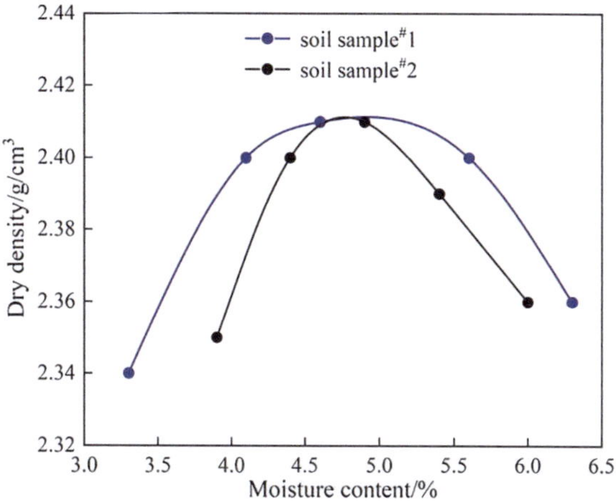

Fig. 2.11 Compaction curve of cement-stabilized gravel base layer

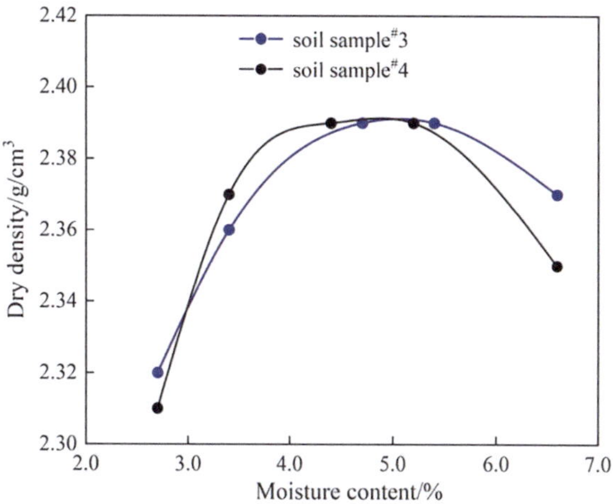

Fig. 2.12 Compaction curve of cement-stabilized gravel sub-base layer

of the gravelly low liquid limit silt in the subgrade. The compaction curve for the gravelly low liquid limit silt obtained from the compaction test is illustrated in Fig. 2.13. From the figure, it can be observed that the optimum moisture content and maximum dry density for Soil Sample 5 are 17.2% and 1.73 g/cm^3, respectively, while for Soil Sample 6, these values are 17.0% and 1.73 g/cm^3. Consequently, the optimum moisture content and maximum dry density for the gravelly low liquid limit silt in the field area are determined to be 17.1% and 1.73 g/cm^3, respectively. Based on these experimental results, the fundamental material properties of the low liquid limit silt in the testing area are summarized in Table 2.4.

Fig. 2.13 Compaction curve of gravel-containing low liquid limit silt

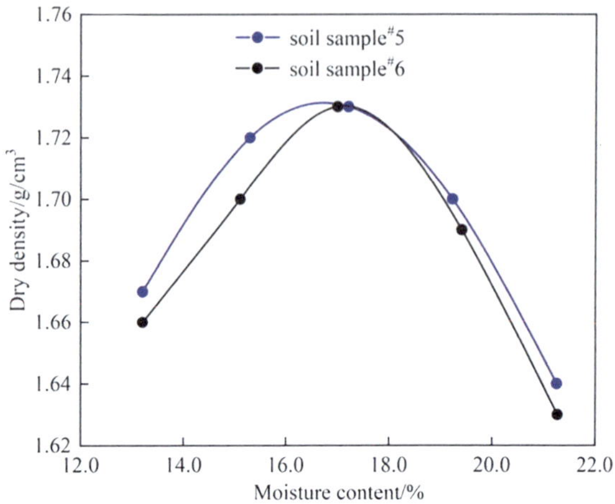

Table 2.4 Basic material properties of gravel-containing low liquid limit silt

Natural moisture content ω_n/%	Optimum moisture content ω_o/%	Maximum dry density ρ_{max}/(g/cm^3)	Liquid limit ω_L/%	Plastic limit ω_P/%	Plasticity index I_P
25.4	17.1	1.73	31.8	21.6	10.2

2.3.3 Testing Plan

(1) Compaction test plan for the water-stabilized layer of the pavement

To evaluate the efficacy of the RICM index in characterizing the compaction quality of the water-stabilized layer of the pavement and to investigate the relationship between the RICM index and the number of passes, this section outlines a corresponding compaction test plan. In practical engineering, compaction parameters typically include the number of passes, driving speed, and excitation force conditions. However, during the construction of the water-stabilized layer, the roller only operates in forward and reverse gears, with speed controlled solely by acceleration and deceleration via the throttle, and excitation force states available are high-frequency low-amplitude and low-frequency high-amplitude. According to construction specifications and on-site requirements, the compaction of the water-stabilized layer is executed solely under "low vibration" operations. The speed cannot be controlled in stages, relying heavily on the operator's skill, leading to significant variability. Therefore, the compaction conditions include only the number of passes, maintaining a speed of 1.43 m/s. The specific test plan is outlined in Table 2.5, which includes two static passes and between one to eight vibratory passes.

The testing area is divided into distinct test strips, each with an actual width of 3 m, suitable for single-pass rolling by the compactor, with an effective length of 10 m. Additional lengths are provided at both ends to mitigate the effects of vibration

Table 2.5 On-site compaction test plan for the water-stabilized layer of the pavement

Working conditions	Excitation force state	Driving speed/ (m/s)	Static rolling for number of passes	Vibration rolling for number of passes
–	High frequency and low amplitude	1.43	2	1–8

initiation and cessation on the vibration signal waveform, ensuring stable acceleration signals can be obtained. Within the centerline of the short edge of each strip, three test pits for in-situ testing are sequentially arranged at distances of 2, 5, and 8 m, along with one randomly placed test pit. The specific layout of the testing site is illustrated in Fig. 2.14. Additionally, adequate adjustment areas between the strips are designated to facilitate the maneuvering of the roller.

The tests are conducted using a SANY SSR260AC-8H model 26-ton single-drum vibratory roller, with a vibrating drum width and diameter of 2170 mm and 1700 mm, respectively. The excitation force state employed features a vibration frequency of 32 Hz and an excitation force of 295 kN, corresponding to a nominal amplitude of 0.95 mm.

The specific steps for the on-site compaction test of the water-stabilized layer of the pavement are divided into three phases: before vibration compaction, during compaction, and after compaction. The detailed testing process is illustrated in Fig. 2.15. In Fig. 2.15a, the cement-stabilized gravel is spread and leveled prior to vibration compaction, with a base layer thickness of 26 cm. The base layer is

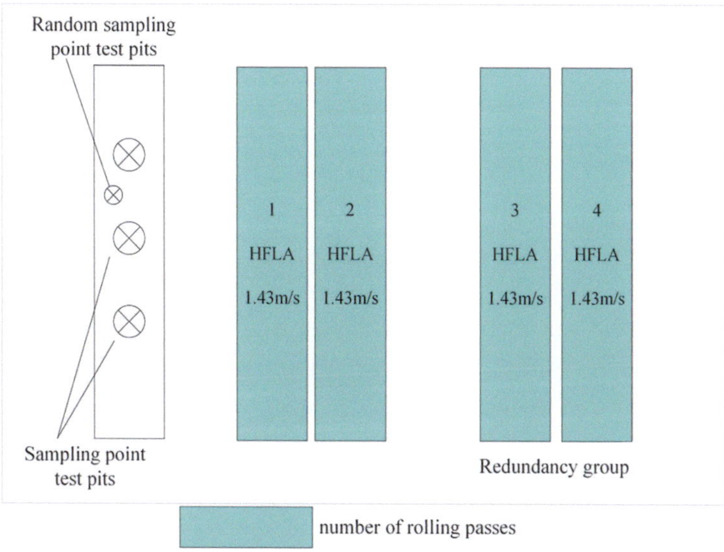

Fig. 2.14 Layout of the compaction test site for the water-stabilized layer of the pavement

Fig. 2.15 On-site testing process for the water-stabilized layer

(a) Before vibratory compaction

(b) During vibratory compaction

(c) After vibratory compaction

constructed in two stages, with a single-layer thickness of 24 cm. This phase also involves marking out the test strips and preliminarily designating the locations of test pits on the leveled filling surface.

Figure 2.15b depicts the compaction process, where appropriate compaction parameters are selected based on the testing plan. Real-time data is collected, including RICM index values, compaction parameters, and other relevant information (such as the real-time spatial position of the roller). Figure 2.15c illustrates the measurement of dry density and other data at the test pit locations following vibration compaction. The dry density of the cement-stabilized gravel is measured using the sand replacement method. This measurement, combined with results from laboratory tests for maximum and minimum dry densities, allows for the calculation of relative density, which serves as an indicator of the compaction quality of the cement-stabilized gravel pavement. The specific formula for calculating relative density is as follows:

$$D_r = \frac{(\rho - \rho_{\min})/\rho_{\max}}{(\rho_{\max} - \rho_{\min})\rho} \times 100\% \tag{2.19}$$

In the equation, (D_r) represents the relative density in percentage (%), and ($\rho_{\min}$) denotes the minimum dry density in g/cm^3.

(2) Compaction test plan for the subgrade layer

Prior to finalizing the test plan, tests were conducted using different compaction parameters of the on-site roller. The results indicated that the number of passes, excitation force state, and drive direction could be controlled to analyze the variations in the RICM index. However, when operating at mid to high speeds in weak vibration mode (HFLA), the roller experienced "bouncing," resulting in "waviness" in the subgrade and inadequate compaction effectiveness. Consequently, the on-site tests were conducted at slow speeds, utilizing low-frequency high-amplitude and high-frequency low-amplitude states for vibration compaction over a specified number of passes. The various experimental conditions are outlined in Table 2.6, which includes two static passes and between one to eight vibratory passes.

The testing area is divided into several test strips, each with a practical width of 3 m and an effective length of 10 m, along with designated passing zones of appropriate size. Additionally, along the length of each strip, four test pits for in-situ testing

Table 2.6 Compaction test plan for subgrade

Working conditions	Excitation force state	Driving speed (m/s)	Number of passes (static rolling)	Number of passes (vibration rolling)
Number of passes	HFLA	1.43	2	1–8
Forward			2	1–8
Reverse			2	1–8
Number of passes	LFHA		2	1–8
Forward			2	1–8
Reverse			2	1–8

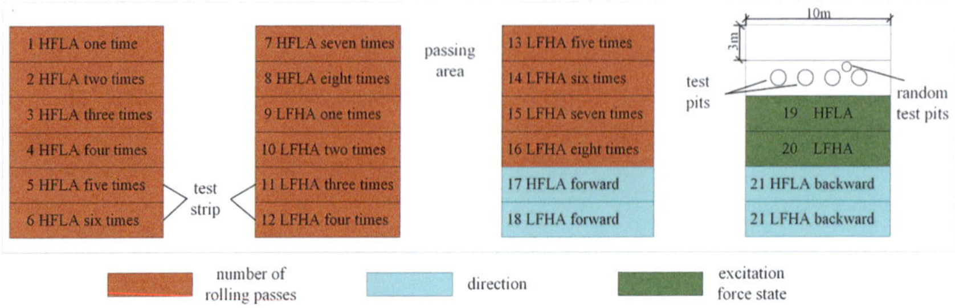

Fig. 2.16 Layout of the compaction test site for subgrade

are sequentially arranged at distances of 2, 4, 6, and 8 m, along with one randomly positioned test pit. The specific layout of the testing site is illustrated in Fig. 2.16.

The tests are conducted using a XCMG XS223J model 22-ton single-drum vibratory roller, with the vibrating drum width and diameter measuring 2130 mm and 1600 mm, respectively. The vibration frequency and excitation forces are set at 28/33 Hz and 353/245 kN, with a nominal amplitude of 1.65/0.83 mm.

The on-site compaction test for low liquid limit silty soil subgrade is divided into three phases: before vibration compaction, during compaction, and after compaction. The specific testing process is illustrated in Fig. 2.17. In Fig. 2.17a, a layer of low liquid limit silty soil containing gravel, with a thickness of 29 cm, is spread prior to vibration compaction. This phase also includes marking out the test strips and preliminarily designating the locations of test pits on the leveled fill surface.

Figure 2.17b depicts the compaction process, where appropriate compaction parameters are selected based on the testing plan. Real-time data is collected, including RICM index values, compaction parameters, and other relevant information (such as the real-time spatial position of the roller). Figure 2.17c illustrates the measurement of dry density and other data at the test pit locations following vibration compaction. The dry density of the low liquid limit silty soil is measured using the sand replacement method. This measurement, combined with results from laboratory tests, allows for the calculation of the compactness, which serves as an index of the compaction quality of the low liquid limit silty soil. The specific formula for calculating the compactness is as follows:

$$K = \frac{\rho}{\rho_{\max}} \times 100\% \tag{2.20}$$

In the equation, (K) represents the compactness in percentage (%), $(\rho_{\max})$ denotes the maximum dry density in g/cm^3, and (ρ) signifies the dry density in g/cm^3.

(a) Before vibratory compaction

(b) During vibratory compaction

(c) After vibratory compaction

Fig. 2.17 On-site testing process for the subgrade layer

2.3.4 *Data Processing*

(1) Preprocessing of vibration acceleration signals

① Filtering

The key to roller-integrated compaction monitoring technology lies in the interaction between the vibratory drum and the embankment material, as well as the monitoring of their respective motion states. Through a self-developed compaction monitoring system, real-time data is collected on the motion state of the vibratory drum, including spatial position information of the drum, excitation force state, and acceleration signals, which indirectly represent the compaction quality of the filling material.

Notably, the spatial position information based on the BeiDou differential positioning system module exhibits high accuracy and can be directly utilized for analysis. However, the acceleration signals and the excitation force state reflected through these signals may be affected by noise and other disturbances from the surrounding environment of the compaction area. Consequently, it is imperative to filter the raw acceleration signals before calculating the RICM index.

Filtering typically encompasses two methods: digital filtering and analog filtering. The former operates on the principle of applying specific numerical computations using corresponding algorithms or devices to achieve filtering. In contrast, the latter involves limiting the minimum or maximum frequency components of the original signal for continuous filtering, which requires constant adjustment of the limiting element's parameters in practical applications, often proving to be time-consuming and labor-intensive. Therefore, for this on-site test, digital filtering has been employed to process the data effectively.

In the realm of digital filtering, Finite Impulse Response (FIR) filters and Infinite Impulse Response (IIR) filters are the most widely utilized, both exhibiting excellent filtering performance, high signal accuracy, and rapid response times. The advantages and disadvantages of these two types of filters are summarized in Table 2.7.

Table 2.7 Comparison of advantages and disadvantages of two types of digital filters

	FIR filter	IIR filter
Advantages	Finite-length sequence, no analytical formula Precise computation of amplitude-frequency characteristics and linear phase signals High accuracy and good stability	Infinite-length sequence, with existing formulas Fewer computational storage units required Lower order of filtering needed Relatively low computational cost
Disadvantages	Higher order of filtering required More storage units needed Higher computational cost	Stability issues present Poor applicability of filter design

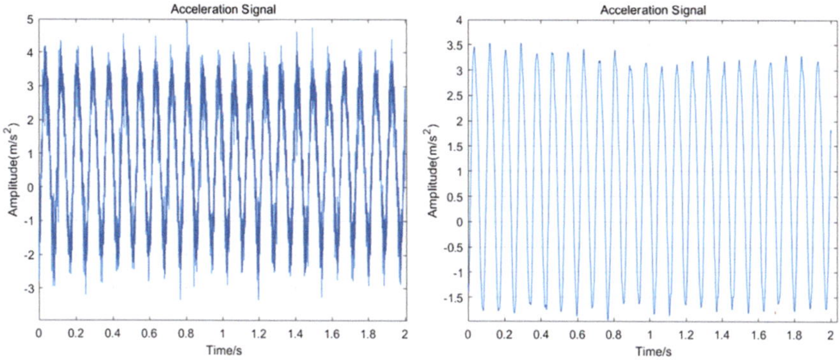

(a) Before filtering (b) After filtering

Fig. 2.18 Time domain representation of vibration drum acceleration signals

The FIR filter can be designed using the window method, Chebyshev approximation method, and frequency sampling method. However, compared to the other two methods, the window method is simpler and more straightforward to implement. Consequently, this experiment opted for the window method to design the required FIR filter.

Among the various window methods available, the Hamming window method is particularly suitable for filtering periodic signals. It effectively minimizes the interference of unwanted frequency signals while causing minimal disruption to the desired signal, thereby ensuring the integrity of the filtered signal. Therefore, the Hamming window was selected for filter design, with the cutoff frequency set to correspond to four times the fundamental frequency. The time-domain representations before and after filtering are illustrated in Fig. 2.18.

② Time-frequency analysis

To calculate the (E) index, the actual excitation frequency (f) during the operation of the vibration drum is required, which can be obtained from the frequency domain representation of the vibration signal. Thus, a spectral transformation of the acceleration signal is necessary. The commonly used method for spectral transformation is the Fourier transform, where the Discrete Fourier Transform (DFT) simplifies complex signals into numerical representations, providing an intuitive display of spectral characteristics and signal variation patterns.

However, DFT is suitable for finite-length discrete data, and applying it to nearly infinite-length original acceleration signal data would require considerable time and computational resources. Therefore, the improved form of DFT, known as the Fast Fourier Transform (FFT), exhibits strong periodicity and significantly reduces the computation time for large periodic sequence data. The FFT equation for the acceleration signal $(x(n))$ is as follows:

$$X(k) = \frac{1}{N} \sum_{n=0}^{N-1} x(n) e^{-jk\frac{2\pi n}{N}} \tag{2.21}$$

In the equation, (N) represents the length of the time-domain signal. $(X(k))$ can also be expressed in terms of amplitude (A (k)) and phase $(\phi(k))$, as follows:

$$X(k) = X_i(k) + X_r(k) \tag{2.22}$$

$$\phi(k) = tg^{-1}\frac{X_i(k)}{X_r(k)} \tag{2.23}$$

$$X(k) = A(k)e^{j\phi(k)} \tag{2.24}$$

In the equation, $(X_i(k))$ represents the imaginary part of $(X(k))$, while $(X_r(k))$ denotes the real part of $(X(k))$. The frequency domain representations before and after filtering are illustrated in Fig. 2.19. Based on the frequency domain representation and the sampling frequency, the frequency at any given point can be determined.

(2) Index calculation

① Unit volume compaction work (E)

As established in Sect. 2.1, the unit volume compaction work (E) of the compacted filling after (n) passes can be calculated using Formula (2.4). In this context, the width of the vibratory drum (B), the thickness of the filling layer (h), and the driving speed (v) are known. The working amplitude (A) can be represented by the average displacement amplitude of the vibratory drum for each pass. The radial position of the vibratory drum can be calculated using the following formula:

$$W = (m_d + m_f)g - m_d a_d - m_f a_f \tag{2.25}$$

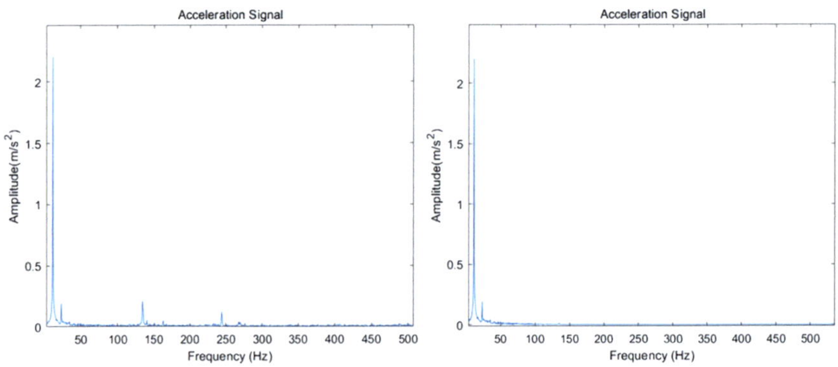

(a) Before filtering

(b) After filtering

Fig. 2.19 Frequency domain representation of acceleration signals

In the equation, (m_d) and (m_f) represent the masses of the vibratory drum and the roller frame, respectively; (a_d) and (a_f) denote the vertical accelerations of the vibratory drum and the roller frame, respectively. The acceleration (a_d) can be expressed as the average amplitude during each compaction cycle, while (a_f) can be considered as zero processing; (g) stands for gravitational acceleration. The actual excitation force (F) can be calculated using Formula (2.9). The frequency (f) of the actual excitation during the operation of the vibratory drum can be obtained from the frequency domain representation of the vibration signal.

For this field test, the sampling frequency of the acceleration sensor is set at 5000 Hz. Based on the FFT method, the frequency and amplitude of the fundamental frequency of the acceleration signal can be determined, corresponding to the operating frequency of the vibratory drum and the mean acceleration voltage value. By integrating this information with Formula (2.4), the value of the (E) index can be calculated.

① Kinetic Energy Compaction Value ($KECV$)

In this experiment, an adaptive peak detection algorithm is employed to obtain the velocity of the vibratory drum at position 2, specifically the maximum vertical velocity. The detailed steps are as follows: First, the acceleration sensor collects the vertical acceleration time-domain signal of the vibratory drum in real time. Next, the data synchronization acquisition device saves the acceleration time-domain signal and transmits it to the on-board controller. Following this, the on-board controller filters the acceleration time-domain signal.

Then, the adaptive peak detection algorithm identifies the transition points ($kT/2$) within each cycle of the filtered acceleration time-domain signal, where ($k = 1, 3, 5, \ldots, 2m + 1, \ldots$); ($m$) is a positive integer greater than 3, and (T) represents one cycle of the acceleration signal. Subsequently, the area under the positive half-cycle prior to each transition point is integrated and summed (as indicated by the orange area in Fig. 2.20), resulting in the peak velocity and maximum kinetic energy increment for each cycle. Finally, the captured peak values are extracted sequentially, and based on the trial pit position, the corresponding Kinetic Energy Compaction Value ($KECV$) is obtained.

2.3.5 *Analysis of Compaction Quality and Correlation*

The factors influencing the compaction quality of filling materials include the physical and mechanical properties of the materials themselves, the design of the compaction process, compaction parameters, as well as human factors and environmental conditions. The degree of correlation between the compaction parameters and the RICM index reflects, to some extent, the quality of compaction. By analyzing the relationship between compaction parameters and the RICM index across different filling materials, a qualitative and quantitative assessment of the correlation between

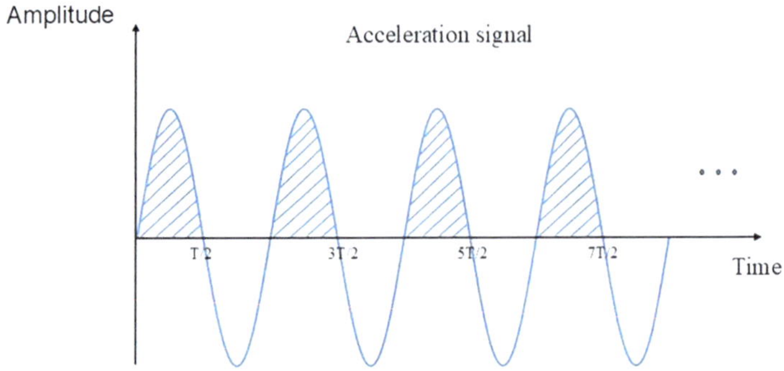

Fig. 2.20 Schematic diagram of filtered acceleration time-domain signal of the vibratory drum

these parameters and the RICM index can be conducted, thereby providing guidance and support for subsequent construction activities.

(1) Analysis of the correlation between the number of compaction passes and the RICM index

During the compaction process of filling materials, vibratory compaction utilizes periodic excitation forces to overcome the friction between the particles, reducing the internal voids within the soil and enhancing the mass of the unit volume of soil, thereby forming a stable and robust filling body. The number of compaction passes represents the number of work cycles the roller exerts on the filling materials. Excessive compaction passes can disrupt the existing particle structure of the filling material due to excessive compaction work, leading to a decrease in compaction quality and resulting in over-compaction of the materials. Conversely, too few compaction passes result in insufficient compaction work to expel the air within the soil pores, leaving a certain proportion of voids and preventing optimal compaction quality, which leads to under-compaction of the materials. Therefore, the number of compaction passes is one of the critical control parameters in compaction quality management and significantly impacts the compaction quality of filling materials.

To minimize errors caused by differences in moisture content and gradation, a series of test strips with similar soil properties were selected from the filling area for vibratory compaction. Following the experimental protocol outlined in Sect. 2.3.3, different filling materials for various filling layers were laid according to construction specifications and site requirements, with the corresponding thickness established. After achieving a smooth surface through static rolling, the actual width was defined as 3 m and the effective length as 10 m for the test strips. The in-situ testing points were marked, and the "small vibration" mode (high frequency, low amplitude) was activated to perform 1–8 compaction passes. During the compaction process, the driving speed was maintained at a stable value of 1.43 m/s, and the acceleration signals of the vibratory drum, along with the corresponding time, were collected and saved, allowing for the acquisition of real-time RICM values during compaction.

The test strip was then divided into several smaller sections using a gridding method, and the average RICM value around the testing points was calculated to serve as the RICM index value for analysis. Based on this, the correlation between different RICM index and the number of compaction passes was analyzed.

① Correlation between *KECV* and the number of Compaction Passes

The relationship between the number of compaction passes and the Kinetic Energy Compaction Value (*KECV*) at various measurement points for different filling materials is illustrated in Fig. 2.21. It is evident from the figure that the *KECV* values for all three types of filling materials at the same measurement point increase with the number of compaction passes.

According to the theoretical analysis in Sect. 2.1, at the beginning of compaction (when the number of passes is relatively low), the soil transitions from a loose state to an initial state of compaction, with deformation primarily occurring in a plastic manner. The energy required for soil compression is substantial, and since the vibratory drum outputs a fixed amount of energy throughout the compaction process, the maximum kinetic energy increment, or Kinetic Energy Compaction Value (*KECV*), is relatively low. As the number of compaction passes increases,

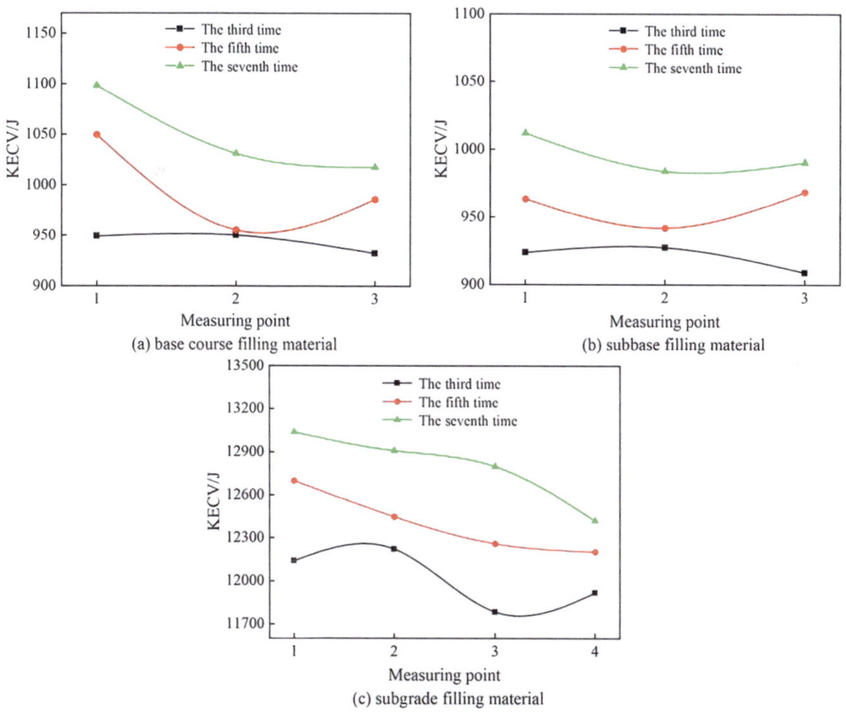

Fig. 2.21 Relationship between *KECV* and number of compaction passes at different measurement points

the soil becomes progressively compacted, resulting in reduced deformation and decreased energy requirements for soil compression, which leads to an increase in the *KECV* value.

Moreover, due to the uniformity of the material source and the challenges of maintaining consistent compaction parameters (such as driving speed) during the compaction process, there are variations in *KECV* values at different measurement points along the same strip under the same number of compaction passes. Comparing Fig. 2.21a–c, it becomes evident that, at a certain number of compaction passes, the *KECV* value for the cement-stabilized gravel base layer is the highest, while the *KECV* value for the gravelly low-plasticity silt is the lowest.

To minimize the impact of compaction process and material source errors during the compaction, the variation patterns of the average *KECV* across different compaction passes were analyzed. This analysis reveals the correlation between the *KECV* indexes of the three filling materials and the number of compaction passes, as illustrated in Fig. 2.22. It is evident from the figure that, with an increase in the number of compaction passes, the *KECV* values for all three types of filling materials rise accordingly.

Based on this, a regression model (either linear or quadratic) is established to relate the number of compaction passes to the *KECV*. The coefficient of determination, (R^2), can be calculated using the following formula:

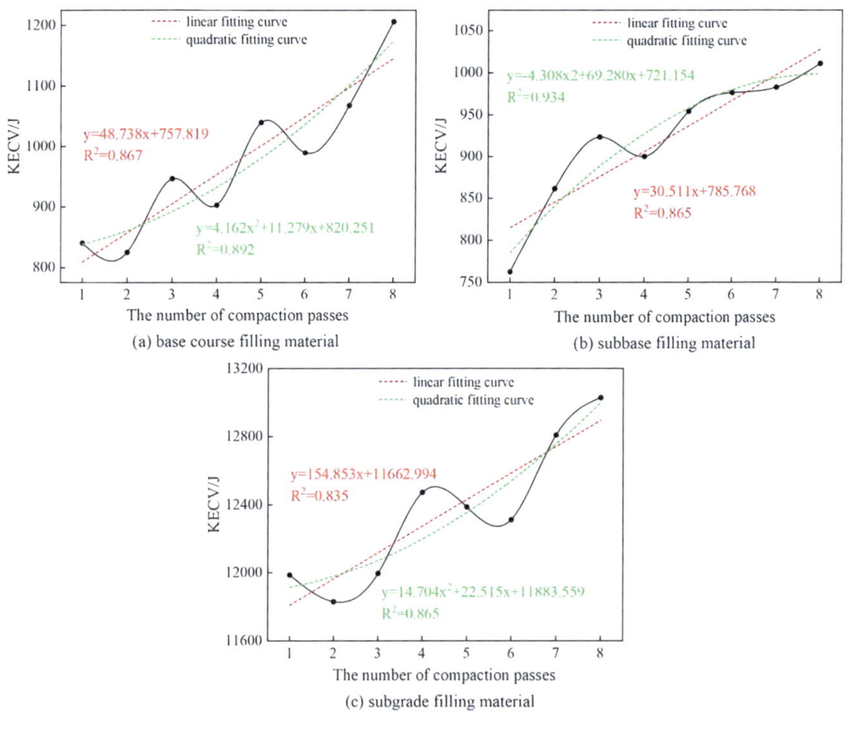

(a) base course filling material

(b) subbase filling material

(c) subgrade filling material

Fig. 2.22 Variation pattern of average *KECV* with number of compaction passes

$$R = \frac{\sum (x_i - \overline{x})(y_i - \overline{y})}{\sqrt{\sum (x_i - \overline{x})^2 (y_i - \overline{y})^2}} \tag{2.26}$$

In the formula, (x_i) and (y_i) represent the sample data for the number of compaction passes and the RICM index (*KECV* or *E* value), respectively, while $(\overline{x})$ and $(\overline{y})$ denote the sample means for the number of compaction passes and the RICM index.

The specific results are presented in Table 2.8. Combining Fig. 2.22 and Table 2.8, it is evident that there exists a significant positive correlation between the *KECV* values of the three filling materials and the number of compaction passes. Among them, the correlation for the subbase material is the strongest, with a linear coefficient of determination (R^2) of 0.865 and a nonlinear coefficient of determination (R^2) reaching 0.934, along with a coefficient of variation ($C.V$) of 8.708%. The correlation for the base filling material follows, with a linear coefficient of determination (R^2) of 0.867 and a nonlinear coefficient of determination (R^2) of 0.892; however, it has the highest coefficient of variation at 13.121%. The correlation for the subgrade filling material is the weakest, with a linear coefficient of determination (R^2) of 0.835 and a nonlinear coefficient of determination (R^2) of 0.865, yet it exhibits the lowest coefficient of variation at 3.359%.

Furthermore, the regression model necessitates a significance test to demonstrate its statistical validity. In this experiment, an F-test is employed to evaluate the model, with the core equation presented as follows:

$$F = \frac{ESS/n}{RSS/(m - n - 1)} \tag{2.27}$$

In the equation, $(ESS = \sum (Yi' - \overline{Y})^2)$ represents the regression sum of squares for the model, while $(RSS = \sum (Yi - Yi')^2)$ denotes the residual sum of squares for the model. Here, (k) signifies the number of independent variables, and (m) indicates the sample size. This expression follows an F-distribution with degrees of freedom $(k - m - 1)$. The test results demonstrate that, at a significance level of $(\alpha = 0.05)$, the significance values for the aforementioned model are all less than 0.05, indicating that the constructed regression model exhibits a high degree of fit and low spatial variability, thus confirming the statistical significance of the experimental results.

Table 2.8 Regression analysis results of *KECV* and number of compaction passes

Filling layer	Regression model	R^2	$C.V/\%$
Pavement base layer	y = 48.738x + 757.819	0.867	13.121
	y = 4.162x^2 + 11.279x + 820.251	0.892	
Pavement subbase layer	y = 30.511x + 785.768	0.865	8.708
	y = −4.308x^2 + 69.280x + 721.154	0.934	
Subgrade	y = 154.853x + 11,662.994	0.835	3.359
	y = 14.704x^2 + 22.515x + 11,883.559	0.865	

② Correlation between E value and number of compaction passes

Figure 2.23 illustrates the relationship between the number of compaction passes and the E value at various measurement points for different filling materials. As depicted, the variation pattern of the E value aligns with that of the $KECV$, indicating that at the same location, the E value increases with the number of compaction passes. According to Eq. (2.4), the E value reflects the accumulation of compaction work per unit volume of filling material, thus as the number of compaction passes increases, the filling material receives more compaction work, leading to an increase in the E value.

Furthermore, at the same number of passes, the E values at different measurement points along the same strip exhibit only minor differences, consistent with the conclusion regarding the energy output of the vibratory drum during actual compaction. To eliminate the impact of compaction processes and material source errors, the average E value's variation pattern across different compaction passes was analyzed, revealing the correlation between the average E value and the number of compaction passes, as shown in Fig. 2.24. It is evident from the figure that there is a strong positive correlation between the average E value and the number of compaction passes. Based on this, a regression model relating the number of compaction passes to the

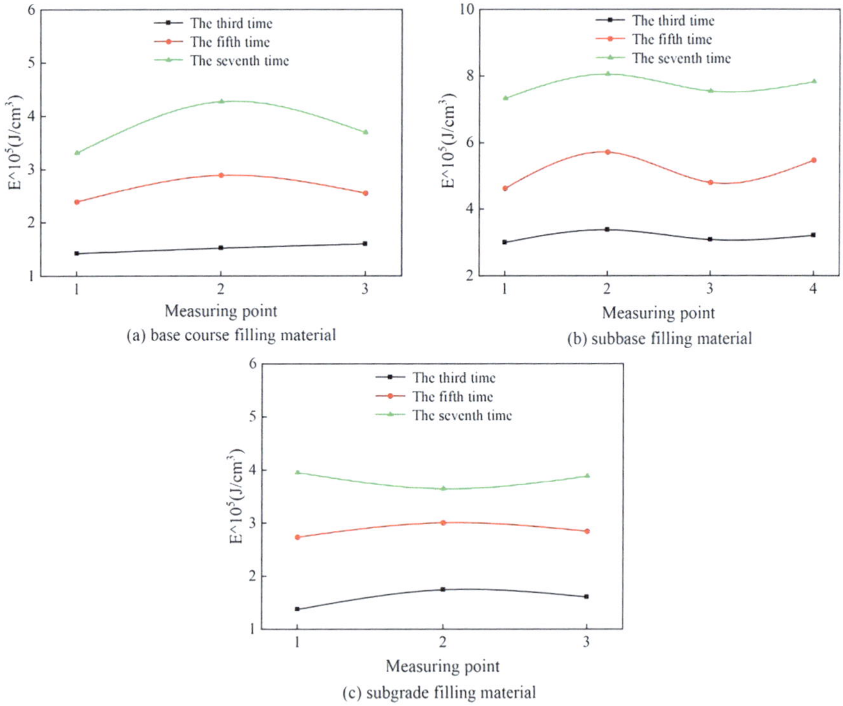

Fig. 2.23 Relationship between E values and number of compaction passes at different measurement points

average E value was constructed, and the coefficient of determination between the two was calculated. The results are summarized in Table 2.9.

The mean values of the average E index data for the three filling materials are 2.241 J/cm^3, 2.453 J/cm^3, and 4.888 J/cm^3, with standard deviations of 1.312 J/cm^3, 1.501 J/cm^3, and 2.821 J/cm^3, respectively. Due to the relatively small means and insufficient precision, the coefficients of variation for each are not calculated here. The correlation between the average E index and the number of compaction passes

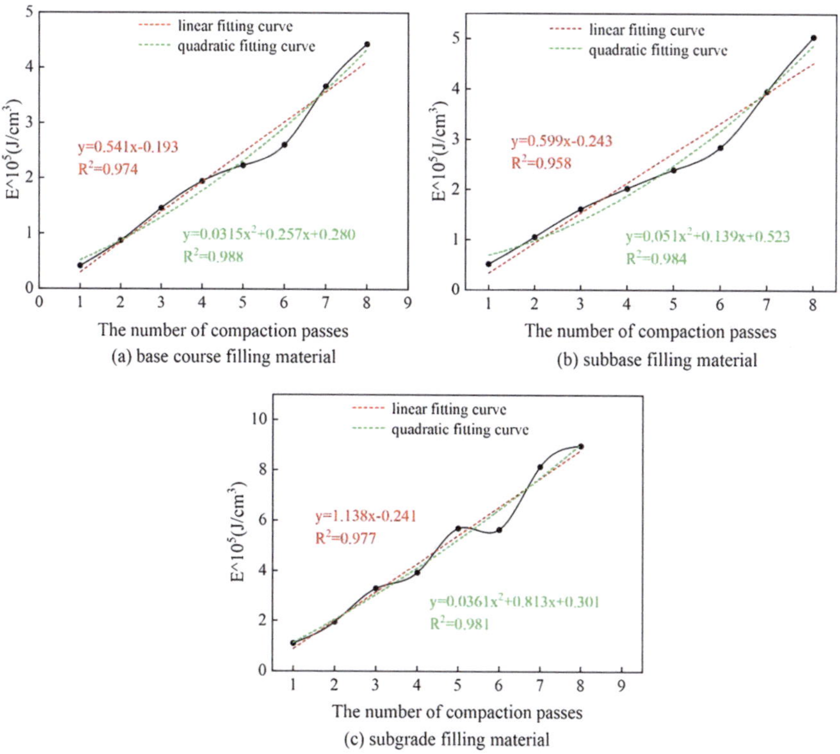

Fig. 2.24 Variation pattern of average E value with number of compaction passes

Table 2.9 Regression analysis results of E value and number of compaction passes

Filling layer	Regression model	R^2
Pavement base layer	$y = 0.541x - 0.193$	0.974
	$y = 0.0315x^2 + 0.257x + 0.280$	0.988
Pavement subbase layer	$y = 0.599x - 0.243$	0.958
	$y = 0.051x^2 + 0.139x + 0.523$	0.984
Subgrade	$y = 1.138x - 0.241$	0.977
	$y = 0.0361x^2 + 0.813x + 0.301$	0.981

for the three filling materials is stronger than that of the average $KECV$. Furthermore, an F-test was conducted on the regression model, and the results indicate that, at a significance level of ($\alpha = 0.05$), the significance values for the aforementioned model are all less than 0.05, confirming the statistical significance of the experimental results.

(2) Correlation analysis between exciting force state and RICM index

The number of compaction passes reflects the accumulation of compaction work exerted by the vibratory roller on the filling materials, while the exciting force state indicates the magnitude of compaction work during a single pass. Therefore, the exciting force state is one of the critical parameters for controlling the quality of compaction in filling materials.

The exciting force state typically includes both exciting force and vibration frequency, which influence each other; high exciting force often necessitates low vibration frequency to be effectively transmitted, while exciting force tends to be lower at high vibration frequencies. Consequently, various exciting force states can be devised. However, in practical construction, the rollers employed are configured with two integrated exciting force states: Low-Frequency High-Amplitude (LFHA) and High-Frequency Low-Amplitude (HFLA). This design allows for adjustments based on the type of filling material and the required level of compaction in the construction area.

Given that the vibration frequency and amplitude differ between these two states, the resulting vibrational energy also varies, leading to differences in the compaction effect of the roller on the filling materials under each state.

According to the experimental scheme outlined in Sect. 2.3.3, test strips with similar material properties were selected and subjected to compaction in LFHA and HFLA states for 1–8 passes, with the driving speed maintained at a stable rate of approximately 1.43 m/s. The processing of the experimental data was conducted in the same manner as that for the number of compaction passes.

Figure 2.25 illustrates the trend of the RICM index for the subgrade filling materials after the eighth compaction pass under different exciting force states. It is evident from the figure that, at a constant number of compaction passes, the $KECV$ and E values in the LFHA state are both greater than those in the HFLA state. This discrepancy arises because, in the LFHA state, the output energy of the vibratory roller is higher, resulting in greater compaction work on the filling materials (i.e., higher E values), thereby providing more energy required for the compression of the filling materials to achieve a dense state in a shorter duration. Consequently, this leads to an increase in the vertical maximum kinetic energy increment of the vibratory roller, signifying an increase in the $KECV$ value. This finding indicates a clear correlation between the exciting force state and the RICM indexes ($KECV$ and E values).

(3) Correlation analysis between driving direction and RICM index

The principle of vibratory compaction in a roller relies on the eccentric block within the vibratory roller undergoing eccentric rotation to generate exciting force, facilitating the vertical movement of the roller to compact the filling materials. The motion

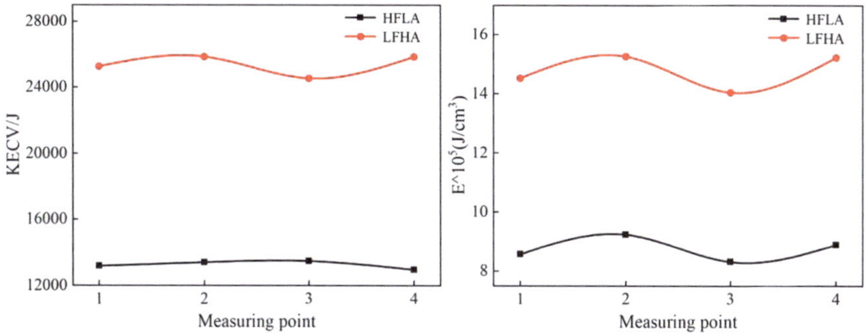

(a) Trend of *KECV* (b) Trend of *E*

Fig. 2.25 Variation trend of RICM index of subgrade filling materials with exciting force state

of the eccentric block is driven by the vibratory motor of the roller, which, in turn, operates on hydraulic oil supplied by the main engine of the roller. Consequently, the oil flow rate of the roller influences the rotational motion of the eccentric block.

In practical engineering applications, the forward and backward movement of the roller affects the rotational behavior of the vibratory roller. Additionally, variations in the hydraulic oil flow supplied to the vibratory motor may occur, resulting in differing vibratory states of the roller. These variations can lead to distinct compaction effects on the filling materials.

Following the experimental scheme outlined in Sect. 2.3.3, test strips with similar material properties were selected, maintaining consistent exciting force states and driving speed for the roller. The roller was operated in both forward and reverse directions for 1–8 passes. The processing of the resulting experimental data was conducted in the same manner as that for the number of compaction passes.

Figure 2.26 illustrates the trend of the RICM index for three different filling materials after the eighth compaction pass under varying driving directions. It is evident from the figure that, at a constant number of compaction passes, the *KECV* and *E* values during forward drive are consistently greater than those during reverse drive, regardless of the exciting force state. This indicates a correlation between drive direction and the RICM index (*KECV* and *E*). However, in comparison to the effects of the number of compaction passes and exciting force state, the impact of drive direction on the RICM indexes is relatively minor.

(4) Correlation analysis between RICM index and compaction quality of filling materials

Based on the aforementioned analysis, there exists a significant correlation between the *KECV* index, the *E* index, and the compaction parameters. Thus, it is imperative to verify the correlation between the RICM index and compaction quality to assess the effectiveness of the RICM index as an index for characterizing the compaction quality of filling materials.

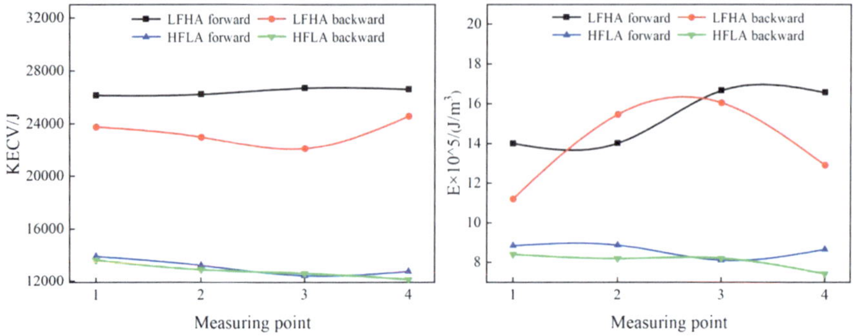

(a) Trend of *KECV* (b) Trend of *E*

Fig. 2.26 Trend of RICM index of subgrade filling materials with driving direction

According to the experimental scheme outlined in Sect. 2.3.3, after conducting on-site vibratory compaction tests for subgrade filling materials, the in-situ measurements of the wet density at test pit locations were performed using the sand replacement method. This data, combined with laboratory test results, allowed for the determination of the compaction quality of the filling materials (including compactness and relative density).

Based on the experimental data, a correlation analysis between the RICM index and the compaction quality of filling materials was conducted, as illustrated in Fig. 2.27. Additionally, a linear regression model was established to depict the relationship between the RICM index and the compaction quality of filling materials, and the corresponding coefficient of determination (R^2) was calculated. The specific results of the regression analysis are presented in Table 2.10.

For base fill materials, the correlation between the *KECV* index and relative density is notably strong, with a coefficient of determination (R^2) reaching 0.766, whereas the correlation between the *E* index and relative density is weaker, with an (R^2) of 0.558 and a coefficient of variation ($C.V$) of 8.392%. For sub-base filling materials, the correlation between the *KECV* index and relative density is stronger than that of the *E* index, with the former achieving an (R^2) of 0.809 and a $C.V$ of 4.907%, while the latter's (R^2) is 0.609. For all subgrade filling materials, the correlation between the *KECV* index and relative density remains strong, with an (R^2) of 0.759 and a $C.V$ of 4.079%, while the *E* index exhibits a weaker correlation, resulting in an (R^2) of 0.653 and a $C.V$ of 8.554%.

This analysis suggests a consistent correlation between the RICM index and compaction quality, with the *KECV* index exhibiting a stronger correlation with compaction quality across all three types of filling materials compared to the *E* index. The relatively low coefficients of variation further indicate that the *KECV* index possesses higher stability and lower data dispersion, thereby providing a more effective representation of the compaction quality of filling materials.

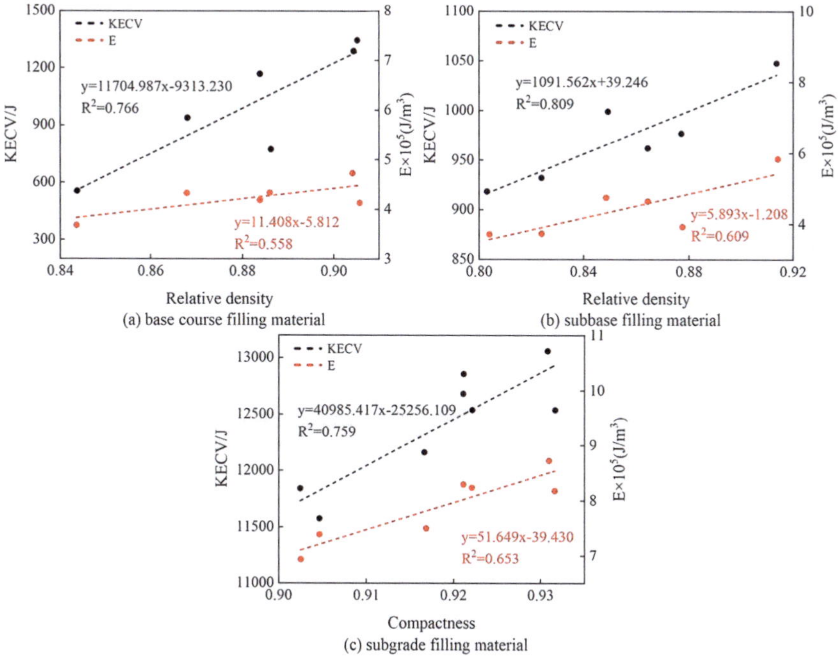

Fig. 2.27 Correlation between compaction quality of filling materials and RICM index

Table 2.10 Regression analysis results of RICM index and compaction quality of filling materials

Filling layer	RICM index	Regression model	R^2	$C.V/\%$
Pavement base	*KECV*	y = 11,704.987x − 9313.230	0.766	–
	E	y = 11.408x − 5.812	0.558	8.392
Pavement subbase	*KECV*	y = 1091.562x + 39.246	0.809	4.907
	E	y = 5.893x − 1.208	0.609	–
Subgrade	*KECV*	y = 40,985.417x − 25256.109	0.759	4.079
	E	y = 51.649x − 39.430	0.653	8.554

References

1. Liu DH, Lin M, Li S (2016) Real-time quality monitoring and control of highway compaction. Autom Constr 62:114–123
2. Zhang JL, Xu GH, Cai Y (2015) An investigation on quality inspection and control for continuously compacting subgrade. Rock Soil Mech 36(04):1141–1146
3. White DJ, Vennapusa PKR, Gieselman HH (2011) Field assessment and specification review for roller-integrated compaction monitoring technologies. Adv Civ Eng 2011(1):783836
4. White JD, Thompson JM (2008) Relationships between in situ and roller-integrated compaction measurements for granular soils. J Geotech Geoenviron Eng 134(12):1763–1770

5. Anderegg R, von Felten DA, Kaufmann K (2006) Compaction monitoring using intelligent soil compactors. In: Geocongress 2006: geotechnical engineering in the information technology age, pp 1–6
6. Komandi G (1999) An evaluation of the concept of rolling resistance. J Terrramech 36(3):159–166
7. White DJ, Jaselskis EJ, Schaefer VR et al (2005) Real-time compaction monitoring in cohesive soils from machine response. Transp Res Rec 1936(1):172–180
8. Rinehart RV, Mooney MA (2008) Instrumentation of a roller compactor to monitor vibration behavior during earthwork compaction. Autom Constr 17(2):144–150
9. Xu Q, Chang KG (2016) Adaptive quality control and acceptance of pavement material density for intelligent road construction. Autom Constr 6278–6288
10. Zhang QL, Liu TY, Zhang ZS et al (2019) Compaction quality assessment of rockfill materials using roller-integrated acoustic wave detection technique. Autom Constru 97110–97121
11. Liu D, Li Z, Lian Z (2014) Compaction quality assessment of earth-rock dam materials using roller-integrated compaction monitoring technology. Autom Constr 44:234–246
12. Hua T, Yang Z H, Yang X et al (2020) Assessment of geomaterial compaction using the pressure-wave fundamental frequency. Transp Geotech 22100318–100318
13. Mooney MA, Rinehart RV (2007) Field monitoring of roller vibration during compaction of subgrade soil. J Geotech Geoenviron Eng 133(3):257–265
14. Li QB, Zhang ZS, Liu TY, Huangfu ZH, Zhang QL, Ma WM, Li CS, An ZZ (2020) Review of vibration compaction model of soil and rock materials. Yellow River 42:105–111
15. Ye YS, Zhu HW, Yao JK et al (2021) Review on the vibration compaction theory and intelligent compaction technology of high-speed railway subgrades. China Railw Sci 42(05):1–11
16. Liu DH, Li ZL, Wang AG (2014) Real-time monitoring and rapid evaluation of the compaction quality of rockfill dams based on the work done by the compactor. J Hydraul Eng 45(10):1223–1230
17. Li D (2016) Simulation analysis of vibrating compaction process of asphalt pavement. Chongqing Jiaotong University
18. Guo DZ, Ren RB (2001) Mechanics of layered viscoelastic systems. Harbin Institute of Technology Press
19. Cai DG, Su K, An ZZ et al (2022) Analysis of the dynamic response during the vibration compaction process of the model trough for high-speed railway subgrades. Railw Eng 62(06):130–133
20. Meehan LC, Cacciola VD, Tehrani SF et al (2017) Assessing soil compaction using continuous compaction control and location-specific in situ tests. Autom Constr 73:31–44
21. Wu L L, Teng J, Ren Z Y et al (2022) Research on identification model of continuous compaction based on energy dissipation. Shock Vibr
22. Zhang ZX, Kou SQ, Jiang LG et al (2000) Effects of loading rate on rock fracture: fracture characteristics and energy partitioning. Int J Rock Mech Min Sci 37(5):745–762
23. Ministry of Transport of the People's Republic of China (2015) JTG/T F20-2015 technical details for the construction of highway pavement base courses. China Communications Press, Beijing
24. Ministry of Transport of the People's Republic of China (2020) JTG 3430-2020 code for geotechnical tests of highways. China Communications Press, Beijing

Chapter 3
Roller-Integrated Compaction Monitoring Technology Based on Positive Maximum Velocity

The subgrade serves as the cornerstone for primary engineering structures and the pavement of highways, with its integrity being paramount for ensuring safety and extending the service life of these infrastructures [1, 2]. QC and QA of compacted layers are essential components in the construction of subgrade layers [3]. A robust foundation is crucial for supporting the upper pavement layers and bearing the vehicular loads transmitted from these surfaces. Specifically, a reliable quality assessment is vital for attaining commendable QC/QA for subgrades. A well-compacted and high-quality subgrade enhances load-bearing capacity, reduces the risk of road damage, and prolongs the lifespan of highways [4]. Conversely, inadequate compaction quality can lead to a multitude of infrastructural problems, with severe cases potentially resulting in significant casualties and economic repercussions.

Current highway construction guidelines predominantly rely on manual oversight of compaction parameters during the construction phase, complemented by random checks post-construction [5]. However, traditional QC/QA approaches for subgrades often depend on human judgment and sporadic on-site tests that measure metrics such as density, moisture content, strength, and modulus. These tests typically employ tools such as the nuclear density gauge (NDG) [6] and the falling-weight deflectometer (FWD) [7]. Despite their widespread usage, these testing methods cover less than 1% of the total subgrade area. Furthermore, they are labor-intensive, time-consuming, and generally conducted only after the compaction process. Since assessments occur after the stabilization and curing of the subgrade, addressing any identified deficiencies becomes increasingly challenging. As a result, conventional methodologies fall short in accurately monitoring the compaction quality of highway subgrades, potentially leading to delays in the construction timeline.

A promising technique in QC/QA is RICM. RICM provides continuous evaluation of compaction conditions by analyzing the interactions between the ground and the roller. This method involves equipping a roller with accelerometers to monitor roller-soil dynamics, allowing for the real-time analysis of captured acceleration data throughout the compaction process [8, 9]. Over the past four decades, the integration

Q. Zhang, *Compaction Quality Nondestructive Testing and Intelligent Control Technology for Infrastructure*, https://doi.org/10.1007/978-981-95-1264-5_3

of RICM technology into QC/QA practices for subgrade construction has gained significant traction in regions such as the United States and Europe [10, 11]. Consequently, a multitude of RICM specifications has been established [8]. A critical issue that arises is the correlation between RICM measurement values (RICMMVs) and on-site measurements [12].

Presently, various RICMMVs have been developed to function as indirect quality indexes for an array of pavement materials, including the compaction meter value (CMV) [13], stiffness index (k_s) [14], machine drive power (MDP) [15], vibration modulus (E_{vib}) [16], total harmonic distortion (THD) [17], compaction control value (CCV) [18], and total compaction power per unit volume (E) [19], among others. Nonetheless, employing existing RICMMVs to assess the compaction quality of silt presents challenges such as low accuracy, complex interpretation, and an inability to identify spatial variability across different rolling areas or shifts in energy during the soil compaction process. To address these issues, a real-time RICMMV, referred to as the maximum velocity compaction value ($MVCV$), has been introduced to directly characterize the compaction status of low-liquid-limit silt in this context.

3.1 Theoretical Analysis

The dynamic conversion of energy has been recognized as the fundamental interaction between the vibration wheel and the soil, as highlighted in various studies. During a specific vibration cycle, the roller's vibration wheel undergoes three distinct downward movement states: transitioning from its highest point to its lowest, and subsequently returning to its original elevation. Beginning at its initial position (position 0, as referenced in Fig. 3.1) and assuming a positive direction for downward motion, the wheel's first state of motion is governed by the activity of the exciting force and the radial load, resulting in an increased positive velocity of the vibration wheel.

Upon reaching the second state of motion, the vibration wheel makes contact with the soil (position 1 in Fig. 3.1) and continues its downward trajectory. This interaction between the wheel and the soil induces soil compression, leading to a decrease in the resultant vertical downward force exerted by the vibration wheel. In the third state of motion, the descent of the vibration wheel persists, producing a resultant force directed vertically upwards. This movement causes the speed of the vibration wheel to diminish, ultimately halting at zero, followed by an observable upward rebound of the vibration wheel.

During the primary rolling phase, a low soil stiffness is noted, accompanied by a relatively smooth curve in the acceleration signal. Due to the pliability of the soil at this stage, energy absorption is substantial, resulting in minimal positive velocity for the vibration wheel, as illustrated in the corresponding figure. However, as compaction passes increase, a marked rise in both soil stiffness and compactness becomes evident. The emergence of high-harmonic components is observed, with their frequency increasing, indicating heightened challenges in soil compression.

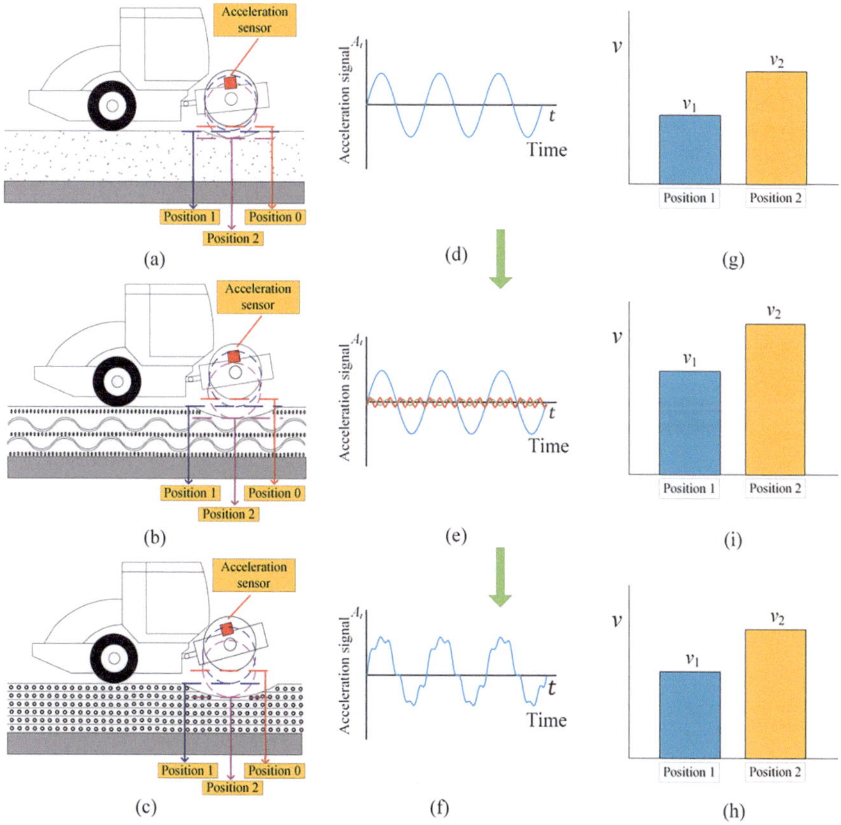

Fig. 3.1 Illustration of changes in the velocity of the vibration wheel with increasing ground stiffness

As amplitude rises, both the radial load on the vibration wheel and the work exerted by the exciting force experience an increment. This dynamic results in a diminishing proportion of energy being absorbed during soil compression, while the kinetic energy fraction of the vibration wheel amplifies, leading to an increased velocity. Subsequent compaction passes yield denser soil, and concurrently, the high-harmonic components approach a steady state. The equilibrium between the energy applied for soil compression and the subsequent surge in vibration wheel velocity becomes apparent, resulting in the stabilization of the vibration wheel's velocity.

(1) Compaction work E_1

In a single vibration cycle, the vibration wheel descends from its highest point, position 0, until it reaches position 2, as illustrated by the compaction function E_1 of the force in the vertical direction.

According to vibration theory [20], the force acting in the vertical direction on the vibration wheel during the vibration process is expressed as follows:

$$P = W + F \sin \omega t \tag{3.1}$$

where W represents the radial load of the vibration wheel; F signifies the exciting force during vibration; and ω denotes the angular velocity of the vibration wheel.

It can be observed from Fig. 3.1 that during the vibration rolling process, a phase difference φ exists between the vertical displacement of the vibration wheel and the eccentric block of the roller due to the damping effect of the soil [19]. When the vibration wheel is in motion state one at time t_1, it transitions to state two at time t_2. Additionally, time t_3 marks the critical moment at which the vibration wheel shifts from state 2 to state 3. At this point, the resultant vertical force of the vibration wheel is zero, and the velocity reaches its maximum. Thus, the expression for E_1 is as follows:

$$\begin{aligned} E_1 = &-\int_{t_1}^{t_2} (W + F \sin \omega t) A\omega \cos(\omega t - \varphi) dt \\ &+ \int_{t_2}^{t_3} (W + F \sin \omega t) A\omega \cos(\omega t - \varphi) dt = WA(2 \sin \varphi - 1) \\ &+ \frac{FA}{4}(4\varphi \sin \varphi - \pi \sin \varphi + 2 \cos \varphi) \end{aligned} \tag{3.2}$$

where A represents the amplitude of the vibration wheel. To mitigate the impact of the roller's initial start, the transition from $t_1 = (3/2\pi)/\omega$ to $t_2 = 2\pi/\omega$ and from $t_2 = 2\pi/\omega$ to $t_3 = (2\pi + \varphi)/\omega$ is considered. During the downward movement of the vibration wheel from its highest position, the speed increases until it reaches the maximum positive velocity. Following time t_3, the speed gradually decreases to 0 as the vibration wheel attains its vertical lowest point.

Previous studies have demonstrated that when the excitation frequency of the roller is close to the natural frequency of the packed material, the phase difference φ can be approximated as $\pi/2$ [20]. Substituting $\varphi = \pi/2$ into Eq. (3.2) yields an expression for the compaction work performed by the vertical force of the vibration wheel during the interval from time t_1 to t_3, as follows:

$$E_1 = A\left(W + \frac{\pi F}{4}\right) \tag{3.3}$$

Furthermore, studies [20] have indicated that as the stiffness of the soil increases, both the amplitude of the vibration acceleration and the displacement of the vibration wheel progressively rise; that is, with an increase in the number of compaction passes, both A and W gradually escalate. The magnitude of the excitation force can be computed [19]. The discrepancy between the actual operating frequency and the rated frequency is minimal, and the difference between the calculated actual exciting force F is also slight, resulting in a negligible impact on the variations in E_1.

Based on the preceding analysis, E_1 steadily increases with an increasing number of compaction passes.

(2) Energy absorbed by soil when compressed E_2

The energy absorbed during the compression of soil under the influence of $\sigma(t) = \sigma_s \cos(\omega_s t)$ can be simplified as the viscous energy of the Kelvin model [21], expressed as follows:

$$E_2 = \frac{\sigma_s^2}{2} \frac{\eta\omega_s^2}{E^2 + \eta^2\omega_s^2} V_s \tag{3.4}$$

where σ_s denotes the stress amplitude, ω_s denotes the angular velocity of the compacted soil, and E represents the composite elastic modulus of the soil. Furthermore, η denotes the composite viscosity coefficient of the soil, and V_s signifies the volume of the soil engaged in the vibration.

From relevant research, it is observed that moisture content and compactness have a minimal impact on the soil viscosity coefficient η. Therefore, η can be regarded as a constant during the compaction process [11, 22]. The stress amplitude σ_s is linked to the roller's output energy throughout the compaction phase. Given that the output energy typically remains stable during this phase, σ_s is consequently consistent. As compactness increases, there is a corresponding rise in the soil's stiffness k, leading to a continued enhancement of its natural frequency $\sqrt{k/m}$ (where m denotes the mass of the soil) [23]. Furthermore, studies indicate that with an increase in the number of compaction passes, there is a rise in the elastic modulus E, while the volume of a consistent fill layer progressively decreases [24].

Based on the aforementioned analysis, as the number of compaction passes increases, E^2/ω_s^2 increases gradually while E_2 decreases progressively.

$$E_2 = \frac{\sigma_s^2}{2} \frac{\eta}{E^2/\omega_s^2 + \eta^2} V_s \tag{3.5}$$

(3) Positive maximum kinetic energy of the vibration wheel during vibration E_3

In a single vibration cycle, when the vertical resultant force of the vibration wheel reaches zero, the positive kinetic energy of the vibration wheel ceases to increase; in other words, the vertical velocity is at its maximum.

$$E_3 = \frac{1}{2} m v_2^2 \tag{3.6}$$

where v_2 represents the positive velocity of the vibration wheel at position 2.

(4) Maximum velocity compaction value ($MVCV$)

The energy consumption of sound, light, and heat generated by primary geological materials, such as soil and rock, under the influence of a dynamic load can be disregarded. Consequently, the compaction work E_1 during the rolling process is converted into the energy E_2 absorbed by the soil and the kinetic energy of the vibration wheel. According to the law of energy conservation, when the positive velocity of the vibration wheel reaches its maximum in a single vibration cycle, the quantitative relationship among the three parameters can be articulated by Eq. (3.7):

$$E_1 = E_2 + E_3 \tag{3.7}$$

As the number of compaction passes increases, E_1 gradually rises, E_2 progressively decreases, and E_3 correspondingly increases; that is, the velocity of the vibration wheel gradually ascends, as illustrated in Fig. 3.1g–i. Specifically, v_1 denotes the velocity of the vibration wheel at position 1, while v_2 represents the velocity of the vibration wheel at position 2. Consequently, the variation in the velocity of the vibration wheel serves as an index of changes in soil stiffness.

Based on the principles outlined above, the maximum velocity compaction value ($MVCV$) is introduced as an index for characterizing the compaction quality of low liquid-limit silt, defined as follows:

$$MVCV = k \times \overline{v_{k_{\max}}} \tag{3.8}$$

where k represents a constant valued at 400, and $\overline{v_{k_{\max}}}$ denotes the average of the positive maximum velocity of the vibration wheel over N vibration cycles, calculated as follows:

$$\overline{v_{k_{\max}}} = \frac{v_{2_1} + v_{2_2} + \cdots + v_{2_i} + \cdots + v_{2_N}}{N} \tag{3.9}$$

where v_{2_i} denotes the positive maximum velocity of the i-th vibration cycle, $i = 1, 2, \cdots, N$.

3.2 The Roller-Integrated Positive Maximum Velocity Detection System

As the compaction process progresses into the middle and later stages, the degree of looseness of the low liquid-limit silt diminishes, while compactness increases and the stiffness of the soil gradually rises, resulting in an escalating difficulty in compressing the soil. Consequently, the portion of compaction work, E_1, allocated for soil compaction demonstrates a declining trend. From an energy conservation standpoint, it is apparent that the fraction of compaction work E_1 devoted to enhancing the positive kinetic energy of the vibration wheel experiences a significant increase, leading to a gradual rise in the positive maximum velocity. In an effort to enable real-time, uninterrupted detection and assessment of highway subgrade compaction quality, RTK-BDS was integrated with a roller-mounted positive maximum velocity detection system, as illustrated in Fig. 3.2.

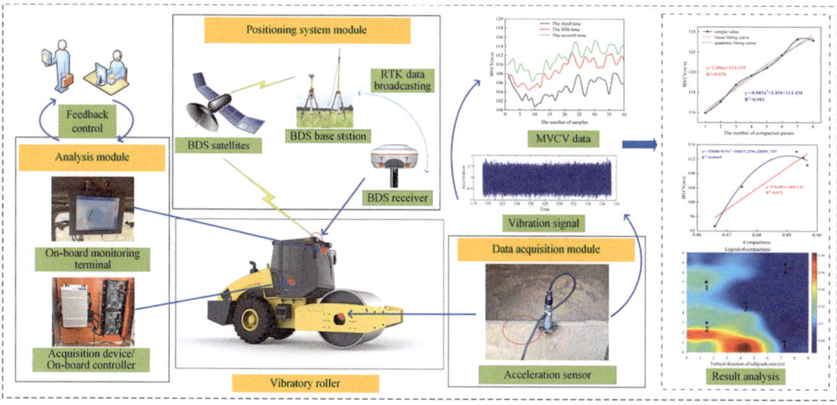

Fig. 3.2 Diagram of the roller-integrated positive maximum velocity detection system

This system primarily consists of three components: a positioning system, a data capture apparatus, and an analytical module. The positioning segment is comprised of BeiDou satellites, BeiDou reference stations, and BeiDou receivers, which utilize differential algorithms to determine the roller's high-precision dynamic differential positioning. This capability facilitates the computation of parameters such as driving velocity and compaction pass count. The data capture apparatus mainly includes an acceleration sensor and a contemporaneous data acquisition instrument. The former, located on the inner frame and directly connected to the vibration wheel, is designed to instantaneously capture the vibration acceleration signal of the wheel. Meanwhile, the latter transmits the roller's acceleration signal and spatial positioning information to the analytical module without delay. The analytical module is divided into two primary components: an integrated controller and an integrated display panel. The integrated controller is responsible for processing, analyzing, and computing the transmitted data related to the compaction process, while the integrated display panel illustrates and showcases the specifics of compaction quality (e.g., *MVCV*) and compaction metric data (e.g., compaction pass count) for the current construction zone. Furthermore, when combined with the on-board controller's embedded feedback control segment, both the roller operator and the field supervisor receive immediate feedback regarding the rolling activity, thereby enabling real-time feedback management.

By integrating the roller-mounted positive maximum velocity detection system with RTK-BDS, the spatial position of the roller and the acceleration signal of the vibration wheel were collected in real time. Building upon this data, the maximum velocity compaction value (*MVCV*) was calculated. Subsequently, correlation analysis was performed between the compaction parameters, *MVCVs*, and in situ measurements. Following this, various correlation models were established using *MVCVs*, in situ measurements, and characteristic parameters (such as moisture content) of the low liquid-limit silt as independent variables. These models aimed to estimate the compaction quality of the low liquid-limit silt subgrade.

3.3 Experimental Test

3.3.1 Test Site and Materials

The Hengyong Highway serves as a crucial component of the 'six vertical and seven horizontal' expressway network, initiated by Hunan Province during its '13th Five-Year Plan.' Extending from Hengyang to Yongzhou, the highway covers a distance of 106.227 km and features a subgrade width of 26.5 m. By linking the existing four highways in southwestern Hunan, the Hengyong Highway enhances the transportation efficiency of the previously established expressways, thereby bolstering economic returns.

A field rolling test is scheduled to be conducted between the K52+800 and K53+030 markers of the fifth section of the Hengyong Highway, which is currently under construction. The filling material for the subgrade in this segment is characterized by low-liquid-limit silt. Figure 3.3 illustrates the distribution of particle sizes. The particles of this material are primarily composed of silt and sand, with only a minor presence of clay. This unique composition imparts distinctive physical attributes to the silt, distinguishing it from conventional sand and clay, while still retaining certain similarities with them.

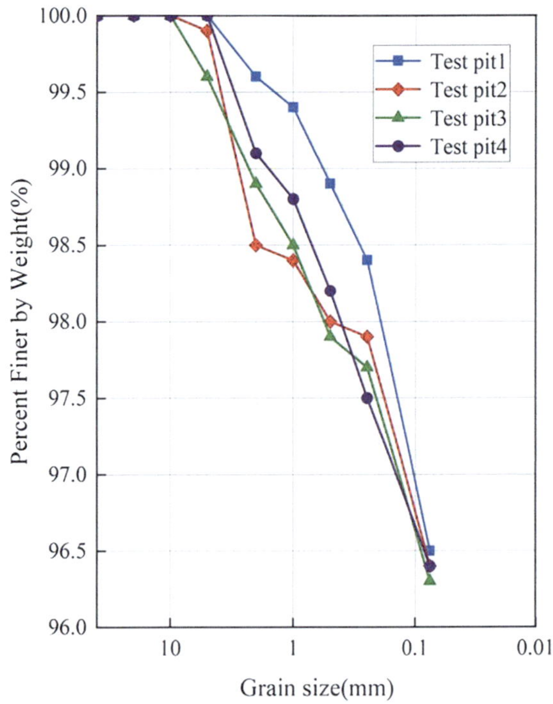

Fig. 3.3 Particle size distribution of low liquid limit silt at the construction site

Table 3.1 Key performance indexes of the filling materials

Filling materials	Maximum dry density (g/cm^3)	Natural moisture content ω_n (%)	Optimum moisture content ω_o(%)	Liquid limit ω_L(%)	Plastic limit ω_P (%)	Plasticity index I_P
Low liquid-limit silt	1.7	27	19	49	33	17

In accordance with the standards established by the specifications (Test Methods of Soils for Highway Engineering, JTG/T 3430–2020), the essential performance indexes of this filling material have been determined and are presented in Table 3.1.

3.3.2 Testing Programme

To collect *MVCV* data under varying operational conditions, the low liquid-limit silt subgrade was divided into distinct test strips. Each test strip was configured as a rectangle with a functional width of 3 m and a length of 10 m. Along its length, five evenly spaced sampling test pits were designated, as illustrated in Fig. 3.4. The compaction parameters for the rolling experiment included two excitation force modes: LFHA, and HFLA. Additionally, three distinct driving speeds—0.56 m/s (low speed), 0.83 m/s (medium speed), and 1.43 m/s (high speed)—were considered, resulting in six unique operational conditions detailed in Table 3.2. Furthermore, during the rolling test, a 22-ton single-drum vibratory roller, specifically the Sany Heavy Industry SSR220AC, was utilized. This roller was equipped with two distinct excitation forces: 410 kN and 300 kN, and operated at vibration frequencies of 29 Hz and 35 Hz, respectively.

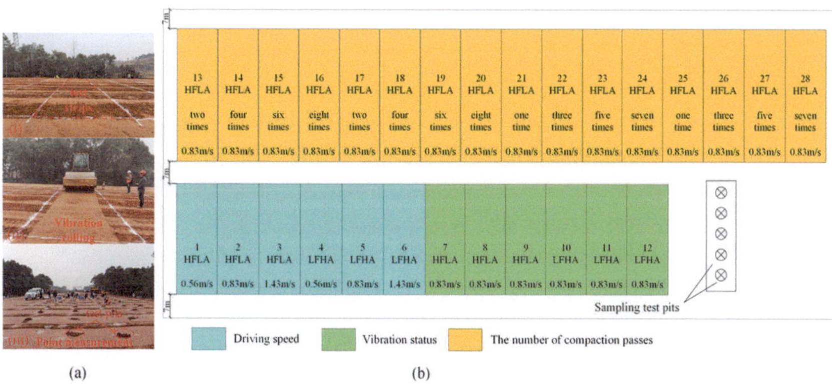

Fig. 3.4 Field rolling test: **a** test procedure; and **b** layout of the test site

Table 3.2 Experiment scheme

Working condition	Exciting force state	Driving speed/(m/s)	Passes (static)	Passes (dynamic)
1	LFHA	0.56	2	1–8
2		0.83	2	1–8
3		1.43	2	1–8
4	HFLA	0.56	2	1–8
5		0.83	2	1–8
6		1.43	2	1–8

To ascertain the influence of the excitation force mode on the compaction quality of the low liquid-limit silt, the roller's driving speed was held constant while alternating between the two excitation force states, LFHA and HFLA, during the rolling. Conversely, to evaluate the impact of driving speed on compaction efficacy, the roller's excitation force was consistently maintained throughout the test.

Based on the field test data, which includes compaction parameters, *MVCVs*, and in situ measurements (compactness and moisture content), the compactness can be calculated as follows:

$$K = \frac{\rho}{\rho_{\max}} \times 100\% \tag{3.10}$$

where K denotes compactness, %; ρ denotes dry density, g/cm^3; $\rho_{\max}$ denotes maximum dry density, g/cm^3.

3.3.3 Data Processing

The acceleration signal captured from the vibration wheel, measured by the acceleration sensor, exhibited an aperiodic nature. In addition to the principal harmonic components, such as the fundamental frequency and the second harmonic, which were identified to elucidate the dynamic characteristics of the vibration wheel, noise signals were also found to interfere with the acceleration signals. To achieve a comprehensive understanding of the dynamics of the 'wheel-soil' system, it was necessary to analyze the vibration acceleration signal within the frequency domain.

The FFT was utilized to explore the frequency domain of the acceleration signal. The fundamental principle involved the decomposition of the signal into sine waves, enabling the identification of the distinct frequency components present within it. The raw acceleration signal was found to be overwhelmed by higher-frequency noise, which significantly affected its amplitude and the uniformity of the acceleration signal. Consequently, the original acceleration signal required filtering prior to the FFT operation. Based on the vibration acceleration spectrum, an FIR low-pass filter

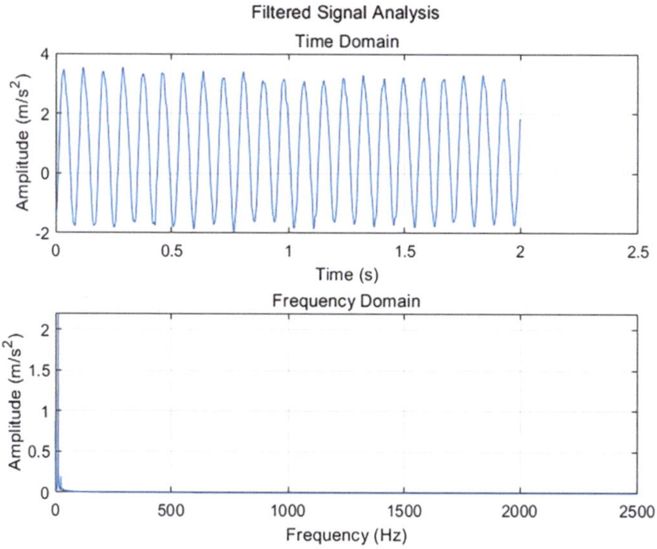

Fig. 3.5 Comparison of time–frequency domain analysis of filtered acceleration signals: **a** time-domain diagram after filtering; and **b** frequency-domain diagram after filtering

was implemented, designed to eliminate frequencies exceeding four times the base wave magnitude [25]. This process yielded filtered acceleration signals in both the time and frequency domains, as depicted in Fig. 3.5a, b, respectively. From these figures, it was evident that post-filtering, the acceleration curve in the time domain exhibited greater fluidity. Furthermore, the frequency domain representation showed that the high-frequency signals had been effectively removed.

Through the time–frequency analysis of the data from the field rolling test, the amplitudes of the fundamental wave component and the second harmonic component of the vibration acceleration were identified, allowing for the determination of the *CMV* value for this rolling test. Utilizing the filtered time-domain representation, each strip acceleration signal was divided into 40 segments. The area defined by the positive values of each segment of the acceleration signal and the zero-acceleration line constituted an *MVCV* sample point. In total, 40 *MVCVs* were identified, as illustrated in Fig. 3.6.

Disregarding the influence of variations in soil material, non-uniformity, and the compaction process on the compaction effectiveness, it was observed that as the number of compaction passes increased, the *MVCVs* for the four operational conditions also exhibited an upward trend. This indicated a direct relationship between *MVCV* and the number of compaction passes. Furthermore, differing states of excitation force and driving speed resulted in varying *MVCVs*, highlighting the impact of driving speed and exciting force conditions on *MVCV*. In the subsequent section, an analysis was conducted to explore the correlation between the RICMMVs (*MVCV* and *CMV*) and the compaction parameters.

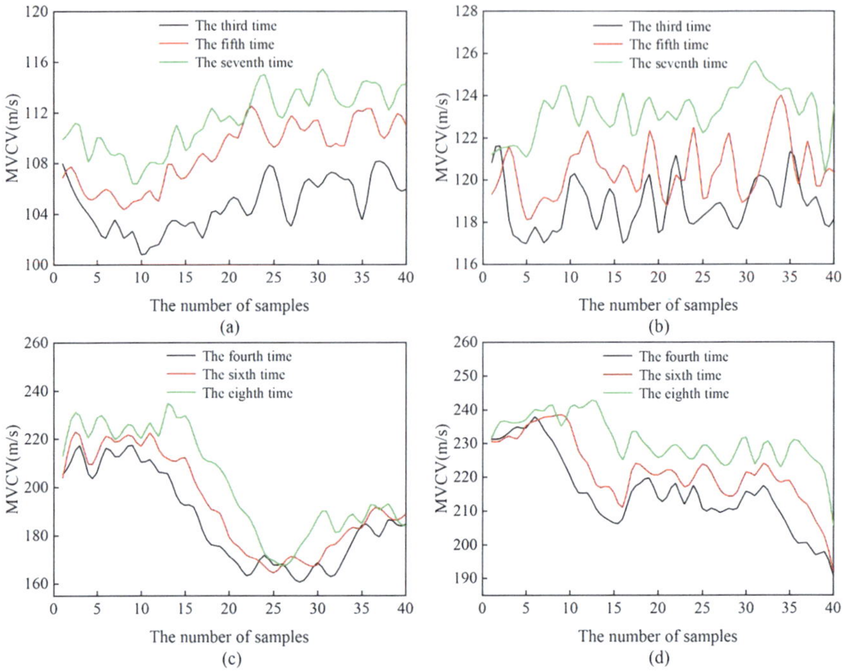

Fig. 3.6 Variation diagram of *MVCV*: **a** strip 2; **b** strip 3; **c** strip 4; and **d** strip 5

3.3.4 Correlation and Compaction Quality Analysis

(1) Correlations between RICM indexes and compaction parameters

① Correlations between RICM indexes and the number of compaction passes

The QC/QA of highway subgrades significantly depends on the number of compaction passes. If the number of compaction passes is not appropriately calibrated, it may lead to either insufficient strength of the filling material or suboptimal compaction quality, both of which can critically undermine the integrity of the subgrade fillings. For the rolling test, test strips with consistent soil material properties were selected to mitigate potential discrepancies arising from factors such as gradation and moisture content. In accordance with the test pits delineated in the experimental design, the RICMMV was calculated every 2 m, resulting in a total of five calculations. To evaluate the influence of the number of compaction passes on the RICMMVs, both the driving speed of the roller and the state of its exciting force were maintained consistently. During the field test, the roller's driving speed was standardized at 0.83 m/s. Initially, static rolling was performed twice to ensure the soil was level. Subsequently, vibration was applied between 1 and 8 times at a steady speed under two distinct exciting force conditions, thereby deriving the RICMMVs

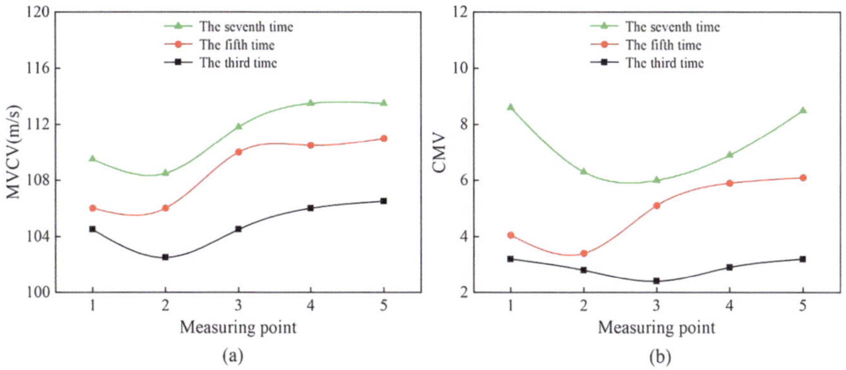

Fig. 3.7 Relationship between the number of compaction passes and RICMMV at different measuring points: **a** *MVCV*; and **b** *CMV*

for two specific sets of working conditions (with the driving speed held constant at 0.83 m/s and the exciting force state set to LFHA and HFLA, respectively).

Figure 3.7 illustrates the correlation between *MVCV* and the number of compaction passes, as well as the relationship between *CMV* and the number of compaction passes across various measurement points. As shown in Fig. 3.7, both *MVCV* and *CMV* of the low liquid-limit silt increase with the number of compaction passes. Additionally, variations in *MVCV* and *CMV* were observed under the same number of compaction passes, attributable to factors such as soil uniformity, the compaction process, and measurement discrepancies.

To gain a clear understanding of the association between the number of compaction passes and the RICMMVs, the coefficient of determination (R^2) was calculated for the number of compaction passes in relation to *MVCV* and *CMV*. A regression equation was formulated to correlate the number of compaction passes with the average *MVCV* and average *CMV*, illustrating the trend of the RICMMVs relative to the number of compaction passes, as depicted in Fig. 3.8. As the number of compaction passes increased, both *MVCV* and *CMV* exhibited consistent growth. The relationship between *MVCV* and the number of compaction passes was notably strong, with an R^2 value of 0.979. The coefficient of variation ($C.V$) was 2.239%, with a mean value of 120.005 m/s and a standard deviation of 2.687 m/s. Conversely, the relationship between *CMV* and the number of compaction passes was less robust, as evidenced by an R^2 value of 0.774. Given that the *CMV* had a mean value of 5.992 and a standard deviation of 1.376, its coefficient of variation was deemed imprecise due to the minimal mean value, and thus was not computed. Based on this analysis, both *MVCV* and *CMV* of the low liquid-limit silt exhibit a pronounced linear positive association with the number of compaction passes.

To ascertain the spatial variability and the association between *MVCV* and the number of compaction passes, as well as between *CMV* and the number of compaction passes, an F-test was applied to the data. Utilizing Eq. (3.11), the significance levels of both regression models, at a confidence level of $\alpha = 0.05$, were

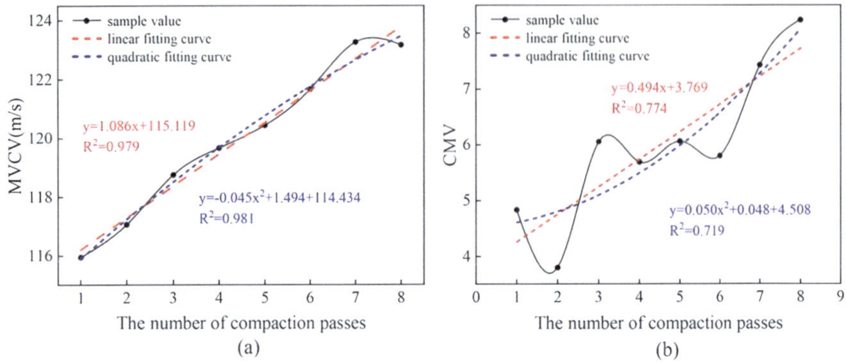

Fig. 3.8 Variation trend of RICMMV with respect to the number of compaction passes: **a** *MVCV*; and **b** *CMV*

found to be below 0.05. This indicates that the models derived from this experiment are statistically significant and have successfully passed the verification test. Given that the coefficient of variation for the *MVCV* model was below 2.3%, the model demonstrated a strong fit and minimal spatial variability.

$$F = \frac{ESS/k}{RSS/(m-k-1)} \tag{3.11}$$

where k denotes the number of independent variables, m represents the sample size, ESS signifies the explained sum of squares of the model, RSS denotes the residual sum of squares of the model, and the expression follows the F distribution with degrees of freedom $(k, m-k-1)$.

② Correlations between RICM indexes and driving speed

To mitigate errors, test strips with nearly identical soil material properties were selected for the experiment. According to the experimental design, vibration rolling was conducted at driving speed of 0.56, 0.83, and 1.43 m/s, with the RICMMV calculated every 2 m to analyze the correlation. Figure 3.9 illustrates the variations in *MVCV* and *CMV* in relation to driving speed. As depicted in Fig. 3.9, as the driving speed increases, *MVCV* and *CMV* display divergent trends: *MVCV* rises while *CMV* decreases. This phenomenon can be attributed to the fact that as driving speed declines, the energy absorbed by the unit volume of soil compression intensifies, resulting in an increase in *CMV*. Viewed from an energy conservation perspective, the positive maximum kinetic energy of the vibratory wheel diminishes during the vibration cycle, leading to a decrease in *MVCV*. In summary, a notable positive relationship exists between *MVCV* and driving speed, whereas a marked negative relationship is observed between *CMV* and driving speed.

③ Correlations between RICM indexes and exciting force state

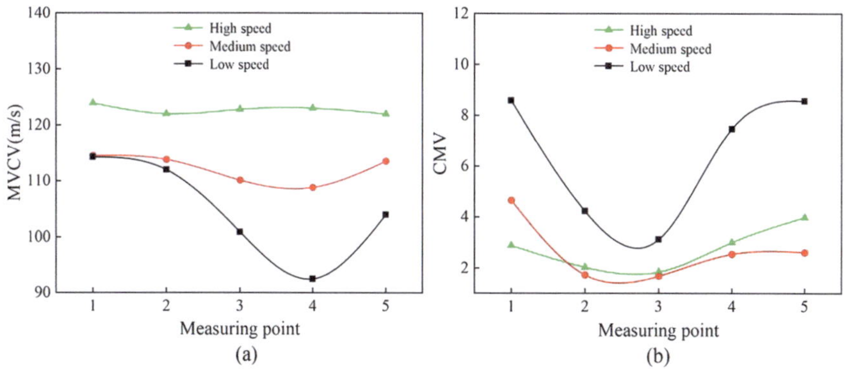

Fig. 3.9 Variation trend of RICMMV with respect to driving speed **a** *MVCV*; and **b** *CMV*

The roller operated under two exciting states, LFHA and HFLA, with vibration frequency and amplitude corresponding to each state. Due to the differences in vibration frequencies and amplitudes associated with LFHA and HFLA, the resulting vibration energies and depths of action also vary. Consequently, there are disparities in the compaction effects of the roller on low liquid-limit silt under the two exciting force states, necessitating the selection of the appropriate exciting force state for rolling based on the compaction condition of the filling material. To eliminate errors, test strips with nearly identical soil material properties were chosen. According to the experimental design, working conditions 2 and 5 were conducted, and RICMMV was calculated every 2 m to analyze the correlation. Figure 3.10 illustrates the trends of *MVCV* and *CMV* with respect to the exciting force states after the eighth pass of compaction. As shown in Fig. 3.10, the *MVCV* and *CMV* for LFHA at the corresponding measurement points were greater than those for HFLA, indicating a significant correlation between the exciting force state and both *MVCV* and *CMV*.

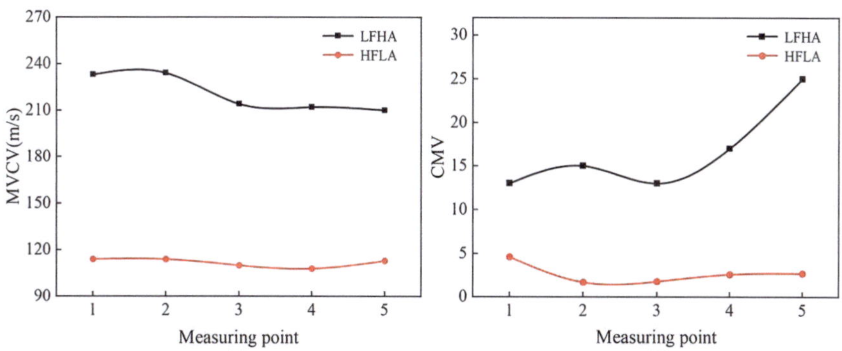

Fig. 3.10 Variation trend of RICMMV with respect to the exciting force state: **a** *MVCV*; and **b** *CMV*

(2) Correlations between RICM indexes and compaction quality of low liquid-limit silt

Based on the preceding analysis, a significant correlation exists between the compaction parameters and both *MVCV* and *CMV*. To validate the characterization efficacy of RICMMV, it is essential to examine its correlation with the compaction quality of low-liquid-limit silt. Following the completion of the initial phase of the rolling test in accordance with the experimental design, the subsequent field test was conducted, wherein the compactness (K) at the test pit point was measured using the sand replacement technique. The relationship between RICMMV (encompassing both *MVCV* and *CMV*) and the compactness of the test pit samples was assessed. A linear regression model was subsequently established to juxtapose the characterization impacts of *MVCV* and *CMV* on the compaction quality of the low-liquid-limit silt.

Figure 3.10 presents the linear regression relationship between RICMMV and the on-site measurements taken at the sampling locations. Analyzing both Fig. 3.10, a clear linear relationship between *MVCV* and compactness (K) emerges, with the coefficient of determination (R^2) reaching a value of 0.872, and the coefficient of variation ($C.V$) at 8.064% (with a mean value of 106.788 m/s and a standard deviation of 8.612 m/s). Conversely, the linear relationship between *CMV* and compactness (K) is weaker, recording a coefficient of determination (R^2) of 0.728, with a mean of 6.826 and a standard deviation of 1.600. Due to the minimal mean of *CMV* samples, which potentially undermines the precision of the coefficient of variation, the coefficient of variation for *CMV* remains uncalculated. Overall, *MVCV* demonstrated a stronger correlation with compactness (K) compared to *CMV*.

In conclusion, consistent relationships were observed between the RICMMVs and on-site measurements, underscoring the utility of RICM technology in the quality control and assurance of highway subgrades. The association between *MVCV* and compactness (K) was notably more pronounced than that between *CMV* and compactness (K). Furthermore, the coefficient of variation ($C.V$) for *MVCV* was lower, indicating that *MVCV* exhibits reduced spatial variability and provides a more precise representation of the compaction quality for low liquid-limit silt (Fig. 3.11).

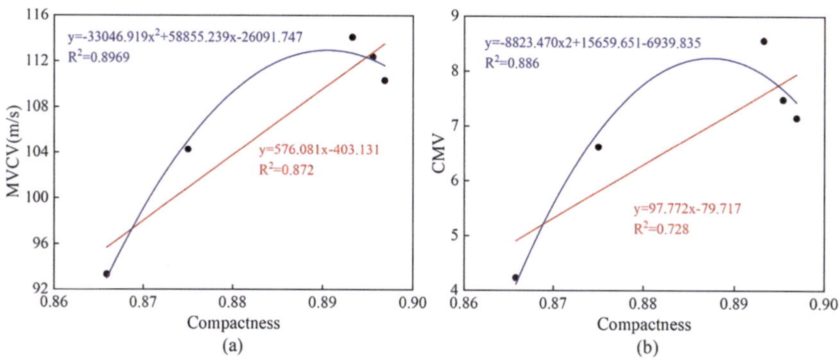

Fig. 3.11 Correlation between RICMMV and compactness: **a** *MVCV*; and **b** *CMV*

References

1. Hu W, Jia X, Zhu X et al (2019) Investigating key factors of intelligent compaction for asphalt paving: a comparative case study. Constr Build Mater 229:116876
2. Wang X, Cheng C, Zhang J et al (2023) Real-time monitoring and quality assessment of subgrade compaction: key factors and ANN model. Acta Geotech 18(6):3349–3366
3. Ganju E, Kim H, Prezzi M et al (2018) Quality assurance and quality control of subgrade compaction using the dynamic cone penetrometer. Int J Pavement Eng 19(11):966–975
4. Asif Imran S, Barman M, Commuri S et al (2018) Artificial neural network–based intelligent compaction analyzer for real-time estimation of subgrade quality. Int J Geomech 18(6):04018048
5. PRC M O T (2019) Technical specifications for construction of highway subgrades
6. Rahman F, Hossain M, Romanoschi S (2008) Intelligent compaction control of highway embankment soil in Kansas. Kansas. Department of Transportation. Bureau of Materials & Research
7. Russell HS, Hossain M (2000) Design resilient modulus of subgrade soils from FWD tests. Pavement Subgrade, Unbound Mater Nondestr Test 87–103.
8. Vennapusa RKP, White JD, Morris DM (2010) Geostatistical analysis for spatially referenced roller-integrated compaction measurements. J Geotech Geoenviron Eng 136(6):813–822
9. Facas N, Mooney M (2010) Position reporting error of intelligent compaction and continuous compaction control roller-measured soil properties. J Test Eval 38(1):1
10. Thompson JM, White JD (2008) Estimating compaction of cohesive soils from machine drive power. J Geotech Geoenviron Eng 134(12):1771–1777
11. Meehan LC, Cacciola VD, Tehrani SF et al (2017) Assessing soil compaction using continuous compaction control and location-specific in situ tests. Autom Constr 73:31–44
12. Wang X, Cheng C, Li J et al (2023) Automated monitoring and evaluation of highway subgrade compaction quality using artificial neural networks. Autom Constr 145:104663
13. White JD, Thompson JM (2008) Relationships between in situ and roller-integrated compaction measurements for granular soils. J Geotech Geoenviron Eng 134(12):1763–1770
14. Anderegg R, von Felten DA, Kaufmann K (2006) Compaction monitoring using intelligent soil compactors. In: Geocongress 2006: geotechnical engineering in the information technology age 1–6
15. Komandi G (1999) An evaluation of the concept of rolling resistance. J Terramech 36(3):159–166
16. White DJ, Jaselskis EJ, Schaefer VR et al (2005) Real-time compaction monitoring in cohesive soils from machine response. Transp Res Rec 1936(1):172–180

17. Rinehart RV, Mooney MA (2008) Instrumentation of a roller compactor to monitor vibration behavior during earthwork compaction. Autom Constr 17(2):144–150
18. Xu Q, Chang KG (2016) Adaptive quality control and acceptance of pavement material density for intelligent road construction. Autom Constr 6278–6288
19. Liu DH, Lin M, Li S (2016) Real-time quality monitoring and control of highway compaction. Autom Constr 62:114–123
20. Mooney MA, Rinehart RV (2007) Field monitoring of roller vibration during compaction of subgrade soil. J Geotech Geoenviron Eng 133(3):257–265
21. Guo DZ, Ren RB (2001) Mechanics of layered viscoelastic systems. Harbin Institute of Technology Press
22. Wu LL, Teng J, Ren ZY et al (2022) Research on identification model of continuous compaction based on energy dissipation. Shock Vib
23. Ye YS, Zhu HW, Yao JK et al (2021) Review on the vibration compaction theory and intelligent compaction technology of high-speed railway subgrades J. China Railw Sci 42(05):1–11
24. Li D (2016) Simulation analysis of vibrating compaction process of asphalt pavement. Chongqing Jiaotong University
25. An ZZ, Liu TY, Huangfu ZH, Zhang QL, Li CS, Li QB (2020) Neural network model for evaluating compaction quality of rockfill materials by compaction meter value. J Hydroelectr Eng 39:110–120

Chapter 4
Non Contact-Type Roller-Integrated Acoustic Wave Detection Technique

Effective control of compaction quality is essential to ensure the safety of rockfill dams [1]. Numerous researchers have endeavored to develop techniques and methods for detecting compaction in soil. These detection techniques can be categorized into contact-type and non contact-type methods, based on whether the detection equipment makes contact with the material. Contact-type methods include the water replacement method, the falling weight deflectometer (FWD), the lightweight deflectometer (LWD), the dynamic cone penetrometer (DCP), the California bearing ratio (CBR), the nuclear density gauge, and the resistivity test method, among others. Non contact-type techniques encompass ground penetrating radar (GPR), Rayleigh wave methods, near-infrared techniques, and others (see Sect. 4.1.2).

Contact-type techniques and methods are employed to assess the compactness of rockfill materials by measuring their density and moisture content. Krebs et al. [2] studied the compaction control of highway materials using the sand cone method, which is notably time-consuming. To address this limitation, Rahman et al. [3] and Meehan et al. [4] further explored compaction control of pavement layer materials and coarse-grained soil using lightweight deflectometers (LWD), falling weight deflectometers (FWD), dynamic cone penetrometers (DCP), and soil stiffness gauges (SSG). Melbouci [5] and Zabilska-Adamska et al. [6] examined compaction control of recycled aggregates, cohesive soils, and fly ash employing the California bearing ratio (CBR) test. Resende et al. [7] utilized the DCP to investigate quality control in mining tailing dams. Kongkitku et al. [8] developed the LWD for estimating the surface stiffness of lateritic soil. Subsequently, Umashankar et al. [9] employed the LWD to study compaction control in pavement layers. To further streamline the operational process and enhance efficiency, Swinford et al. [10] and White et al. [11] investigated soil compaction using nuclear density gauges. While contact-type techniques and methods offer high accuracy and a broad range of applications, they also present significant disadvantages, such as point sampling, destructive measurements, low efficiency, and higher operational requirements.

Q. Zhang, *Compaction Quality Nondestructive Testing and Intelligent Control Technology for Infrastructure*, https://doi.org/10.1007/978-981-95-1264-5_4

To circumvent the disadvantages associated with contact-type techniques and methods, numerous researchers have developed non contact-type alternatives. Thurner et al. [12] and Forssblad [13] noted that the vibration acceleration amplitude and harmonic amplitude correlate with soil density and underlying stiffness. Anderegg et al. [14] extracted soil stiffness based on vibration acceleration and phase difference. Through extensive research, a relative measurement index value (RICMMV) based on the Relative Impact Compaction Method (RICM) was proposed to assess soil compactness [15–18]. RICMMV is defined in various forms, including compaction meter value (*CMV*) as used by Geodynamic [16], k_b by Ammann [19], machine drive power (*MDP*) by Caterpillar [15, 17, 20], and OMEGA by Bomag [18]. Furthermore, Rinehart et al. [21, 22] identified total harmonic distortion (*THD*) as a highly sensitive index for evaluating soil compaction status, while White et al. [20] introduced E_{vib} as a vibration modulus. Related studies have shown that these indexes closely correlate with the stiffness or modulus of compacted materials, suggesting that non contact-type techniques based on RICMMV can effectively detect compactness and evaluate soil compaction quality.

By integrating automated feedback control (AFC) techniques, some researchers have developed IC systems for road compaction control, which are now widely utilized in Europe, Japan, and the United States [23–25]. Additionally, wave techniques have been employed to detect subgrade compactness [26, 27]. However, the laws and mechanisms governing wave propagation in various soil types have yet to be thoroughly studied, leading to larger detection errors and higher operator requirements. For pavement assessment, ground penetrating radar (GPR) techniques can gauge moisture content and compactness of the subgrade; however, the impact of surface jet moisture on GPR signals during compaction remains unclear [28]. Austin et al. [29] employed near-infrared and microwave resonance sensing techniques to monitor continuous roller compaction processes; however, this method involves complex application and data processing procedures. Overall, while non contact-type techniques and methods offer advantages such as being nondestructive, rapid, and efficient, they also present challenges including lower accuracy, complexity in interpreting compaction conditions, significant variability, greater potential for errors, and susceptibility to material factors.

In non contact-type techniques and methods, the *CMV* is frequently employed to characterize the compactness of granular soils and assess compaction quality [16, 29]. Adam [18], Sandström [30], and Vennapusa et al. [31]. indicated that *CMV* measurements should be interpreted alongside resonant meter value (*RMV*) measurements; however, this necessity has created challenges in the practical application of *CMV* [31]. Moreover, research by White and Thompson [16] demonstrated that *CMV* exhibited significant variability and measurement errors; the discreteness of *CMV* values tended to increase with larger particle sizes, and the variability-induced measurement errors rendered *CMV* inadequate for practical applications. Consequently, *CMV* is deemed unsuitable for rockfill materials with large particle sizes exceeding 200 mm. Liu et al. [29] employed RICM techniques to evaluate the compaction quality of earth-rock dam, observing that the grain size of the material was less than 120 mm. They noted that the variability of *CV* values escalated with increasing particle size, and Liu's *CV* method and regression model were not appropriate for materials with a particle size range of 0.1–400 mm. To address the issues associated with current detection techniques, we propose a novel compaction detection method characterized by non contact-type, high efficiency, high precision, continuity, and convenience, making it suitable for rockfill materials.

4.1 Theoretical Analysis

In terms of the theoretical model, the interaction between the soil and the vibration wheel is conceptualized as Lamb's problem, which pertains to the propagation of vibration across the surface of a semi-infinite isotropic elastic solid. The acoustic field resulting from this interaction is interpreted as the infinite baffle piston radiation acoustic field. The new technique is predicated on a vocal mechanism as follows: when the vibration wheel engages with the soil, the soil experiences a vertical contact force alongside horizontal friction. The combined effects of these two forces induce the corresponding soil particles to vibrate both vertically and horizontally in a compound manner, simultaneously generating sound. Concurrently, the fluctuations caused by the vibration of these particles lead to alterations in the density and pressure of the air medium at the interface between the soil and the gas, resulting in changes in the state of the air medium at the solid-air interface. Consequently, acoustic waves are produced at this interface and radiate through the air medium in a specific manner.

Taking into account the mode of interaction between the vibration wheel and the soil, as well as its boundary conditions, the acoustic wave radiation field formed at the solid–gas interface can be regarded as a piston-type radiation field on an infinite plate. Specifically, as the vibration wave propagates across the surface of the solid medium, a defined area on this surface can be treated as a circular piston radiator on an infinite baffle. The piston vibrates vertically at a speed (u), radiating acoustic waves into the semi-infinite space in front of the baffle.

4.1.1 Lamb's Problem

The elastic fluctuation problem induced by dynamic line or point loads on the surface or within a semi-space is a matter of considerable importance in the study of fluctuations in elastic solids. The foundational research can be traced back to the seminal work of Lamb [32]. In the early 1930s, Cagniard developed a generalized method for addressing Lamb's problems, employing the displacement field of the Fourier transform, with the inverse transformation derived using the Laplace transform. Subsequently, De Hoop [33] made further enhancements to this methodology. Reissner [34] analyzed the behavior of a flexible circular plate subjected to a uniformly distributed load on an elastic semi-space, demonstrating that the problem could be resolved through the integration of Rahm's solution for point load scenarios. In 1953, Quinlan [35] and Sung [36] published two papers that expanded upon Reissner's solution by considering the effects of pressure distribution on the circular contact area at the surface of the semi-space.

Regarding the dynamic response of saturated soil, Biot [37] was the first to establish a phenomenological theory addressing the fluctuations in two-phase media, which paved the way for a genuine investigation into the vibrational characteristics of foundations situated on saturated soil. Following Biot's formulation of the wave equation for saturated porous media in 1956, Paul [38] and Philippacopoulos et al. [39] conducted thorough examinations of Lamb's problem in the context of horizontal and vertical forces acting on the surface of saturated soil. Paul, neglecting the viscous fluid component in saturated soil, utilized the Hankel transform and Cagniard method, employing the Helmholtz decomposition proposed by Deresiewicz, thus becoming the first to explore Lamb's problem within a saturated semi-space. Wang et al. [40], hypothesizing that viscosity was incorporated into the dynamic permeability coefficient, employed the integral transformation method to solve the coupled equations, circumventing the need for potential functions, and provided an integral solution for saturated elastic semi-space under low-frequency harmonic concentrated forces. Cai [41] investigated the steady-state response of axisymmetric soil subjected to vertical loads, developing the amplitude curve for surface displacement. Building upon these foundational solutions, extensive research has also been conducted on the dynamic interaction between saturated soil and various types of foundations [42, 43]. Lamb's problem has garnered significant international attention, leading to marked advancements in analytical methods, resulting in increasingly refined solutions.

In this section, the interaction between the soil and the vibratory wheel is framed as Lamb's problem for saturated soil. To address this issue, an innovative differential pulse excitation method (DPEM) has been proposed, building upon the vertical displacement solutions applicable to normal concentrated loads acting on the surface of the space. This method aims to determine the vertical displacement within the semi-space under simple harmonic loading. The isotropic and homogeneous semi-space is illustrated in Fig. 4.1, with its surface represented as the (X–Y) plane.

The coordinate axis (z) extends into the interior of the semi-space. At the origin (o), there exists a normal concentrated load ($QH(t)$). Due to the symmetry of the load about the (z) axis, the wave motion problem within the infinite elastic semi-space exhibits axial symmetry. Utilizing a cylindrical coordinate system (r, θ, z), the vertical displacement (u_z) can be expressed in terms of the displacement potentials (φ) and ($\chi(\psi = 0)$) in the symmetric configuration of the cylindrical coordinate system, as follows:

$$u_z = \frac{\partial \varphi}{\partial z} - \frac{1}{r}\frac{\partial}{\partial r}\left(r\frac{\partial \chi}{\partial r}\right) \tag{4.1}$$

Drawing upon the research of Lamb [32] and other scholars [35, 38, 40], the vertical displacement at the surface is described as follows:

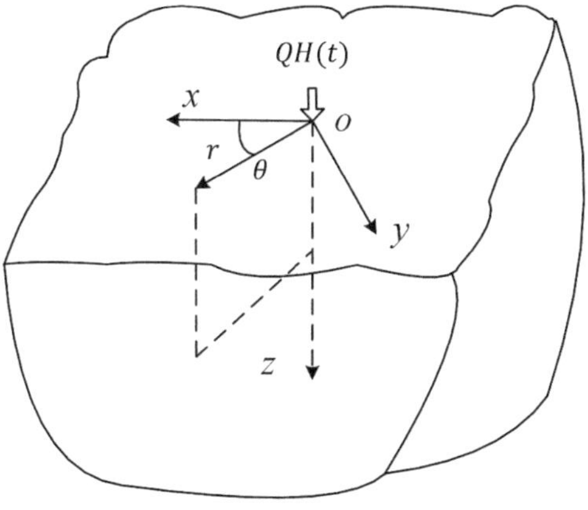

Fig. 4.1 Isotropic and homogeneous semi-space

$$u_z(r,0,t) = \begin{cases} 0 & (\tau < c_T/c_L) \\ -\frac{Q}{\pi^2 Gr} G_1(\tau) & (c_T/c_L < \tau < 1) \\ -\frac{Q}{\pi^2 Gr[G_1(\tau)+G_2(\tau)]} & (\tau > 1) \end{cases} \tag{4.2}$$

Following additional analysis, Pekeris [44] obtained the analytical results of the integral through the partial fraction decomposition of the integrand functions $G_1(\tau)$ and $G_2(\tau)$. Figure 4.2 illustrates the alternative relationship between the surface vertical displacement $u_z(r, 0, t)$ and the variation of τ.

Given that a normal concentrated load $QH(t)$ is applied at the origin o (as depicted in Fig. 4.1), and considering the symmetry of the load about the z-axis, Q can be treated as a constant force. For the vibratory roller, the exciting force Q_o is introduced by the centrifugal force generated when the eccentric oscillator rotates at high speed. This force is solely dependent on the static eccentricity M_e and the angular frequency

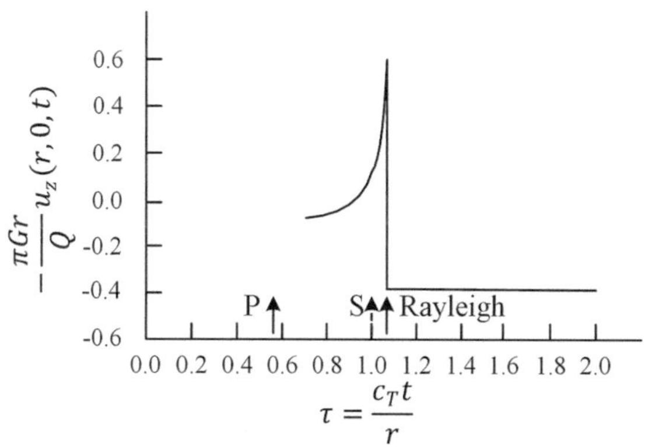

Fig. 4.2 Relation between the surface vertical displacement $u_z(r, 0, t)$ and τ

ω of the vibrator. In a damped vibration system, a periodic force must be exerted by an external source to sustain the vibration. Typically, harmonic excitation serves as the primary force, and the vibration wheel of the roller operates under simple harmonic excitation. Assuming that the vibration system is subjected to a simple harmonic load defined as $Q(t) = Q_0\cos\omega t$, it becomes necessary to determine the vertical displacement u_z of the soil at the surface of the semi-space under the influence of this harmonic load. Consequently, an innovative differential pulse excitation method has been proposed, employing differential pulse excitation to simulate the harmonic load and facilitate the resolution of vertical displacement. The specific methodology is outlined as follows:

(1) Initially, the periodic harmonic load excitation is discretized into N parts, whereby the N-th excitation pulse is denoted as Q_n (where $(n \in [1, N])$, as illustrated in Fig. 4.3.
(2) Subsequently, we assume that $Q_n = Q^n((n-1)\nabla\tau)H(t+(n-1)\nabla\tau) - Q^n(t+(n-1)\nabla\tau)H(t+n\nabla\tau))$, where $\nabla\tau$ represents the delay time, and $Q^n((n-1)\nabla\tau) = Q_0\cos(\omega(n-1)\nabla\tau)$, as depicted in Fig. 4.4.
(3) Impulse excitations $(Q_1, Q_2, Q_3, Q_4, \ldots, Q_k, \ldots, Q_N,)$ are sequentially applied at the origin o, and the vertical displacement u^k (where $(k \in [1,N])$) at a distance r from the origin is determined as each distinct pulse load interacts with the semi-space surface. Over the interval $[0, t]$ (where $(t \in [0, 2\pi])$), the true vertical displacement u_z of the soil under the simple harmonic load $Q(t) = Q_0\cos\omega t$ at any moment within a cycle can be acquired by integrating the vertical displacement response induced by the pulse excitation. The displacement u_z is expressed as follows:

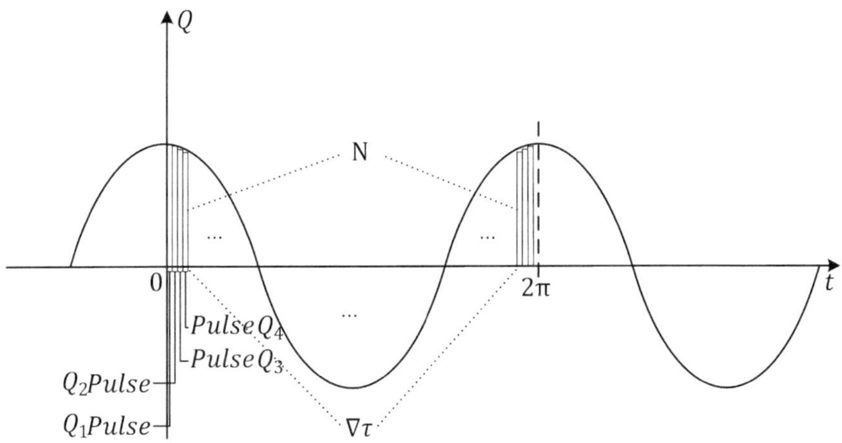

Fig. 4.3 Schematic diagram of pule load formed by differential concentrated load

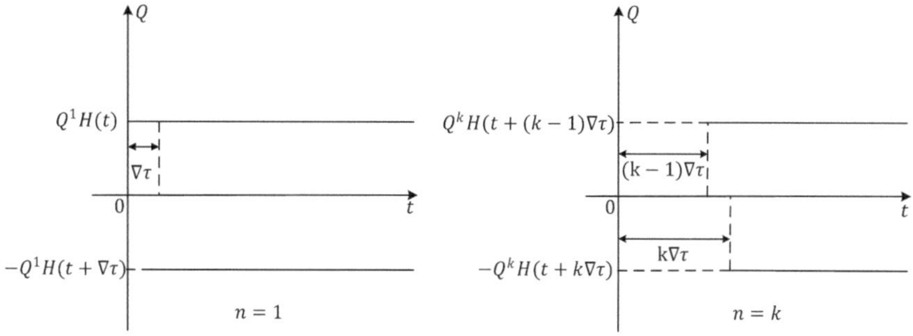

Fig. 4.4 Periodic simple harmonic load excitation discretization

$$u_z = \int_0^t u^k dt \tag{4.3}$$

where $k \in [1, t/\Delta t]$ and $t/\Delta t$ is an integer.

To ensure the accuracy of the waveform and displacement depicted in Fig. 4.3, let $\nabla\tau = 0.001$s, resulting in $N = 2\pi/\nabla\tau \approx 6283$. By employing the aforementioned method to resolve the vertical displacement under a simple harmonic load, the requisite vertical displacement can be derived by integrating the research findings of Pekeris with pertinent data. Figure 4.5 illustrates the variations in vertical displacement at $t = 500\nabla\tau$, $1000\nabla\tau$, $4000\nabla\tau$ and 6283 $\nabla\tau$, respectively.

By resolving the problem, the true vertical displacement u_z of the soil at the surface of the semi-space at any moment within a cycle under the harmonic load $Q(t) = Q_0\cos\omega t$ can be expressed as follows:

$$u_z(r, 0, t) = \frac{Q_0}{A_0\pi Gr}\cos\omega_0 t \tag{4.4}$$

where $A_0 = 0.03605$ and $\omega_0 = 2\pi/T_0$, $T_0 = 0.036s$.

To determine the real-time vertical displacement over multiple cycles, the displacement function from Eq. (4.4) simply requires a horizontal shift.

4.1.2 *Infinite Baffle Piston Radiation Acoustic Field*

Figure 4.6a illustrates a circular piston radiator positioned on an infinite baffle. When the piston vibrates with a velocity $u = u_A e^{iwt}$ in the z direction, it radiates acoustic waves into the semi-infinite space in front of the baffle. In Fig. 4.6b, a surface element with an area of dS and located at a distance q from the center of the piston is selected. Since the focus of this study is on far-field acoustic wave radiation, this surface element can be treated as a point source, with the entire piston radiator

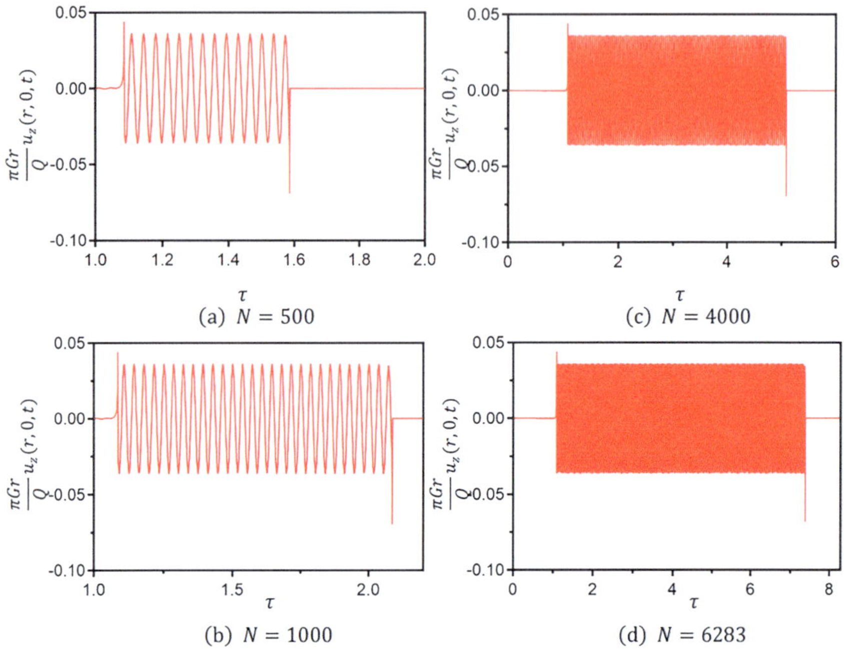

Fig. 4.5 Change curve of vertical displacement at different times

comprising numerous such point sources. Drawing upon educational materials and relevant literature, the far-field radiation sound pressure is expressed as follows:

$$p = i\frac{a\rho_a cu_A J_1(ka\sin\theta)}{2r_P\sin\theta}e^{i(\omega t - kr_P)} \tag{4.5}$$

For an N-th order first-class Bessel function, if n is a non-negative integer, then $\Gamma(\mathrm{n}+\mathrm{m}+1) = (\mathrm{n}+\mathrm{m})!$. Furthermore, by considering the relationship between the zero-order and first-order Bessel functions, and assigning $n = 0$ and $n = 1$, the following formula is derived:

$$J_1(x) = \frac{x}{2} - \frac{x^3}{2^3\cdot 2!} + \cdots + (-1)^k\frac{x^{2k+1}}{2^{2k+1}k!(k+1)!} + \cdots \tag{4.6}$$

Considering the characteristics of the acoustic far field, the following relations are established: $r_p \gg a$ and $r_p \gg a^2/l$. According to the small argument property of the Bessel function, as $x \to 0$, substituting $J_1(x)$ into Eq. (4.5) yields the following formula:

$$p(r_p, t) = i\frac{ka^2\rho_a cu_A\left[\frac{1}{2} - \frac{1}{16}(ka\sin\theta)^2\right]}{2r_p}e^{i(\omega t - kr_p)}$$

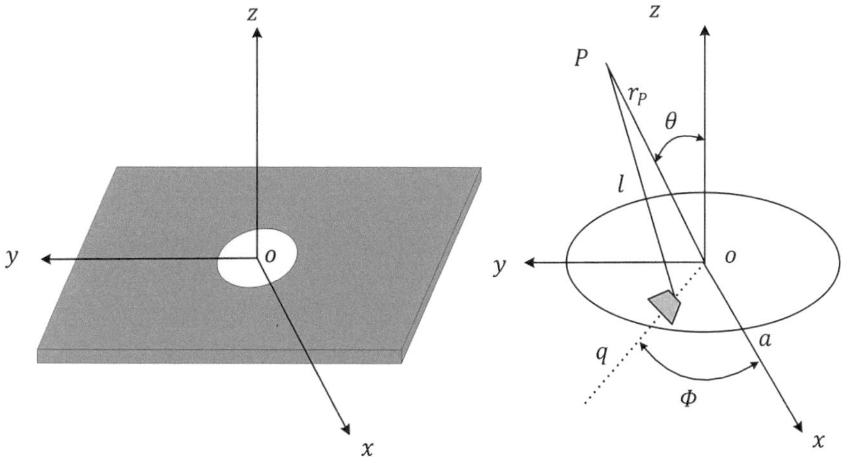

Fig. 4.6 **a** Circular piston on infinite baffle; **b** radiated acoustic field of circular piston

$$\approx i\frac{ka^2\rho_a c u_A}{4r_p}e^{i(\omega t-kr_p)} \tag{4.7}$$

4.1.3 Analytical Solution

In acoustics, it is often necessary to compute sound pressure and particle velocity. For the sake of convenience in calculations, plural forms are frequently employed to represent these physical quantities. The velocity of the piston in the z direction can be expressed as $u = u_A\cos\omega t$, while the displacement of the piston in the z direction is given as follows:

$$\overline{u}_z^0 = \int_0^t u dt = u_A \sin \omega t \tag{4.8}$$

When $\tau = 1.088$, Rayleigh waves will arrive, and since $\tau = c_T t_0/r = 1.088$, it follows that $t_0 = 1.088r/c_T$. Given that $T = \sqrt{G/\rho}$, at time t_0, the boundary condition between the soil and the air medium can be articulated by Eq. (4.9):

$$\overline{u}_z^0\big|_{t=t_0} = u_z(r, 0, t_0) \tag{4.9}$$

Substituting Eqs. (4.4) and (4.8) into Eq. (4.9), we note that $k = \omega/c$, where k represents the wave number, ω is the angular frequency, and c is the wave velocity. Consequently, Eq. (4.10) is derived:

$$k = \frac{1}{c}\arcsin\left(\frac{Q_0}{\pi G r u_A}\right)\cos\left(1.088\omega_0 r\sqrt{\frac{\rho}{G}}\right) \tag{4.10}$$

By substituting Eq. (4.10) into Eq. (4.7), the expression for sound pressure is derived as follows:

$$p(r_p, t) = i\frac{a^2\rho_a u_A}{4r_p}\arcsin\left(\frac{Q_0}{\pi G r u_A}\cos\left(1.088\omega_0 r\sqrt{\frac{\rho}{G}}\right)\right)e^{i(\omega t - kr_p)} \tag{4.11}$$

where $u_A = p_0/\rho_a c_0$, with p_0 representing the standard atmospheric pressure at ambient temperature, c_0 denoting the propagation velocity of the acoustic wave in the air medium at ambient temperature, ρ being the mass density of the soil, and μ signifying the shear modulus of the soil.

After applying the Fast Fourier Transform (FFT) to Eq. (4.11), the root mean square (RMS) amplitude can be determined. In Eq. (4.12), b_0 is a constant:

$$A = \frac{a^2\rho_a u_A}{4r_P}b_0\arcsin\left(\frac{Q_0}{\pi G r u_A}\cos\left(1.088\omega_0 r\sqrt{\frac{\rho}{G}}\right)\right) \tag{4.12}$$

Lade and Nelson [45] formulated a nonlinear isotropic model to describe the elastic behavior of granular materials. In their study, Young's modulus is articulated as shown in Eq. (4.13):

$$E = Mp_a\left[\left(\frac{I_1}{p_a}\right)^2 + R\frac{J_2'}{p_a^2}\right]^2 \tag{4.13}$$

Building upon the work of Lade and Nelson, Liu et al. [46] conducted a study on constitutive modeling of dense gravelly soils subjected to cyclic loading. In their paper, the modulus G is further articulated as shown in Eq. (4.14):

$$G = G_0 p_a \frac{(2.97 - e)^2}{1 + e}\left[\left(\frac{3p}{p_a}\right)^2 + 9\frac{K_0}{G_0}\frac{J_2}{p_a^2}\right]^{m/2} \tag{4.14}$$

where G represents the shear modulus, G_0 and m are material constants, e denotes the current void ratio, and p_a is the atmospheric pressure.

By substituting Eq. (4.14) and $e = 1 - \rho_0/\rho$ into Eqs. (4.12) and (4.15) is derived:

$$A = A_0\arcsin\left(B_0\cos\left(1.088\omega_0 r\frac{\rho}{1.97\rho + \rho_0}\sqrt{M_0(2\rho - \rho_0)}\right)\right) \tag{4.15}$$

where, $M_0 = G_0 p_a\left[(3p/p_a)^2 + 9(K_0/G_0)\left(J_2/p_a^2\right)\right]^{m/2}$, ρ is the mass density, and ρ_0 is the dry density.

A diagrammatic representation of the compacting area during the pulse load duration time $\Delta\tau$ is depicted in Fig. 4.7. The velocity of the roller is denoted as v. As previously mentioned, the harmonic load is divided into N parts (refer to Fig. 4.3). The duration of each pulse load is $\Delta\tau$. The compaction area within the time interval $\Delta\tau$ is illustrated as rectangle ABCD in Fig. 4.7. In this context, the area ABCD can be regarded as consisting of N_c rows and M_c columns of small piston radiators. Consequently, the radiation field generated by the interaction between the vibrating wheel and the soil comprises N_c rows and M_c columns of small piston radiators positioned on an infinite baffle. Each piston has a radius of $a = 0.01$ cm, and given that the length of the vibrating wheel is $L = 2.2$ m, the number of pistons in each column is calculated as $N_c = L/2a = 11000$, while the number of pistons in each row is given by $M_c = v\Delta\tau/2a$, $M_c \in Z$.

When an interaction occurs between the vibratory wheel and the soil, the cumulative effect of the combined piston radiation acoustic fields in the far field at point P is expressed as follows:

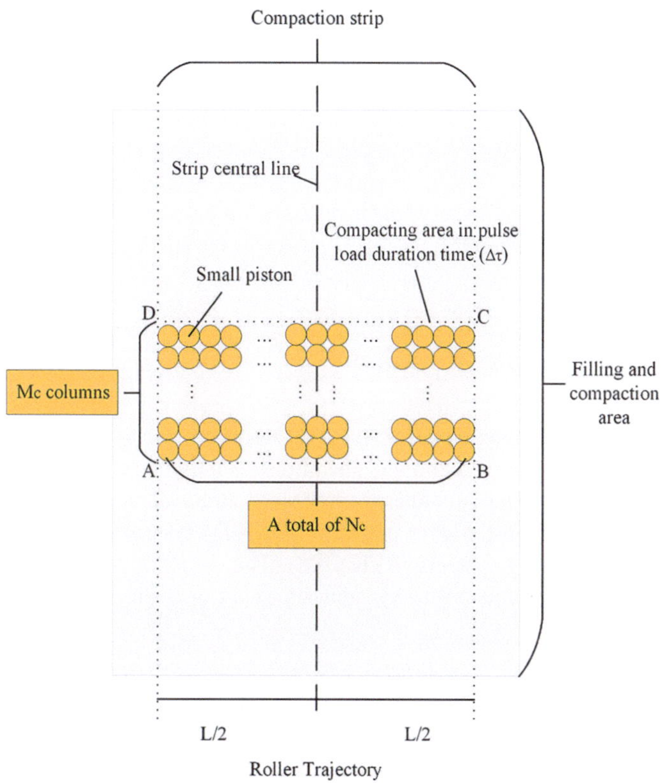

Fig. 4.7 Diagrammatic sketch of compaction area in pulse load duration time $\Delta\tau$

$$A_{all} = 2M_c \int_{r_{p0}}^{r_{pN/2}} A dr_p$$
$$= C_0 \ln\left(\frac{r_{pN/2}}{r_{p_0}}\right) \arcsin\left(B_0 \cos\left(1.088\omega_0 r \frac{\rho}{1.97\rho + \rho_0}\sqrt{M_0(2\rho - \rho_0)}\right)\right) \tag{4.16}$$

Based on Eq. (4.16), a theoretical relational model between the soil's shear wave attenuation (SHA) and its dry density was established, referred to as the A-model.

4.1.4 Numerical Solution

In this section, numerical results are presented for the A-model established in the previous segment. All parameters and theoretical models have been implemented using MATLAB. The numerical solutions related to gravel materials were validated by comparing them with results obtained from field tests based on the new compaction detection technique proposed in this section and traditional point measurements (water replacement method). Table 4.1 illustrates the key parameters necessary for the numerical analysis. Figure 4.8 depicts the relationship curve between the dry density of specific gravel materials and the *SCV*, according to the A-model. The results indicate an approximately linear relationship between the *SCV* and the dry density of the gravel materials.

4.2 Technical Implementation

In terms of technical implementation, the detection principle is outlined as follows: this acoustic wave detection technique encompasses both detection and signal analysis. The compaction data is collected, processed and analyzed by the continuous compaction control acoustic wave detection (CAWD) system with real-time kinematic global positioning systems (RTK-GPS). The detection equipment is affixed to the vibratory roller and primarily consists of an acoustic field microphone, a

Table 4.1 Key parameters required for the numerical analysis

Parameter	Value	Parameter	Value	Parameter	Value
p_0/Pa	10^5	r/m	$\leq 10^{-6}$	a/cm	0.01
ρ_a/(kg/m^{-3})	1.2	π	3.14	ω_0/(rad/s^{-1})	175
c_0/(m/s^{-1})	340	ρ_0/(g/cm^{-3})	2.0 –2.5	ρ/(g/cm^{-3})	2.2
m_0	0.375	r_p/cm	23	μ/(MN/m^{-2})	100–200

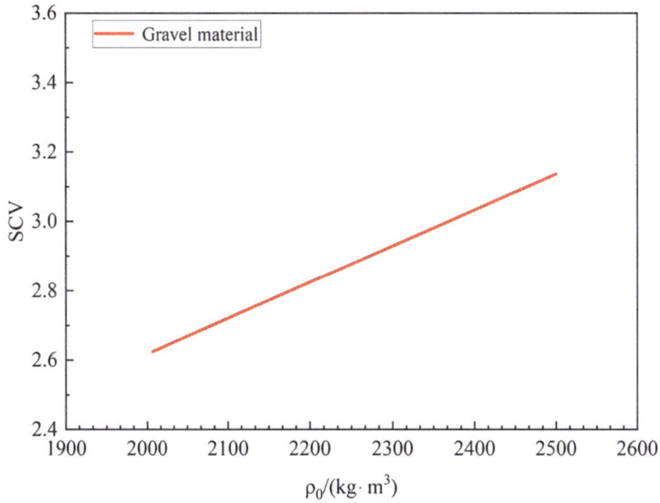

Fig. 4.8 Relation curve between dry density of gravel materials and *SCV* based on A-model

signal acquisition analyzer, a display, and a GPS receiver. As the filling layer is compacted by the roller, the time-domain signal of the acoustic wave field generated near the contact surface between the vibratory wheel and the soil is captured by the acoustic field microphone mounted on the outer frame of the vibrating wheel. This signal is subsequently collected by the analyzer to produce digital signals. The GPS receiver, also installed on the roller, concurrently provides spatial data associated with the roller's location. The signal acquisition analyzer filters and conducts spectrum analysis on the effective acoustic signal to derive the second harmonic amplitude (SHA). Based on the correlation between SHA and the compactness or density of the rockfill material, a CCI is established following calibration. Coupled with spatial location information, a spatiotemporal compactness distribution map of the rolling area is displayed on a roller-integrated screen. Furthermore, in conjunction with the compactness feedback module, intelligent compaction can be easily achieved by adjusting the mechanical parameters.

The schematic diagram of this technique is illustrated in Fig. 4.9. The detection device 2 comprises an acoustic field microphone 3, a microphone mounting apparatus 10, a signal conditioning module 4, a data acquisition module 5, an analysis processing module 6, a display 7, and a GPS receiver 9. Additionally, the system includes a vibratory roller 1 and an onboard battery 11. The acoustic field microphone 3 is primarily employed to capture time-domain signals from the acoustic field. The signal conditioning module 4 converts the signals received by the acoustic field microphone 3 into standard signals through amplification and filtering. These standard signals are then collected by the data acquisition module 5, which simultaneously gathers spatial data associated with the roller's position, as provided by the GPS receiver 9. The two signals are subsequently transmitted to the analysis processing module 6 via wired communication.

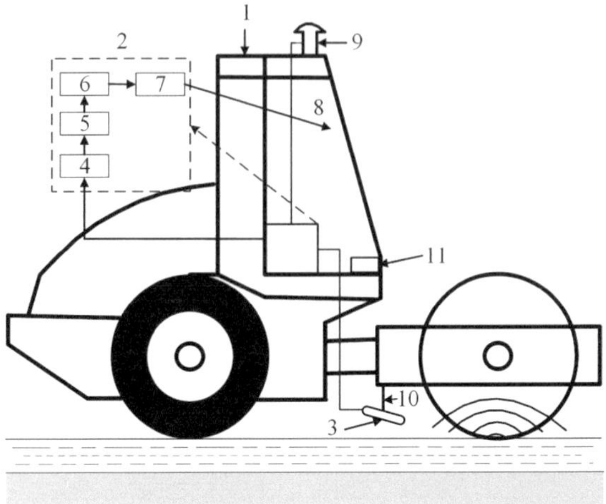

Fig. 4.9 Schematic diagram of acoustic wave detection technology

The analysis processing module 6 performs filtering, spectrum analysis, and logarithmic processing of the acoustic signals to derive the SHA of the effective acoustic signal, calculates the *SCV* values in real time through multiplication by the calibration coefficient, processes the spatial signals related to the roller's position using a differential algorithm, and computes the current position coordinates of the vibratory roller 1 with centimeter-level accuracy. The display 7 is utilized to present a real-time compaction nephogram showcasing spatial location characteristics, along with other pertinent information such as driving speed, compaction monitoring data, and the number of compaction cycles. The GPS receiver 9 is responsible for acquiring spatial signal information related to the roller's position. The RTK-GPS receiver, satellite, and GPS base station collaboratively establish the roller's position. The microphone mounting apparatus 10 secures the acoustic field microphone 3, while vibration isolation between the acoustic field microphone 3 and the mounting device 10, as well as between the mounting device 10 and the vibratory roller 1, is achieved using soft rubber. The onboard battery 11 provides power support for the entire CAWD system.

4.3 *SCV*

During rolling, the compacted layer is influenced by both static pressure and vibrational forces. This combined effect can alter the material's state from static to dynamic, thereby reducing the friction and cohesion among soil (or rock) particles. Consequently, smaller soil (or rock) particles fill the voids between larger particles, resulting in a gradual densification of the material. At the onset of rolling construction, the filling materials are relatively loose (Fig. 4.10a), the sound pressure curve appears smooth (Fig. 4.10d), and the corresponding sound spectrum predominantly features the fundamental frequency f_0, with the amplitude of the *SCV* being relatively low (Fig. 4.10g).

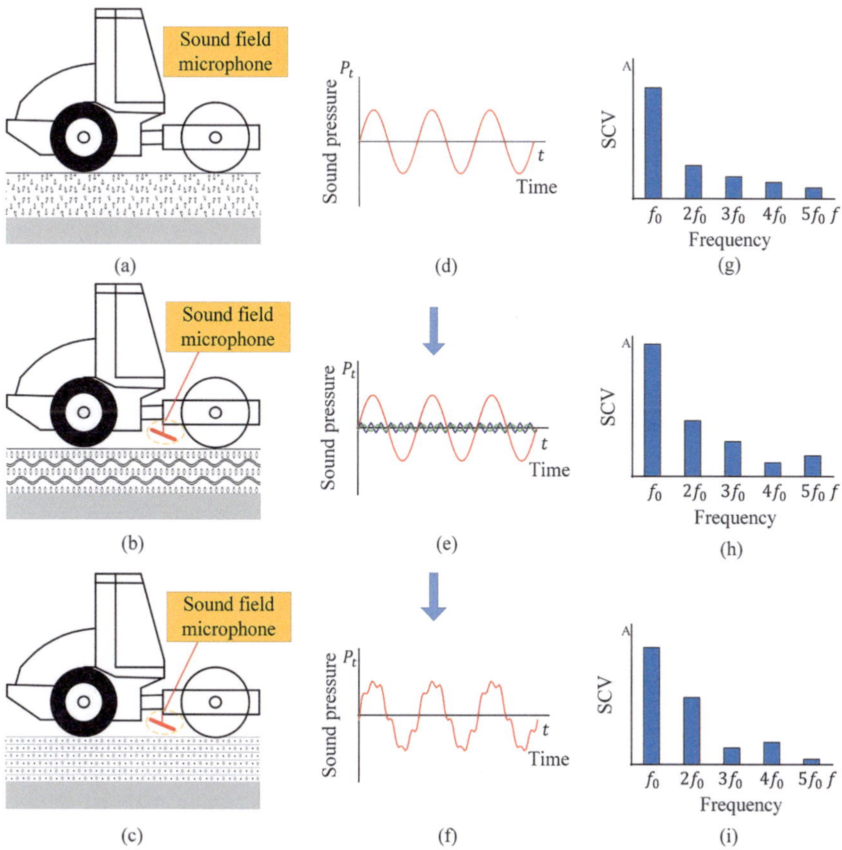

Fig. 4.10 Illustration of changes in sound pressure and *SCV* with increasing ground stiffness

As the number of compactions increases, the compacted material becomes denser (Fig. 4.10b), and the higher harmonic components gradually rise (Fig. 4.10e). Simultaneously, the *SCVs* also undergo gradual changes (Fig. 4.10h). Once the number of compaction cycles reaches a certain threshold, the content of higher harmonic components stabilizes (Fig. 4.10f), and the compacted material attains greater density, approaching a stable state (Fig. 4.10c). The *SCVs* continue to evolve, ultimately tending toward stability (Fig. 4.10i). This section defines a real-time *SCV* as the compaction index for characterizing the compaction status of rockfill materials as follows:

$$SCV = k \times SHA \times 100\% \tag{4.17}$$

where k is the calibration coefficient of the material.

4.4 Case Study

In response to the issues of low efficiency, poor accuracy, lack of timeliness, and insufficient application experience in earth-rock dam engineering associated with existing detection methods, this chapter introduces a novel compaction detection method based on contact-type roller-integrated acoustic wave detection technology for assessing the compaction of shell materials (rockfill) in earth-rock dams. This subsection focuses on the construction of the Qianping Reservoir located in Luoyang, Henan Province, using natural gravel (rockfill) required for the construction of the main dam (a clay core earth-rock dam) as the subject of study to validate the compaction detection method utilizing non contact-type roller-integrated acoustic wave detection technology.

4.4.1 Testing Site and Materials

A reservoir project is situated in Henan Province, China, featuring primary structures that include the main dam, an auxiliary dam, a spillway, a water tunnel, and a power station. The main dam is constructed as a earth-rock dam with a maximum height of 90.3 m. This experimental research concentrated on the near-grain material (NGM) at the upstream cofferdam, which forms a component of the main dam. During the initial phase of the reservoir project, the China Institute of Water Resources and Hydropower Research (IWHR) conducted a rolling test on gravel material. The grading curve of the gravel material sourced from the VI zone material field is depicted in Fig. 4.11. Among five sets of tests, only one group of test materials conformed to the representative grading envelope, while the others fell outside the original grading parameters. Considering the significant variability of gravel material across different material fields, the NGM was conservatively estimated to have a maximum particle size of 200 mm. The test results from IWHR revealed that the proportion of particles less than 5 mm ranged from 8 to 26%, while the proportion of particles exceeding 200 mm varied from 1.3 to 7.1%.

The maximum size of the natural gravel material reached up to 400 mm, with the content of particles smaller than 5 mm maintained at no more than 15%. The design standard for the dry density of the NGM was established to exceed 2.144 g/cm^3, while the required compactness was stipulated to be greater than 75%, in accordance with the design criteria.

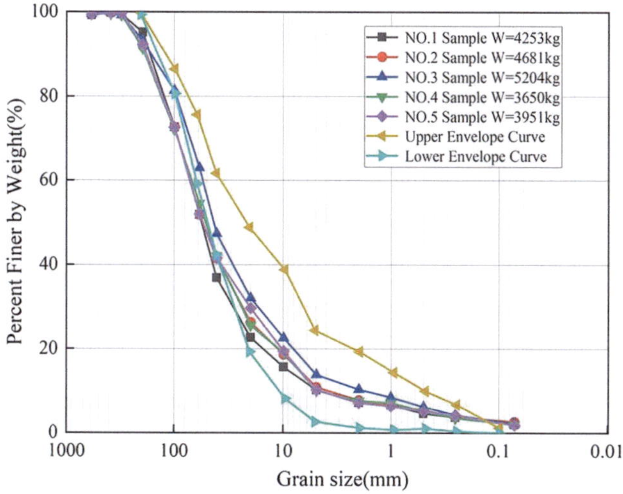

Fig. 4.11 Grading curve of gravel material taken from in situ VI zone material field

4.4.2 *Testing Program*

To evaluate the efficacy of the *SCV* in assessing the compaction quality of rock-fill materials, a two-part field experiment was conducted. The first segment of the experiment examined the correlation between the compaction parameters and the *SCV*. Four testing strips were filled with NGM characterized by consistent moisture content and gradation. Each strip measured 30 m in length and 3 m in width. Figure 4.12 displays images of the field rolling test conducted in the rock-fill zone located at Elevation 1 and Elevation 2 (upstream cofferdam), while the test scheme is outlined in Table 4.2. In this context, NGM refers to natural gravel materials, and HFHA denotes high-frequency/high-amplitude. The second part of the experiment involved traditional point measurements utilizing the water replacement method, aimed at analyzing the relationship between the *SCV* and the number of compaction passes as well as between *SCV* and the compactness.

To assess this technique, a field study was conducted utilizing 30-m test strips (Fig. 4.12) filled with NGM. A comprehensive set of field data was collected, encompassing compaction parameters, *SCVs*, and results from traditional point measurements, all of which were employed for correlation analysis to validate the effectiveness of this technique. The test strips were compacted using a prototype SSR260C-6 vibratory smooth drum roller, with the depth of the filling layer approximately (and not exceeding) 0.8 m. Two test sites situated at different elevations were utilized for this field experiment. Each strip was segmented into a start region, an effective rolling region, and a stop region (Fig. 4.12). The start and stop regions each measured 3 m in length, while the effective rolling region spanned 24 m. In light of the diameter of

Fig. 4.12 Photos of field rolling test in the rockfill zone

Table 4.2 Test scheme for field rolling experiment

Material	Strip	Passes (static)	Passes (vibration)	Vibration frequency (Hz)	Velocity (km/h)	Vibration status	Lift thickness (m)
NGM	1	2	12	31	3	HFLA	About 0.8
NGM	2	2	12	31	3	HFLA	About 0.8
NGM	3	2	12	31	3	HFLA	About 0.8
NGM	4	2	12	31	3	HFLA	About 0.8

the test pit used in the water replacement method and the precision required by this technique, the effective rolling region was further divided into six blocks (Fig. 4.12). Initially, static compaction was conducted across the four strips, followed by vibration compaction (Table 4.2). Static compaction was executed twice, while vibration compaction was carried out twelve times.

4.4.3 Results

(1) Spectrum analysis of detection results

Field tests were conducted using a detection device developed in the previous section. Data from four test strips were collected and subsequently analyzed. Given the extensive volume of data, the analysis process is illustrated using the original acoustic data

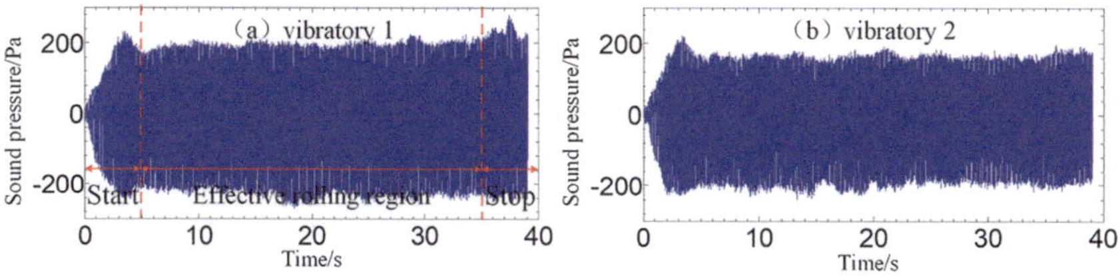

Fig. 4.13 Original acoustic signal data collected from Strip 1 by acoustic field microphone 3

from Strip 1. Figure 4.13 depicts the original acoustic data recorded by the acoustic field microphone from Strip 1 during the initial four cycles of vibration compaction. It is evident from Fig. 4.13 that the amplitude of the sound pressure during vibratory compaction fluctuated between −200 and 200 Pa. The acoustic signal profile when the roller commenced and ceased its rolling operation is illustrated in Fig. 4.13a. The results presented in Fig. 4.13, along with those from the other test strips, indicated no zero drift in the original acoustic signal, and there was no discernible pattern in the variation of the time-domain signal.

To conduct the spectrum analysis and derive the *SCV*, the original acoustic signal for each compaction pass was processed through a series of steps: first, the signals from the start and stop rolling regions were eliminated, and zero drift was corrected; next, the remaining acoustic signals were partitioned into six blocks, each approximately 4 m in length (Fig. 4.12); subsequently, the effective acoustic signals of each block were transformed using the Fast Fourier Transform (FFT), yielding the corresponding frequency spectrum; thereafter, the magnitude of the spectrum data was converted using $\log_{10}$, allowing for the calculation of the Spectral Harmonic Amplitude (SHA) for each acoustic signal block; finally, the *SCV* was determined using Eq. (4.17).

To elucidate the findings, the frequency spectrum of the signal collected from Block 1 in Strip 1 was analyzed (Fig. 4.14). In Fig. 4.14, labels (1)–(8) correspond to the number of compaction cycles from 1 to 8, with the first three peaks representing the fundamental wave amplitude f_0, the second harmonic amplitude $2f_0$, and the third harmonic amplitude $((3f_0))$. The magnitude of the fundamental wave amplitude (A_{f0}) ranged between 2×10^4 and 4×10^4. The results displayed in Fig. 4.14, along with those from the other strips, indicated that the magnitude of the SHA generally increased as the number of compaction cycles rose.

(2) Correlations between compaction parameters and *SCV*

Based on the aforementioned analysis, the *SCV* was determined. To explore the relationship between the number of compaction cycles and the *SCV*, a series of field tests were conducted on the NGM at the main dam. Figure 4.15 illustrates the correlation between *SCV* and the number of compaction cycles in Strip 1. The linear correlation between *SCV* and the number of compaction cycles, ranging from 1 to 8,

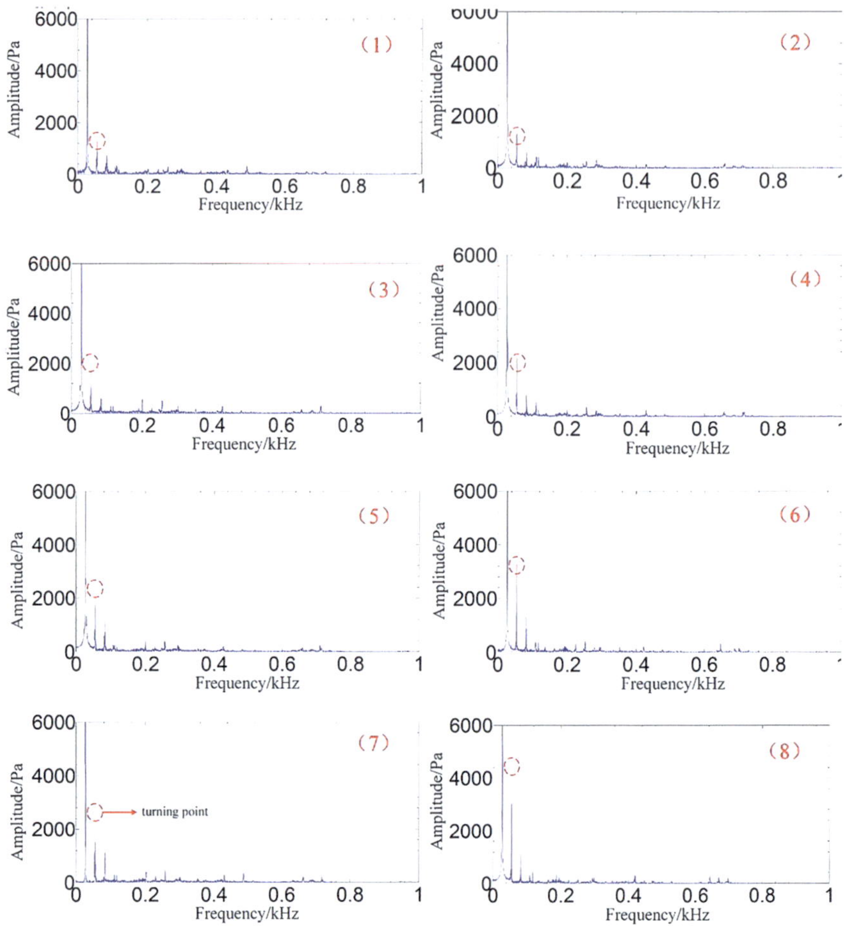

Fig. 4.14 Spectrum analysis of domain signal collected from Block 1 in Strip 1

was analyzed. For each test strip, the *SCV* for each of the six blocks within the strip was calculated for each compaction cycle. This figure demonstrates that the *SCV* for NGM increased with the number of compaction cycles. The *SCV* computed for each block within each strip exhibited a strong linear relationship with the number of compaction cycles, with the determination coefficient R^2 ranging from 0.7803 to 0.9696. The specific determination coefficients R^2 are presented in Table 4.3.

(3) Correlations between *SCV* and compaction quality of rockfill materials

In the construction area of the upstream cofferdam, field tests were conducted across four strips to ascertain the dry density, compactness, and various other property parameters of the rockfill materials (including P5 content, minimum dry density, and

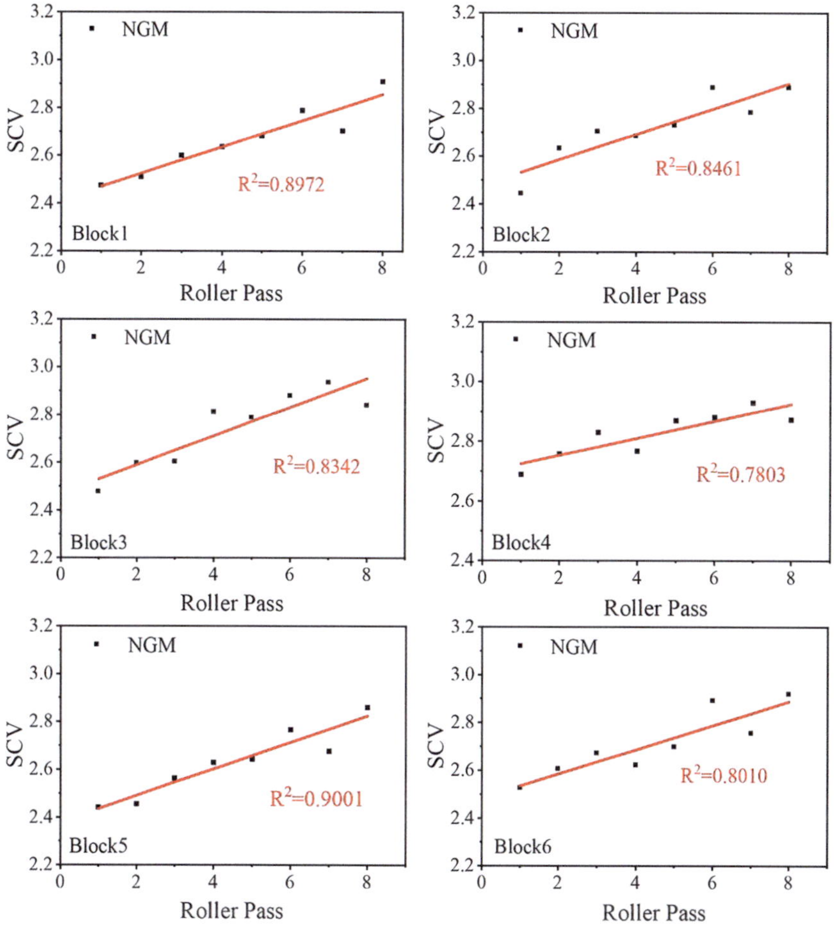

Fig. 4.15 Correlation between the *SCV* and the number of compaction passes in Strip 1

Table 4.3 Specific determined coefficient R^2 between the *SCV* and the number of compaction passes for NGM

Block	Strip			
	1	2	3	4
1	0.897	0.809	0.815	0.910
2	0.846	0.832	0.824	0.920
3	0.834	0.876	0.970	0.842
4	0.780	0.863	0.923	0.806
5	0.900	0.796	0.8680	0.871
6	0.801	0.862	0.796	0.897

maximum dry density). A total of 16 groups of data were collected for correlation analysis. Figure 4.16 illustrates the relationship between the *SCV* and dry density, as well as between the *SCV* and compactness for compaction passes of 6, 8, 10, and 12, respectively, in Strip 1. The correlations between the *SCV* and both dry density and compactness exhibited similar trends; as the *SCV* increased, both dry density and compactness also rose, and vice versa. The correlation between *SCV* and compactness is depicted in Fig. 4.17, clearly demonstrating a strong linear correlation with a determination coefficient R^2 ranging from 0.7371 to 0.8064. Given the strong correlation between *SCV* and compactness, the *SCV* serves as an effective index that reflects the compaction status of rockfill materials, while compactness remains a critical index for assessing the compaction quality of the dam foundation and the dam body.

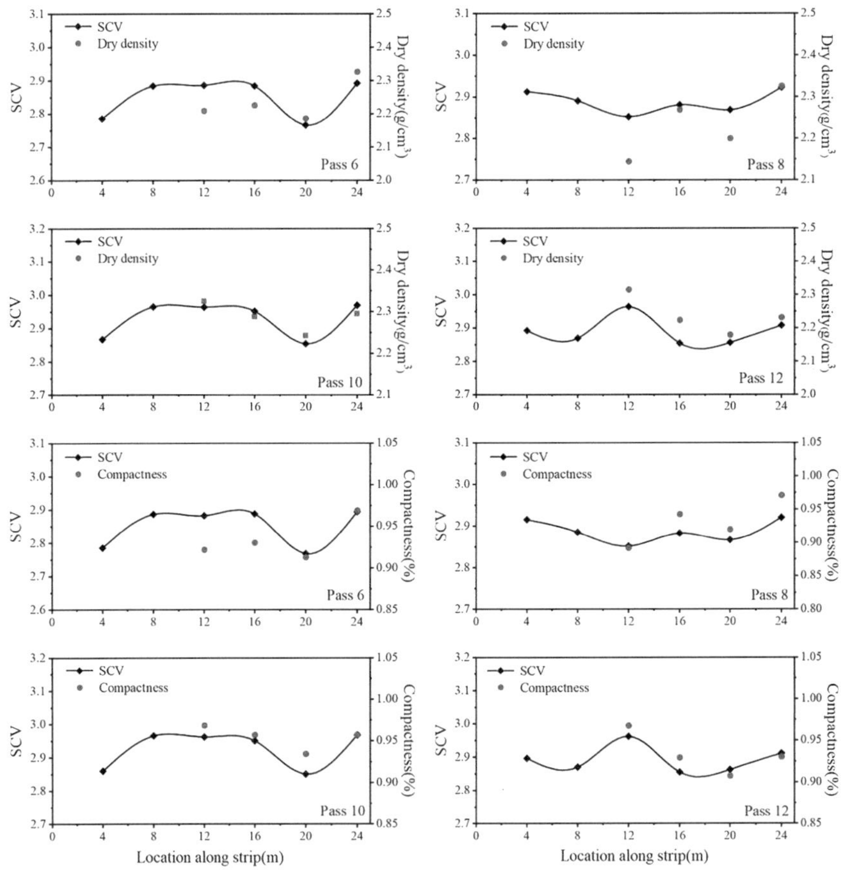

Fig. 4.16 Relationship between the *SCV* and dry density, as well as between the *SCV* and compactness at each block along Strip 1

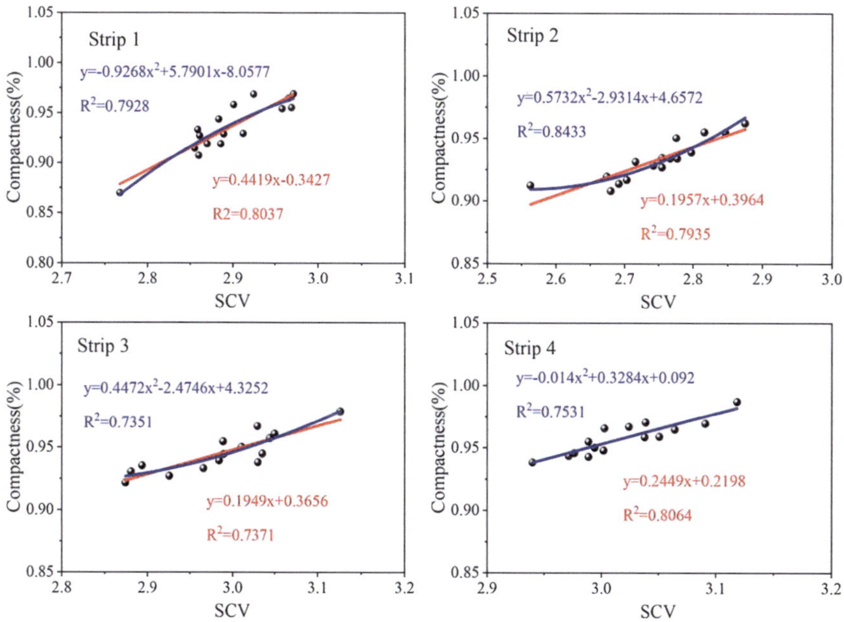

Fig. 4.17 Linear correlations between *SCV* and compactness computed from all test strips

(4) Relations between the theoretical model and in situ measurements

Figures 4.16 and 4.17 illustrate the compaction quality by analyzing the correlations between the *SCVs* and the measured values obtained from field tests. To a certain extent, these analyses reflect the efficacy of the new technique proposed in the previous section. To further validate the effectiveness of this technique, the relationships between the actual values from in situ measurements and the calculated values derived from the A-model were examined. The measured values from the field tests were averaged for compaction passes of 6, 8, 10, and 12, respectively, in each strip. Figure 4.18 presents a comparison of the dry density between the actual values and the calculated values based on the A-model. The relative errors between the measured dry density and the calculated values from the A-model are detailed in Table 4.4. The average relative errors of the A-model are less than 5%, indicating a high level of accuracy. Figure 4.19 shows the variation of the SCV as the number of compaction times in Strip 2. The overall compaction performance covering 100% of the work area can be clearly determined by the nephogram.

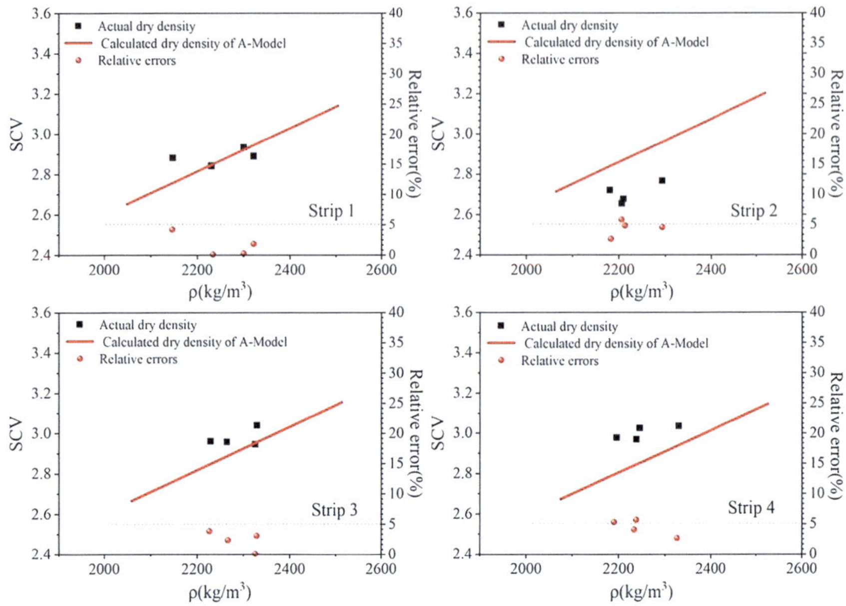

Fig. 4.18 Comparison between actual dry density values and calculated valued computed by A-model

Table 4.4 Relative errors between measured and calculated dry density

Strip	Maximum relative error (%)	Minimum relative error (%)	Average relative error (%)
1	4.32	0.05	1.64
2	5.81	2.92	4.75
3	3.90	0.19	2.40
4	5.67	2.80	4.59

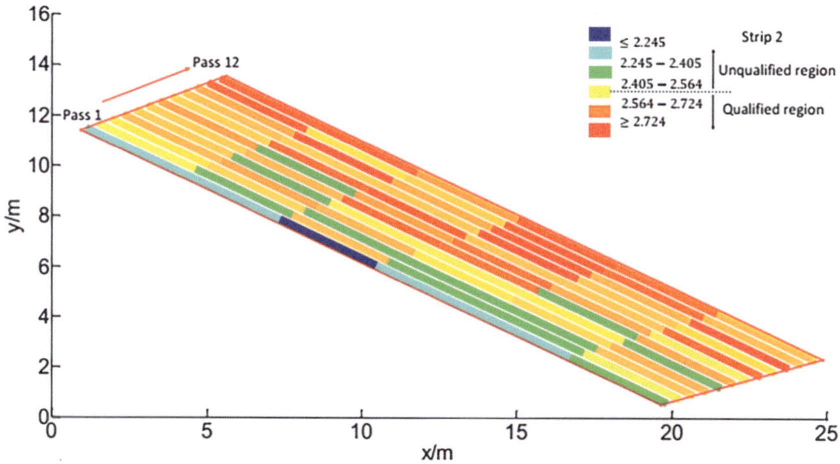

Fig. 4.19 Compaction nephogram of Strip 2 using *SCV*

References

1. Liu DH, Sun J, Zhong DH et al (2012) Compaction quality control of earth-rock dam construction using real-time field operation data. J Constr Eng Manag 138(09):1085–1094
2. Krebs RD, Waller ED (1971) Highway materials. McGraw-Hill, New York
3. Rahman F, Hossain M, Romanoschi S (2008) Intelligent compaction control of highway embankment soil in Kansas. Kansas Department of Transportation. Bureau of Materials & Research
4. Meehan CL, Tehrani FS, Vahedifard F (2012) A comparison of density-based and modulus based in situ test measurements for compaction control. Geotech Test J 35:387–399
5. Melbouci B (2009) Compaction and shearing behavior study of recycled aggregates. Constr Build Mater 23:2723–2730
6. Zabilska-Adamska K, Sulewska MJ (2014) Dynamic CBR test to assess the soil compaction. J Test Eval 43:1028–1036
7. Resende LRM, Oliveira Filho WL, Nogueira CL (2013) Use of the DCP test for compaction control of staged dikes in mining tailings dams. REM Rev Esc Minas 66:493–498
8. Kongkitkul W, Saisawang T, Thitithavoranan P et al (2014) Correlations between the surface stiffness evaluated by light-weight deflectometer and degree of compaction. Tunn Undergr Constr 2014
9. Umashankar B, Hariprasad C, Kumar G (2016) Compaction quality control pavement layers using LWD. J Mater Civ Eng 28
10. Swinford JM, Meyer JH (1985) An evaluation of a nuclear density gauge for measuring infield compaction in soils of the South African sugar industry. S Afr Sugar Technol Assoc 19:218–224
11. White DJ, Morris M, Thompson M (2006) Power-based compaction monitoring using vibratory padfoot roller. In: GeoCongress 2006: geotechnical engineering in the information technology age, pp 1–6
12. Thurner H, Forsblad L (1978) Compaction meter on vibrating rollers; Res Bulletin Nr 8022 Monograph. Swedish National Road and Transport Research Institute (VTI), Linköping, Sweden
13. Forssblad L (1980) Compaction meter on vibrating rollers for improved compaction control. Proc Int Conf Compact 2:541–546

14. Anderegg R, Kaufmann K (2004) Intelligent compaction with vibratory rollers: feedback control systems in automatic compaction and compaction control. Transport Res Rec 1868:124–134
15. Thompson JM, White JD (2008) Estimating compaction of cohesive soils from machine drive power. J Geotech Geoenviron Eng 134(12):1771–1777
16. White JD, Thompson JM (2008) Relationships between in situ and roller-integrated compaction measurements for granular soils. J Geotech Geoenviron Eng 134(12):1763–1770
17. Komandi G (1999) An evaluation of the concept of rolling resistance. J Terrramech 36(3):159–166
18. Adam D (1997) Continuous compaction control (CCC) with vibratory rollers. Environmental geotechnics. In: Proceedings of the 1st Australia-New Zealand conference on environmental geotechnics-geoenvironment 97, Melbourne, Victoria, Australia, 26–28 November 1997. A. A. Balkema: Rotterdam, The Netherlands, pp 245–250
19. Anderegg R, von Felten DA, Kaufmann K (2006) Compaction monitoring using intelligent soil compactors. In: Geocongress 2006: geotechnical engineering in the information technology age pp 1–6
20. White DJ, Jaselskis EJ, Schaefer VR et al (2005) Real-time compaction monitoring in cohesive soils from machine response. Transp Res Rec 1936(1):172–180
21. Mooney MA, Rinehart RV (2007) Field monitoring of roller vibration during compaction of subgrade soil. J Geotech Geoenviron Eng 133(3):257–265
22. Rinehart RV, Mooney MA (2008) Instrumentation of a roller compactor to monitor vibration behavior during earthwork compaction. Autom Constr 17(2):144–150
23. Liu DH, Lin M, Li S (2016) Real-time quality monitoring and control of highway compaction. Autom Constr 62:114–123
24. Xu Q, Chang KG (2016) Adaptive quality control and acceptance of pavement material density for intelligent road construction. Autom Constr 6278–6288
25. Singh D, Zaman M, Commuri S, Zaman M (2015) Application of intelligent compaction technology for estimation of effective modulus for a multilayered asphalt pavement. J Test Eval 43:1–11
26. Karray M, Lefebvre G, Ethier Y, Bigras A (2010) Assessment of deep compaction of the Peribonka dam foundation using modal analysis of surface waves (MASW). Can Geotech J 47:312–326
27. Donohue S, Forristal D, Douohue LA (2013) Detection of soil compaction using seismic surface waves. Soil Tillage Res 128:54–60
28. Shangguan P, Al-Qadi I, Coenen A et al (2016) Algorithm development for the application of ground-penetrating radar on asphalt pavement compaction monitoring. Int J Pavement Eng 17(03):189–200
29. Liu D, Li Z, Lian Z (2014) Compaction quality assessment of earth-rock dam materials using roller-integrated compaction monitoring technology. Autom Constr 44:234–246
30. Sandström Å (1994) Numerical simulation of a vibratory roller on cohesionless soil. In: Internal Report; Geodynamic: Stockholm, Sweden, pp 1–22
31. Vennapusa RKP, White JD, Morris DM (2010) Geostatistical analysis for spatially referenced roller-integrated compaction measurements. J Geotech Geoenviron Eng 136(6):813–822
32. Thurner HF, Sandström A (1989) Compaction meter and compaction documentation system. Unbound Aggreg Roads 46–51
33. Pampagnin LH, Peyret F, Garcia G (1988) Architecture of a GPS-based guiding system for road compaction. In: Proceedings of the 1988 IEEE international conference on robotics and automation, Philadelphia, PA, USA, pp 2422–2427
34. Liu DH, Cui B, Liu YG, Zhong DH (2013) Automatic control and real-time monitoring system for earth-rock dam material truck watering. Autom Constr 30:70–80
35. Liu D, Chen J, Li S (2018) Collaborative operation and real-time control of roller fleet for asphalt pavement compaction. Autom Constr 98:16–29
36. Zhong DH, Li XC, Cui B, Wu BP, Liu YX (2018) Technology and application of real-time compaction quality monitoring for earth-rockfill dam construction in deep narrow valley. Autom Constr 90:23–38

37. Zeng HE, Liu JP (2018) Preliminary experiment study on Georobot/WLAN-based intelligent monitoring system for roller compaction. Water Resour Hydropower Eng 49:141–147
38. Huang S, Zhang W (2018) A fast calculation method of rolling times in the GNSS real-time compaction quality supervisory system. Adv Eng Softw 128:20–33
39. Xu J, Liu FP, Wang JX (2020) Research and application of continuous compaction monitoring system based on BDS. Agric Eng 10:59–63
40. Meehan CL, Khosravi M, Cacciola D (2013) Monitoring field lift thickness using compaction equipment instrumented with global positioning system (GPS) technology. Geotech Test J 36:755–767
41. Junhui Z, Le D, Ling Z et al (2020) Using portable falling weight deflectometer to determine treatment depth of subgrades in highway reconstruction of Southern China. Transp Saf Environ 2(1):18–28
42. Jin B, Liu H (1999) Vertical dynamic response of a disk on a saturated poroelastic half-space. Soil Dyn Earthq Eng 18:437–443
43. Senjuntichai T, Sapsathiarn Y (2003) Forced vertical vibration of circular plate in multilayered poroelastic medium. J Eng Mech 129(11):1330–1341
44. Pekeris CL (1955) The seismic surface pulse. Proc Natl Acad Sci 41(7):469–480
45. Lade PV, Nelson R (2014) Modeling of the elastic behavior of granular materials. Numer Anal Methods Geomech 38:1503–1518
46. Liu H, Zou D, Liu J (2014) Constitutive modeling of dense gravelly soils subjected to cyclic loading. Int J Numer Anal Meth Geomech 38(14):1503–1518

Chapter 5
Contact-Type Roller-Integrated Acoustic Wave Detection Technology

In view of the problems that the mapping relationship between the physical mechanism and the compaction state of the soil is not fully established in the existing compaction quality detection indicators, some detection indicators have low accuracy, it is difficult to accurately reflect the compaction quality, and they are easily disturbed by the construction environment, the authors have constructed an original compaction quality detection method based on acoustic theory, continuum mechanics and soil dynamics, and proposed a new compaction quality detection index, namely contact-type sound compaction value (*CSCV*), to directly characterize the compaction state of gravel soil and weathered slate. On this basis, an intelligent assessment model considering small sample uncertainty is constructed to quantify multi-source uncertainties such as data sample insufficiency, data noise, materials and construction disturbances, so as to achieve a rapid evaluation of the compaction quality of the entire rolling area.

5.1 Theoretical Analysis

5.1.1 Acoustic Theory Derivation

Donnell-Mushtari shell theory is used to describe the vibration of the cylindrical shell, and its geometric theoretical model is shown in Fig. 5.1. The elastic cylindrical shell with simple support at both ends has a length of L, a radius of a, and a shell thickness of h. A cylindrical coordinate system is established as shown in the figure, where x, y, and z are the vibration displacement components along the axial x, circumferential θ and radial directions r [1].

The surface vibration equation of a cylindrical shell is shown below:

Q. Zhang, *Compaction Quality Nondestructive Testing and Intelligent Control Technology for Infrastructure*, https://doi.org/10.1007/978-981-95-1264-5_5

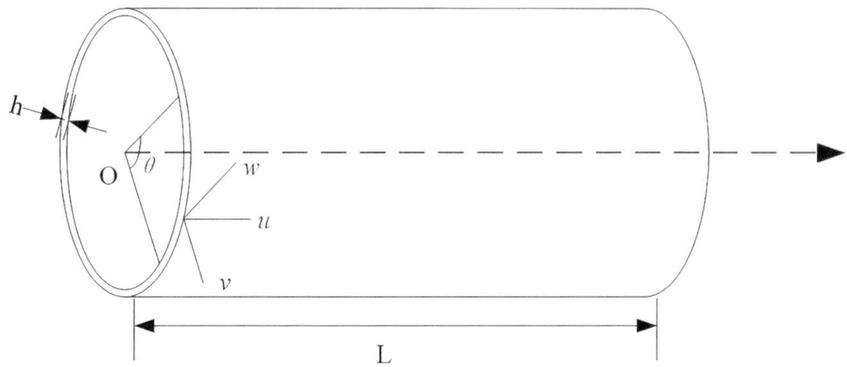

Fig. 5.1 Cylindrical shell geometry model

$$\frac{\partial^2 u}{\partial x^2}+\frac{(1-\sigma)}{2a^2}\frac{\partial^2 u}{\partial \theta^2}+\frac{(1+\sigma)}{2a}\frac{\partial^2 v}{\partial x\partial \theta}+\frac{\sigma}{a}\frac{\partial w}{\partial x}-\frac{\ddot{u}}{c_L^2}=0$$
$$\frac{(1+\sigma)}{2a}\frac{\partial^2 u}{\partial x\partial \theta}+\frac{(1-\sigma)}{2}\frac{\partial^2 v}{\partial x^2}+\frac{1}{a^2}\frac{\partial^2 v}{\partial \theta^2}+\frac{1}{a^2}\frac{\partial w}{\partial \theta}-\frac{\ddot{v}}{c_L^2}=0$$
$$\frac{\sigma}{a}\frac{\partial u}{\partial x}+\frac{1}{a^2}\frac{\partial u}{\partial \theta}+\frac{w}{a^2}+\beta^2\left(a^2\frac{\partial^4 w}{\partial x^4}+2\frac{\partial^4 w}{\partial x^2\partial \theta^2}+\frac{1}{a^2}\frac{\partial^4 w}{\partial \theta^4}\right)+\frac{\ddot{w}}{c_L^2}=\frac{f_r\left(1-\sigma^2\right)}{Eh} \tag{5.1}$$

where σ is Poisson's ratio, E is Young's modulus, and f_r is the amplitude of the radial external force acting on the cylindrical shell.

For the boundary conditions of simply supported ends, the displacements of the cylindrical shell in three directions can be expanded by Fourier series as follows:

$$\begin{aligned} u(x,\theta) &= \sum_{m,n} U_{mn}\cos\left(\frac{m\pi x}{L}\right)\cos(n\theta) \\ v(x,\theta) &= \sum_{m,n} V_{mn}\sin\left(\frac{m\pi x}{L}\right)\sin(n\theta) \\ w(x,\theta) &= \sum_{m,n} W_{mn}\sin\left(\frac{m\pi x}{L}\right)\cos(n\theta) \end{aligned} \tag{5.2}$$

where U_{mn}, V_{mn}, W_{mn} are the displacement amplitudes in the axial, circumferential, and radial directions, respectively, and $m(m=1,2,3,\ldots)$ and $n(n=1,2,3,\ldots)$ are the moduli in the axial and circumferential directions, respectively. Among the three displacement components mentioned above, only the radial displacement has an effect on the vibration and acoustic radiation.

Assume that the radial single-frequency harmonic point force acting on the surface of the cylindrical shell is expanded into the normal displacement order modal series through two-dimensional Fourier transform, that is:

$$f_r(x_0,\theta_0)=\sum_{m=1}^{\infty}\sum_{n=0}^{\infty}\sin\left(\frac{m\pi x_0}{L}\right)\left[F_{mn}^{c}\cos n\theta_0+F_{mn}^{s}\sin n\theta_0\right] \tag{5.3}$$

Substituting Eqs. (5.2) and (5.3) into the vibration Eq. (5.1), we can obtain the following matrix form:

$$\begin{pmatrix} L_{11} & L_{12} & L_{13} \\ L_{21} & L_{22} & L_{23} \\ L_{31} & L_{32} & L_{33} \end{pmatrix}\begin{pmatrix} U_{mn} \\ V_{mn} \\ W_{mn} \end{pmatrix}=\begin{pmatrix} 0 \\ 0 \\ \frac{F_{mn}a^2(1-\sigma^2)}{Eh} \end{pmatrix} \tag{5.4}$$

Therefore, the radial displacement amplitude expression can be obtained:

$$W_{mn}=\frac{(1-\sigma^2)a^2}{Eh}\frac{(L_{11}L_{22}-L_{12}L_{21})}{\Delta}F_{mn} \tag{5.5}$$

where Δ is the determinant of the coefficient matrix.

When expressing the shell vibration displacement function, the superposition of the symmetric and antisymmetric modes of its free vibration is selected. The displacement function that satisfies the simple support at both ends:

$$w(x,\theta)=\sum_{m=1}^{\infty}\sum_{n=0}^{\infty}\sin\left(\frac{m\pi x}{L}\right)\left[W_{mn}^{c}\cos n\theta+W_{mn}^{s}\sin n\theta\right] \tag{5.6}$$

By taking the time derivative based on Eq. (5.6), the velocity response of the shell under excited vibration can be obtained:

$$\begin{aligned}\dot{w}(x,\theta)&=j\omega w(x,\theta)\\ &=j\omega\sum_{m=1}^{\infty}\sum_{n=0}^{\infty}\sin\left(\frac{m\pi x}{L}\right)\left[W_{mn}^{c}\cos n\theta+W_{mn}^{s}\sin n\theta\right]\end{aligned} \tag{5.7}$$

For the acoustic radiation generated by a vibrating structure with a closed structure surface, the meaningful area is usually the fluid medium inside or outside the vibrating structure. The homogeneous acoustic wave equation can be used to analyze this type of radiated sound waves generated by structural vibration. Assuming that the density of the air medium in which the radiating structure is located is ρ and the sound velocity is c, the linear homogeneous acoustic wave equation is:

$$\nabla^2\tilde{p}(x,y,z,t)-\frac{1}{c^2}\frac{\partial^2\tilde{p}(x,y,z,t)}{\partial t^2}=0 \tag{5.8}$$

where $\tilde{p}(x,y,z,t)$ is the sound pressure and c is the sound velocity in the fluid medium.

For simple harmonic vibration, the sound pressure can be set to $\tilde{p}(x, y, z, t) = p(x, y, z)e^{i\omega t}$, and the wave equation of the sound field can be obtained by substituting:

$$\left(\nabla^2 + k^2\right)p(r) = 0 \tag{5.9}$$

In the formula, ∇ is the Laplace operator, $k = \frac{\omega}{c}$ is the sound wave number, ω is the angular frequency, and formula (5.9) is the Helmholtz differential equation.

Since the property of the δ function is that the intensity of a point source is finite and the volume is zero, the δ function can be used to represent the point source in space, and the Helmholtz differential equation can be solved using the Green function.

$$\left(\nabla^2 + k^2\right)G(r/r_\theta) = -\delta(r - r_\theta) \tag{5.10}$$

Integrate both sides of the above equation over the volume domain enclosed by the boundary of arbitrary area S. According to the properties of Green function and δ function, the Helmholtz integral equation can be obtained after further simplification:

$$p(r) = -\iint\limits_{s}\left[G\left(\frac{r}{r_0}\right)\frac{\partial p}{\partial \eta_0}(r_0) - p(r_0)\frac{\partial G}{\partial \eta_0}\left(\frac{r}{r_0}\right)\right]dS(r_0) \tag{5.11}$$

where η_0 is a vector perpendicular to the structure boundary surface and pointing to the fluid medium.

Using the phase stability method, the far-field Green function can finally be simplified to:

$$G(R, \theta, \phi/a, \phi_0, x_0) = -\frac{\exp\left[-jk(R - x_0\cos\theta)\right]}{\pi^2 akR\sin\theta}\sum_{n=0}^{+\infty}\frac{j^{n+1}}{\varepsilon_n}\frac{\cos[n(\phi - \phi_0)]}{H_n^{(2)}(ka\sin\theta)} \tag{5.12}$$

Substituting Eqs. (5.7) and (5.12) into the vibration Eq. (5.10), and using the orthogonality of trigonometric functions, the expression of the far-field sound pressure is finally obtained as:

$$p(R, \theta, \phi) = \frac{\rho\omega^2 e^{-jkR}}{\pi kR\sin\theta}\sum_{n=0}^{\infty}\sum_{m=1}^{\infty}\left[W_{mn}^{c}\cos(n\phi)\alpha_{mn} + W_{mn}^{s}\sin(n\phi)\alpha_{mn}\right] \tag{5.13}$$

5.1.2 Modal Truncation

The far-field sound pressure formula uses a complete set of modes, which is the superposition of all modes, and different modes contribute differently to the response.

For low-frequency responses, the influence of high-order modes is small, so high-order modes are often ignored in actual structures. To simplify the calculation, the first few modes of the structure can be used to represent the vibration of the structure and perform modal truncation. Modal truncation will produce certain errors, but it still has sufficient accuracy to represent the vibration of the structure.

When the roller rolls the backfill layer, the vibration wheel of the vibratory roller works under simple harmonic excitation. The simple harmonic excitation frequency is a low-frequency response relative to the natural frequency of the vibration wheel. Therefore, the far-field sound pressure formula generated when the roller is working is modally truncated. Related literature research shows [2] that under specific excitations of different frequencies, the strain energy of the overall structure varies significantly with the number of modes intercepted. When performing modal truncation, the order of the intercepted mode is related to the excitation frequency, and the mode close to the excitation frequency should be intercepted. And when the N order natural frequency of the structure is greater than the excitation frequency, and the ratio of the sum of the strain energy of the first $N-1$ order mode to the sum of the strain energy of the first N order mode is close to 1, only the first a-order mode needs to be intercepted for structural dynamic characteristics analysis.

The free vibration natural frequency of a finite cylindrical shell can be determined by the following formula:

$$\left(\Omega^2\right)^3 - A_2\left(\Omega^2\right)^2 + A_1\left(\Omega^2\right) - A_0 = 0 \tag{5.14}$$

where A is the coefficient derived from the vibration equation of the cylindrical shell, Ω is the dimensionless frequency, and the three roots of the formula correspond to the three natural frequencies of each mode.

The coefficients A_0, A_1 and A_2 are:

$$A_0 = \frac{(1-\sigma)}{2}\left[k^4a^4\left(1-\sigma^2\right) + \beta^2\left(k^2a^2+n^2\right)^4\right] \tag{5.15}$$

$$A_1 = \frac{(1-\sigma)}{2}\left[\left(k^2a^2+n^2\right)^2 + n^2 + k^2a^2(3+2\sigma)\right] + \frac{(3-\sigma)}{2}\beta^2\left(k^2a^2+n^2\right)^3 \tag{5.16}$$

$$A_2 = 1 + \frac{(3-\sigma)}{2}\left(k^2a^2+n^2\right) + \beta^2\left(k^2a^2+n^2\right)^2 \tag{5.17}$$

where σ is Poisson's ratio, which is 0.3. β is the thickness factor of the cylindrical shell, $\beta^2 = h^2/12a^2$, and k is $m\pi/L$.

The roller is a XCMG XS223J 22t single steel wheel vibratory roller, with a vibration wheel width L and radius a of 2130 mm and 800 mm respectively, and a vibration wheel thickness h of 50 mm. Substituting the above geometric parameters into Eqs. (5.15), (5.16) and (5.17), the $A_0 = 0.616$, $A_1 = 42.689$, and $A_3 = 15.027$ can be obtained.

Substituting the coefficients A into formula (5.14), the three roots of the dimensionless frequency Ω^2 are 0.01451, 3.7816 and 11.2311, where the minimum value corresponds to the radial resonance frequency.

Let ω_r be the frequency of the cylindrical shell angle ring, then:

$$\omega_r = \frac{\sqrt{\frac{E}{\rho(1-\sigma^2)}}}{a} \tag{5.18}$$

$$\Omega = \frac{\omega}{\omega_r} \tag{5.19}$$

Substituting the smallest root of the dimensionless frequency Ω^2 into Eqs. (5.18) and (5.19), we obtain the radial resonance frequency, i.e., the first-order natural frequency of the vibration wheel, which is 818.96 Hz, which is greater than the excitation frequency of the roller when it is working, 28/33 Hz. The ratio of the sum of the strain energies of the first $N-1$ order modes, i.e., the first-order modes, to the sum of the strain energies of the first N order modes, i.e., the first-order modes, is 1, so the first-order mode can be intercepted for structural dynamic characteristics analysis.

By performing modal truncation simplification on the far-field sound pressure formula (5.13), we can obtain:

$$\begin{aligned} \alpha &= \frac{j^{n+1}\widetilde{x_m}(k\cos\theta)}{H_n^{(2)'}(ka\sin\theta)} = \frac{j^4\int_0^L \sin\left(\frac{m\pi x}{L}\right)e^{jk\cos\theta x}dx}{\left[j^3 e^{j\left(\frac{\pi}{4}ka\sin\theta\right)}\sqrt{\frac{2}{\pi ka\sin\theta}}\right]'} \\ &= j\frac{\frac{\frac{\pi}{L}e^{jkL\cos\theta}-1}{(jk\cos\theta)^2+\left(\frac{\pi}{L}\right)^2}}{-e^{j\left(\frac{\pi}{4}-ka\sin\theta\right)}\sqrt{\frac{2}{\pi ka\sin\theta}}\left(\frac{1}{2ka\sin\theta}+j\right)} \end{aligned} \tag{5.20}$$

The W in the original sound pressure formula is derived by Fourier decomposing the external force into multiple modes and then deriving it according to the dynamic equation. In the simplification process of modal truncation, only the first-order mode is considered. To simplify the calculation, the multi-modal W formula is replaced as follows:

$$w(x,\theta) = \sum_{m,n} W_{mn}\sin\left(\frac{m\pi x}{L}\right)\cos(n\theta) \tag{5.21}$$

Substituting Eqs. (5.20) and (5.21) into Eq. (5.13), the radiation sound pressure formula of the first-order mode can be extracted as follows:

$$p(R,\theta,\phi) = j\frac{\rho\omega^2 e^{-jkR}}{\pi kR\sin\theta}\frac{\frac{\frac{\pi}{L}e^{jkL\cos\theta}+\frac{\pi}{L}}{(jk\cos\theta)^2+\left(\frac{\pi}{L}\right)^2}}{-e^{j\left(\frac{\pi}{4}-ka\sin\theta\right)}\sqrt{\frac{2}{\pi ka\sin\theta}}\left(\frac{1}{2ka\sin\theta}+j\right)}W\cos(3\phi) \tag{5.22}$$

5.1.3 Vertical Displacement of Soil Under Simple Harmonic Load

The isotropic, uniform surface of the half-space is shown in Fig. 5.2. The coordinate z points into the half-space. A normal concentrated load of magnitude $QH(t)$ is applied at the origin o. Since the load is symmetric about the axis, the wave problem in the infinite elastic half-space is also axisymmetric. Using the cylindrical coordinate system (r, θ, z), the vertical displacement component u_z can be expressed by the displacement potentials φ and $x(\psi = 0)$ of the symmetric case of the cylindrical coordinate system as follows:

$$u_z = \frac{\partial \varphi}{\partial z} - \frac{1}{r}\frac{\partial}{\partial r}\left(r\frac{\partial \chi}{\partial r}\right) \tag{5.23}$$

For a vibratory roller, in a vibration system with damping, if the vibration is to be maintained, a periodic forcing force must be applied from the outside, usually simple harmonic excitation. The vibration wheel of a vibratory roller works under simple harmonic excitation. The simple harmonic load on the vibration system is:

$$F = f \cos \omega t \tag{5.24}$$

Based on existing research results [3], the true vertical displacement of soil on the half-space surface at any time within a cycle under simple harmonic load is:

$$W = \frac{f}{A_0 \pi G r} \cos \omega_0 t \tag{5.25}$$

Combining Eqs. (5.24) and (5.25), we get the far-field radiation sound pressure formula of a cylindrical shell with a baffle:

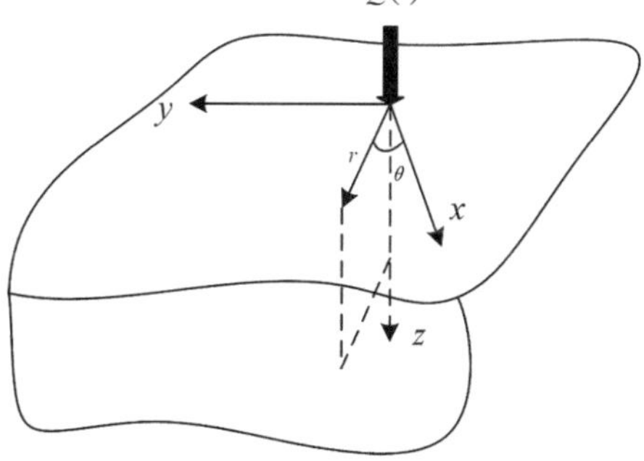

Fig. 5.2 Isotropic and homogeneous half-space

$$p(R,\theta,\phi,t)=j\frac{\rho\omega^2 e^{-jkR}}{\pi kR\sin\theta}\frac{\frac{\frac{\pi}{L}e^{jkL\cos\theta}+\frac{\pi}{L}}{(jk\cos\theta)^2+\left(\frac{\pi}{L}\right)^2}}{-e^{j\left(\frac{\pi}{4}ka\sin\theta\right)}\sqrt{\frac{2}{\pi ka\sin\theta}}\left(\frac{1}{2ka\sin\theta}+j\right)}$$
$$\frac{f}{A_0\pi Gr}\cos\omega_0 t\cos(3\phi)$$
$$=\frac{\frac{\pi}{L}}{\left(k^2\cos^2\theta+\frac{\pi^2}{L^2}\right)\sqrt{\frac{2}{ka\sin\theta}}}\frac{\rho\omega^2}{\pi kR\sin\theta}$$
$$\frac{f}{A_0\pi Gr}\cos\omega_0 t\cos(3\phi)\frac{j\left(e^{jkL\cos\theta}+1\right)}{e^{j\left(\frac{\pi}{4}+kR-ka\sin\theta\right)}\left(\frac{1}{2ka\sin\theta}+j\right)} \tag{5.26}$$

In the formula, $k=\omega/c$. Perform Fourier transform on formula (5.26) and take its root mean square amplitude to obtain:

$$A=a_0\frac{\frac{\pi}{L}}{\left(\frac{\omega^2\cos^2\theta}{c^2}+\frac{\pi^2}{L^2}\right)\sqrt{\frac{2c}{\omega a\sin\theta}}}\frac{\rho\omega c}{\pi R\sin\theta}\frac{f}{A_0\pi Gr}\cos\omega_0 t\cos(3\phi) \tag{5.27}$$

In the formula, a_0 can be taken as 2.

5.1.4 *Low Frequency Response of Cylindrical Shell Under Point Forces at Different Positions*

For a simple harmonic point force $f(x_0,\theta_0)e^{j\omega t}$ acting on (x,θ), the modal coefficients of the force are obtained by modal expansion using Fourier transform. The displacement is given as a superposition of the modes of each order, and the information of the point force position is reflected in the excitation coefficients of each mode. For the convenience of calculation, the multi-mode is truncated and simplified, so the response of the cylindrical shell under the action of point forces at different positions is simplified accordingly. In a multi-modal system, the position of the point force does not significantly affect the displacement peak, the acoustic pressure at the monitoring point, and other responses [4].

As can be seen from Fig. 5.3, in a multi-modal system, the position of the point force does not significantly affect the displacement peak, monitoring point sound pressure and other responses. The point force position mainly affects the number of high-order modal resonance peaks. In a multi-modal system, the point force position is reflected in the modal coefficients of each order mode, and in the previous modal truncation simplification, the multi-modal system was simplified to the first order mode, and the modal coefficients were ignored. Therefore, it can be approximately considered that the point position has no effect. In summary, the low-frequency response of the cylindrical shell under the action of point forces at different positions

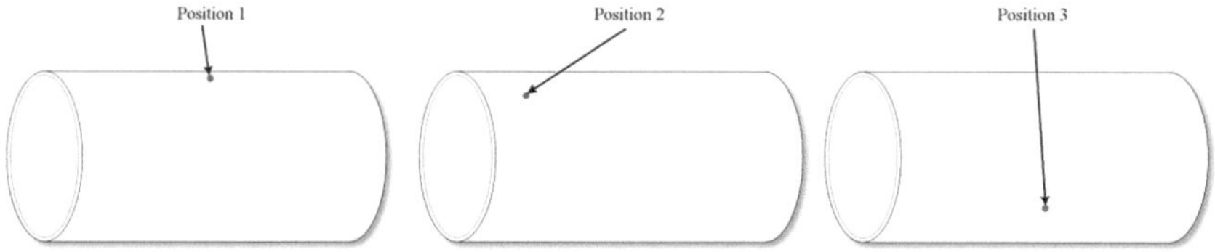

Fig. 5.3 Displacement response of simple harmonic external forces of the same magnitude at different locations

is appropriately simplified, and it is considered that when a simple harmonic external force acts on any point on the cylindrical shell, the sound pressure at a certain point in the far field remains unchanged.

5.1.5 Relationship Model Between Root Mean Square Amplitude of Sound Pressure and Dry Density

When $T = 1.088$ s, the Rayleigh wave arrives, then $T = c_t t/r = 1.088$, we can get:

$$t = \frac{1.088r}{c_T} = 1.088r\sqrt{\frac{G}{\rho}} \tag{5.28}$$

The compaction area is shown in Fig. 5.4. The duration of each pulse load is Δt. The compaction area within the time interval Δt is a rectangle with a length of L and a width of $v\Delta t$. At this time, the force on the vibrating wheel in the rectangle can be equivalent to multiple uniformly distributed simple harmonic point forces. The sound pressure generated by each point force at the monitoring point is the same. Align the integrals and sum them. Substituting formula (5.28) into formula (5.27) yields:

$$\begin{aligned} A &= \frac{(Lv\Delta t)^2}{2\pi aL} \cdot a_0 \frac{\frac{\pi}{L}}{\left(\frac{\omega^2\cos^2\theta}{c^2} + \frac{\pi^2}{L^2}\right)\sqrt{\frac{2c}{\omega a \sin\theta}}} \cdot \frac{\rho\omega c}{\pi R \sin\theta} \cdot \frac{f}{A_0 \pi G r} \cos\omega_0 t \cdot \cos(3\phi) \\ &= \frac{v^2 \Delta t^2}{a\left(\frac{\omega^2 cos^2\theta}{c^2} + \frac{\pi^2}{L^2}\right)\sqrt{\frac{2c}{\omega a \sin\theta}}} \cdot \frac{\rho\omega c}{\pi R \sin\theta} \cdot \frac{f}{A_0 \pi G r} \cos\left(1.088\omega_0 r\sqrt{\frac{\rho}{G}}\right) \cdot \cos(3\phi) \end{aligned} \tag{5.29}$$

The shear modulus G can be further expressed as [5]:

$$G = G_0 p_a \cdot \frac{(2.97 - e)^2}{1 + e} \left[\left(\frac{3p}{p_a}\right)^2 + \frac{9K_0 J_2}{G_0 p_a^2}\right]^2 \tag{5.30}$$

Fig. 5.4 Schematic diagram of compaction area

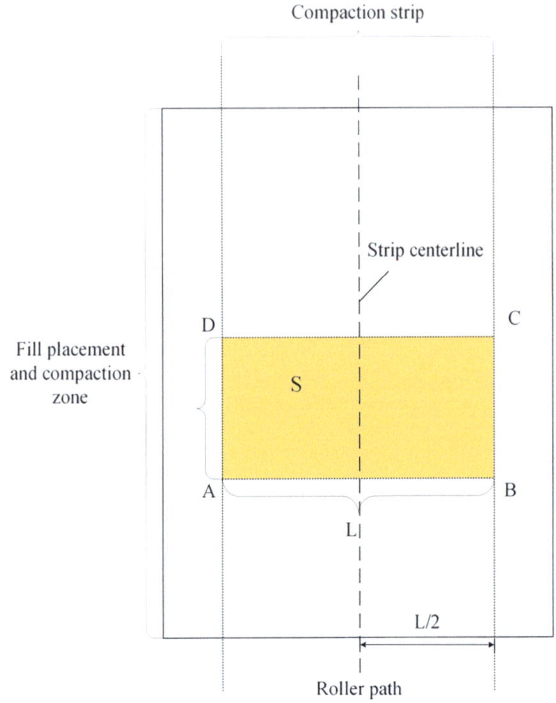

where G_0 and m are material constants, e is the porosity, and p_a is the atmospheric pressure.

Substituting Eq. (5.30) and $e = 1 - \rho_0/\rho$ into Eq. (5.29), the relationship model between the root mean square amplitude of sound pressure and dry density can be obtained:

$$A = \frac{v^2 \Delta t^2}{a\left(\frac{\omega^2 cos^2\theta}{c^2} + \frac{\pi^2}{L^2}\right)\sqrt{\frac{2c}{\omega a \sin\theta}}} \cdot \frac{\rho\omega c \cos(3\phi)}{\pi R \sin\theta} \cdot a_1 \cos\left(1.088\omega_0 r \frac{\rho\sqrt{a_2(2\rho - \rho_0)}}{1.97\rho + \rho_0}\right) \tag{5.31}$$

where $a_1 = \frac{fa_2}{A_0\pi r} \cdot \frac{\rho(2\rho-\rho_0)}{(1.97\rho+\rho_0)^2}$, $a_2 = G_0 p_a\left[\left(\frac{3p}{p_a}\right)^2 + \frac{9K_0 J_2}{G_0 p_a^2}\right]^{\frac{m}{2}}$.

5.2 CSCV

This section introduces contact-type sound compaction value, namely *CSCV*, as a compaction index to characterize the compaction state of gravel soil and weathered slate, as follows:

Table 5.1 Key parameters required for the numerical analysis of gravel soil

Parameter	Value	Parameter	Value
L (m)	2.13	c (m/s)	5900
a (m)	0.8	ρ/(kg/m^3)	2482
v (km/h)	2.95	ρ0/(kg/m^3)	1932
R (m)	2.5	A1 (Pa)	20,000

Table 5.2 Key parameters required for the numerical analysis of weathered slate

Parameter	Value	Parameter	Value
L (m)	2.13	c (m/s)	5900
a (m)	0.8	ρ/(kg/m^3)	2600
v (km/h)	2.95	ρ0/(kg/m^3)	2100
R (m)	2.5	A1 (Pa)	20,000

$$CSCV = C \cdot \frac{A_2}{A_1} \tag{5.32}$$

where A_2 is the amplitude of the second harmonic component of the sound pressure, A_1 is the amplitude of the fundamental frequency of the sound pressure, and C is the calibration constant.

Existing research results show that the amplitude of the fundamental frequency during vibration is about 2 × 104 Pa, so the fundamental frequency amplitude of sound pressure is taken as 2 × 104 Pa. Based on Matlab software, numerical solutions are performed for gravel soil and weathered slate. The important parameters of gravel soil and weathered slate used for model numerical calculation are shown in Tables 5.1 and 5.2. Figures 5.5 and 5.6 show the relationship curves between the dry density of gravel soil [6] and weathered slate[7] and the root mean square amplitude of sound pressure, respectively. The results show that the dry density of gravel soil and weathered slate is approximately positively correlated with the root mean square amplitude of sound pressure.

5.3 Case Study

5.3.1 Testing Site and Materials

The entire Chenglong Highway in Hunan Province, China is located in Chengbu Miao Autonomous County. It is 51.79 km long from Chengbu to Longsheng (the border between Hunan and Guangxi). It is designed as a 4-lane highway with a subgrade width of 26 m, including a central divider, 4 lanes and hard shoulders. The maximum designed vehicle speed is 100 km/h. Chenglong Expressway is part of the S91 Dongcheng Expressway. It has a great role in improving the structure of Hunan's

Fig. 5.5 Relationship curve between dry density of a certain gravel soil and CSCV

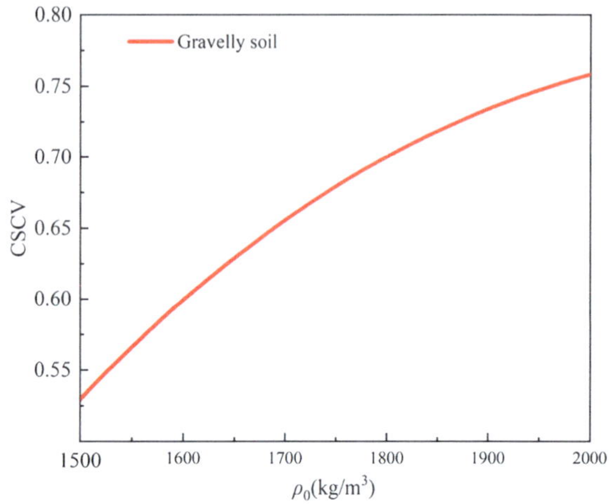

Fig. 5.6 Relationship curve between dry density of a certain weathered slate and CSCV

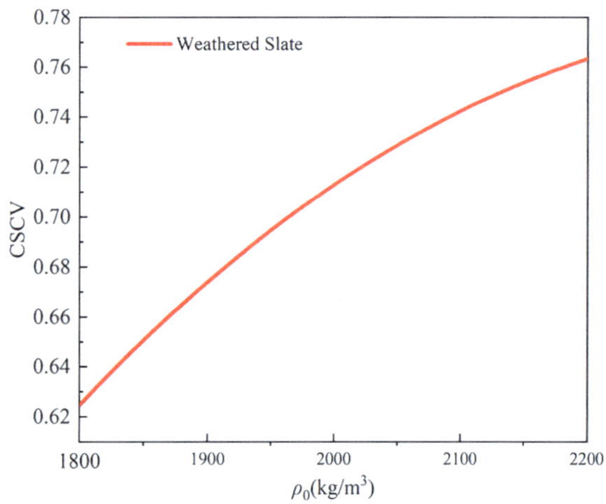

Highway network, opening up the highway out of the province, giving full play to the radiation function of the highway network, and strengthening passenger and freight transportation exchanges in southern Hunan Province. The gradation curve of the on-site filling material is shown in Fig. 5.7.

5.3.2 *Testing Program*

In order to study the changing trend of *CSCV* under different compaction parameters and verify the characterization effect of *CSCV* on the compaction quality of different subgrade filling materials, field rolling tests of gravel soil and weathered slate were

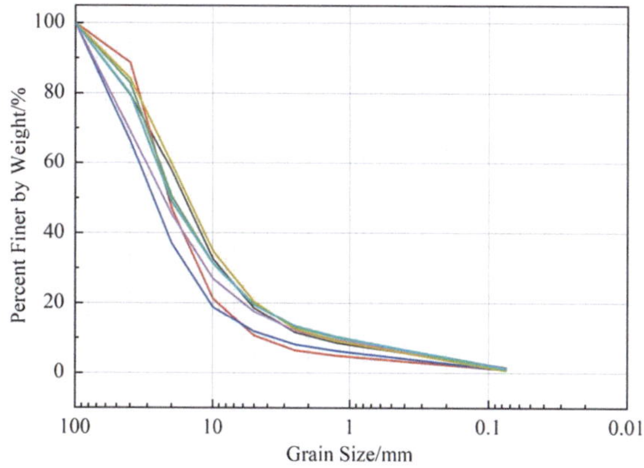

Fig. 5.7 Grading curve of filling materials

carried out. In actual engineering, the number of rolling passes, exciting force state and driving direction are used as compaction parameters to analyze the changing trend of *CSCV*.

The type of roller used in the field test is the XCMG 22t single steel wheel vibratory roller, with a vibration wheel width and diameter of 2130 mm and 1600 mm respectively, a vibration frequency of 28/33 Hz, an exciting force of 353/245kN, and a nominal amplitude of 1.65/0.83 mm. The field test roller travels at a slow speed and vibrates and compacts a certain number of times in two exciting force states: low frequency high amplitude (LFHA) and high frequency low amplitude (HFLA). The field test layout is shown in Fig. 5.8. The rolling test plan is shown in Table. 5.3.

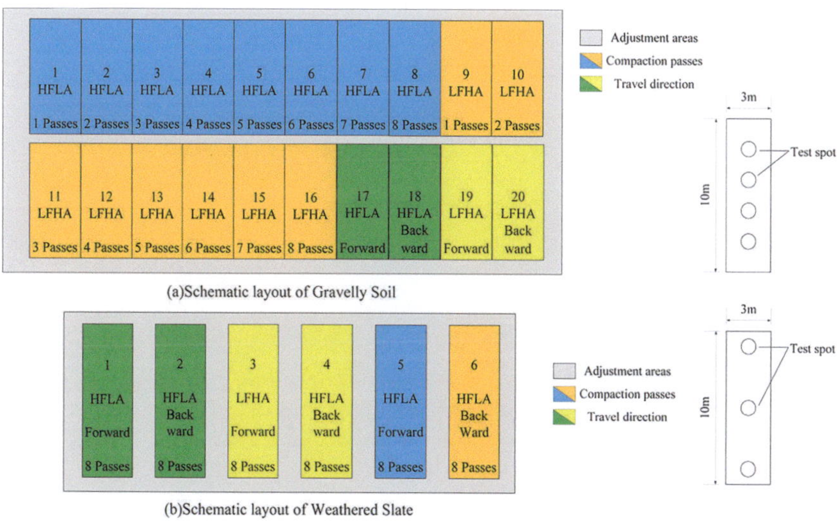

Fig. 5.8 On-site rolling test layout

Table 5.3 Test scheme for field rolling experiment

Material	Operation condition	Vibration status	Passes (static)	Passes (vibration)	Velocity (km/h)	Driving direction
Gravelly soil	1	LFHA	2	1–8	0.83	Bidirectional
	2	HFLA	2		0.83	
	3	LFHA	2	8	0.83	Forward
	4	HFLA	2	8	0.83	
	5	LFHA	2	8	0.83	Backward
	6	HFLA	2	8	0.83	
Weathered slate	1	LFHA	2	8	0.83	Bidirectional
	2	HFLA	2	8	0.83	
	3	LFHA	2	8	0.83	Forward
	4	HFLA	2	8	0.83	
	5	LFHA	2	8	0.83	Backward
	6	HFLA	2	8	0.83	

The test area consists of several test strips with a length of 10 m and a width of 3 m and a certain area for vehicles to pass each other. Rolling operations are carried out on the test strips and data is collected. After that, an on-site in-situ test is carried out to determine the density or settlement. The compaction process first uses a sheep's foot roller to crush the large-size stone by concentrated pressure from the convex blocks to form a good dense structure to enhance the compaction effect. In combination with the use of a smooth wheel, different rolling processes are used to achieve the final compaction quality requirements. The on-site rolling test strip is shown in Fig. 5.9.

Fig. 5.9 On site test diagram

For weathered slate, field tests mainly use settlement difference or cumulative settlement difference as the compaction quality assessment index. Related studies have analyzed the relationship between the mechanical assessment index of compaction quality (settlement difference/settlement rate) and the physical assessment index (density index) [8]. After a certain number of rolling passes, if the soil mass and area remain basically unchanged, then:

$$\rho = \frac{\rho_0}{1-a} = \rho_0\left(1 + a + a^2 + \cdots\right) \approx \rho_0(1 + a) = \rho_0\left(1 + \frac{S}{H_0}\right) \tag{5.33}$$

where ρ_0 is the initial density of the filling material, H_0 is the initial thickness of the compacted layer, ρ is the density, S is the settlement amount, and a is the settlement rate. The results of this study show that the density index has a strong correlation with the settlement difference. For weathered slate, the settlement difference and dry density are also used as compaction quality assessment indexes, which to a certain extent confirms the relationship model between sound pressure and compaction indexes. Similarly, the settlement difference can also be used for corresponding correlation analysis.

5.4 Result Analysis

5.4.1 Correlation Analysis Between CSCV *and Compaction Parameters*

(1) Correlation analysis between *CSCV* and rolling passes

In order to analyze the relationship between the number of rolling passes and *CSCV*, a series of field tests were conducted. For the specific test plan, please refer to Sect. 5.3.2. Figure 5.10 shows the trend and correlation of *CSCV* with the number of rolling passes under different gravel soil and slate test strip rolling conditions. As can be seen from the figure, whether it is gravel soil or weathered slate, *CSCV* increases with the increase of rolling passes, and *CSCV* has a significant correlation with the number of rolling passes.

(2) Correlation analysis between *CSCV* and exciting force state

The exciting force state includes vibration frequency and amplitude. A high exciting force has a lower vibration frequency, and correspondingly, a low exciting force has a higher vibration frequency. In the actual rolling process, the exciting force state of the roller is divided into two types: high frequency low amplitude (HFLA) and low frequency high amplitude (LFHA) to meet the requirements of different filling materials and different compactness. The two different exciting force states have different vibration effects, so the roller has different compaction effects on the

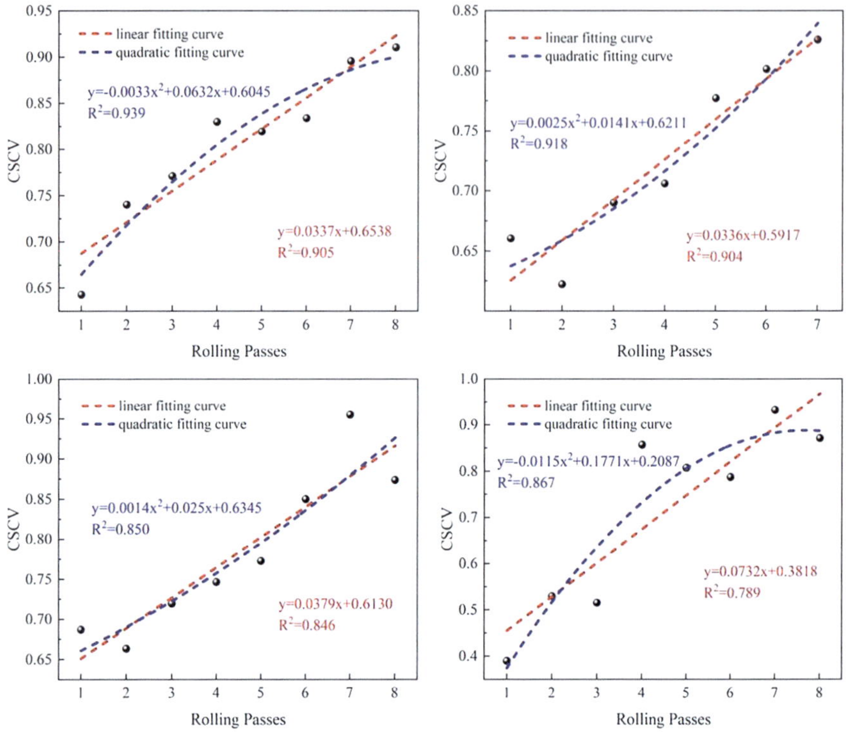

Fig. 5.10 Correlation analysis between *CSCV* and rolling passes

filling material under different exciting force states. According to the test plan in Sect. 5.3.2, at the same driving speed, the test strip was rolled under two exciting force states: HFLA and LFHA [9]. Figure 5.11 shows the change trend of *CSCV* under different exciting force states after 8th rolling. It can be seen from the figure that, under the same driving speed, for gravel soil or weathered, the *CSCV* in the HFLA state is generally greater than the *CSCV* in the LFHA state. It can be concluded that there is a certain correlation between *CSCV* and the exciting force state of the roller.

(3) Correlation analysis between *CSCV* and driving direction

The vibration of the vibration wheel is driven by the exciting force generated by the eccentric rotation of the eccentric block inside it. The hydraulic oil provided by the roller engine makes the vibration motor work, and the vibration motor realizes the rotation of the eccentric block. Therefore, the oil supply of the vibration motor affects the rotation of the eccentric block, and then affects the vibration effect of the roller. In the actual rolling process, the driving direction of the roller affects the rotation of the eccentric block. At the same time, the oil supply of the vibration motor is different, so different driving directions of the roller have different vibration effects. Figure 5.12 shows the changing trend of *CSCV* under different driving directions

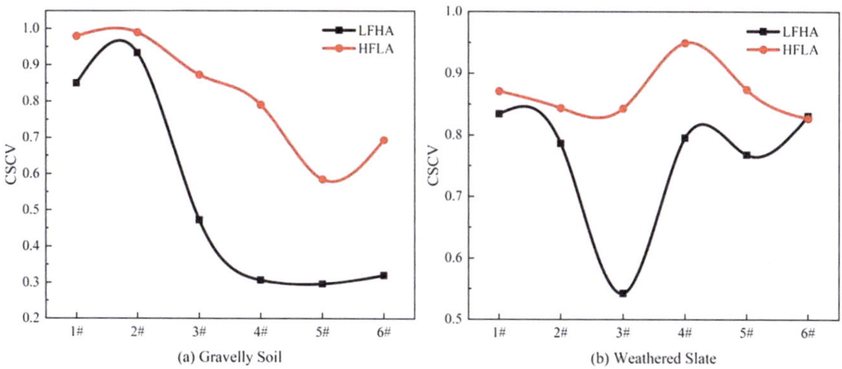

Fig. 5.11 Variation trend of *CSCV* under different exciting force states

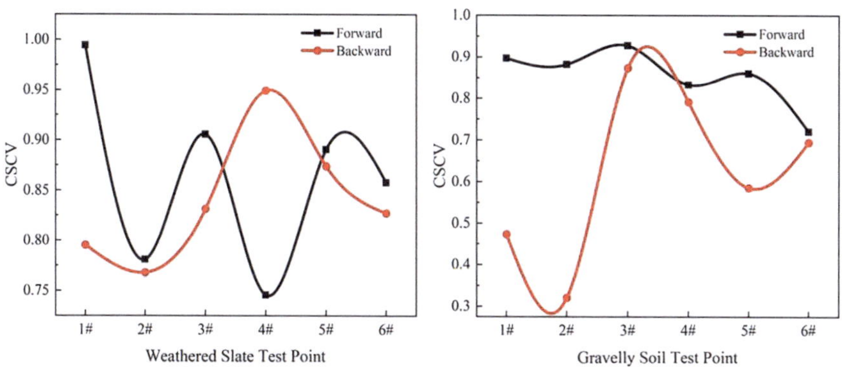

Fig. 5.12 Variation trend of *CSCV* in different driving directions

after the 8th rolling. It can be seen from the figure that under the same exciting force, the *CSCV* of forward driving is always greater than the *CSCV* of reverse driving. *CSCV* has a certain correlation with the driving direction of the roller, but compared with other compaction parameters, the influence of driving direction is small.

5.4.2 *Compaction Quality Intelligent Assessment Considering Uncertainty of Small Samples*

Due to the large number of data samples, this section only takes slate as an example to show how to perform compaction quality intelligent assessment considering the uncertainty of small samples. Based on the 144 slate sample data obtained from the field test, the first 80% of the sample data is used as the training set of the compaction quality assessment model, and the last 20% is used as the test set, that is, there are a

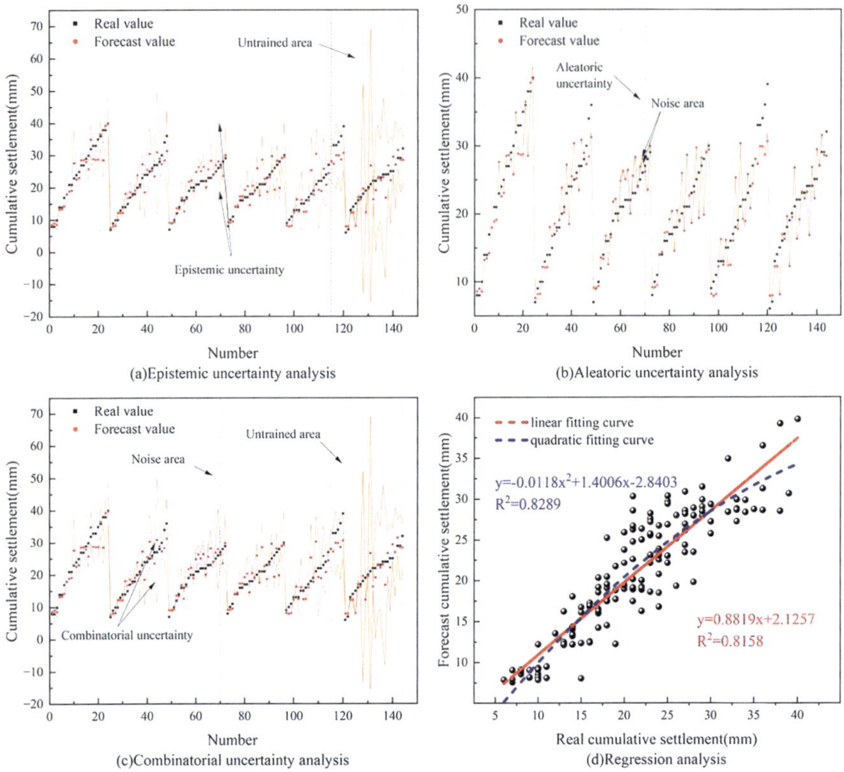

Fig. 5.13 Uncertainty analysis of slate

total of 116 training sample data and 28 test sample data. Considering the uncertainty of small samples, it is substituted into the compaction quality intelligent assessment model. Figure 5.13 shows the compaction quality intelligent assessment results of slate considering the uncertainty of small samples.

References

1. Liu XL (2008) Research on active control of vibration and acoustic radiation of elastic cylindrical shell structures. Harbin Engineering University
2. Lu L, Zhang KY, Jiang H et al (2018) Structural model truncation analysis based on modal strain energy. J Wuhan Univ Technol (Transp Sci Eng) 42(06):1044–1048
3. Zhang QL (2018) Study on the compaction monitoring system of earthwork and its application. Tsinghua University
4. Li MZ, Tian XY (2023) Free and forced vibration characteristics of the cylindrical shell structure of the carrier roller. Mech Electr Eng Technol 52(09):21–24+30
5. Liu HB, Zou DG, Liu JM (2014) Constitutive modeling of dense gravelly soils subjected to cyclic loading. Int J Numer Anal Methods 38:1503–1518

6. Li MW, Lu Z, Tang MX et al Study on mechanical properties and compaction quality evaluation of gravel soil subgrade based on impact load method. J Rock Mech Eng 1–9
7. Zhan YX, Yao HL, Zhang JB et al (2016) Research on the filling technology of embankment with strong to moderate weathered slate in Sansui-Liping Eexpressway. Highway 61(05):8–14
8. Zhao HJ (2015) Study on the evaluation index of roadbed compaction quality. Chang'an University
9. Liu DH, Li ZL, Wang AG (2013) Correlation analysis of rolling parameters and real-time monitoring index for rockfill dam compaction quality evaluation. J Tianjin Univ Sci Technol 46:361–366

Chapter 6
Compaction Quality Interpolation Assessment Methods and Models

Whether the compaction quality of filling materials is qualified directly affects the safety, stability and economic benefits of pavement and subgrade. In addition to the disadvantages of destructive testing, excessive proportion of human factors and post-evaluation, the biggest shortcoming of conventional detection methods is that there are few test samples, lack of continuity, and it is impossible to evaluate the compaction quality of the entire rolling area. Therefore, researchers have gradually shifted their focus to quantitative models based on compaction parameters and compaction quality control indexes that indirectly reflect the compaction quality, so as to achieve real-time compaction quality assessment during rolling, thereby improving the construction quality of earthwork projects and ensuring the safety, stability and service life of infrastructure.

The proposal of roller-integrated compaction monitoring technology and the development of corresponding systems are attempts and innovations of civil engineering personnel in rolling construction. By monitoring compaction information, real-time feedback of the compaction quality of the rolling area is achieved. The compaction information includes compaction parameters, compaction quality control indexes, etc. The compaction quality feedback effect depends on the accuracy of the constructed assessment model. At present, commonly used assessment model construction methods include statistical regression [1, 2], fuzzy prediction [3, 4], random forest [5, 6] and neural network [7, 8], etc. However, due to the sampling frequency of the acquisition system, driving speed, etc., the RICM data is small in discreteness, and the compaction quality of the rolling area represented is still discontinuous. In this regard, this chapter introduces the spatial interpolation method into the roller-integrated compaction monitoring system, aiming to fill the compaction quality information of the unknown discontinuous area with the assessment model constructed by the statistical regression method and the compaction quality data of the adjacent positions, so as to realize the compaction quality assessment of the entire rolling area.

Q. Zhang, *Compaction Quality Nondestructive Testing and Intelligent Control Technology for Infrastructure*, https://doi.org/10.1007/978-981-95-1264-5_6

6.1 Compaction Quality Assessment Method Based on Kriging Interpolation

Chapter 4 of this book introduces the SCV as a novel index for characterizing the compaction quality of filling materials and validates its effectiveness through field experiments. Building upon this foundation, this chapter will employ correlation analysis and statistical regression methods to establish a compaction quality assessment model for filling materials, utilizing multiple sets of experimental sample data collected from field rolling tests. Subsequently, it will explain how to apply the Kriging interpolation method to perform interpolation calculations for compaction quality at any location on the working surface, thereby enabling a comprehensive assessment of the compaction quality across the entire body of the earth-rock dam.

6.1.1 Establishment of Compaction Quality Assessment Model

(1) Multiple groups of on-site rolling tests based on working surface

The experiment employs a modified vibratory roller equipped with GPS positioning, autonomous driving capabilities, and roller-integrated acoustic detection and exciting force measurement devices. The testing is conducted using rockfill materials in a saturated state from the Qianping Reservoir and consists of four test groups (28 Hz—Area A, 28 Hz—Area B, 28 Hz—Area C, and 28 Hz—Area D). The experimental site is depicted in Fig. 6.1.

The lift thickness for the trial is set at 80 cm. After filling, two static passes are conducted, followed by eight vibratory passes. Measurements for settlement, density, gradation, and P5 content are taken after the eighth pass of rolling (which corresponds to the standard number of passes used in field construction), along with continuous monitoring of acoustic signals and excitation forces. The entire experimental area measures 30 m in length and 20 m in width, encompassing a core testing zone as well as a surrounding non-testing area that serves a transitional function. The core testing area is divided into four sections, labeled A, B, C, and D. Taking Area A as an example, the core rolling section measures 24 m in length and 3 m in width. Density measurements are performed using the water replacement method, with each of the four groups featuring six sampling points, resulting in a total of 24 sampling points.

(2) Correlation analysis between dry density and *SCV* of rockfill materials

Based on the field rolling tests, a dataset was obtained that includes *SCV*, dry density (ρ_d), and other parameters of the filling materials (P5 content, moisture content). Previous studies have indicated that the mechanical properties of rockfill materials are significantly influenced by the characteristics of particle gradation, which primarily encompasses fine particles (<5 mm) and coarse particles (200–400 mm).

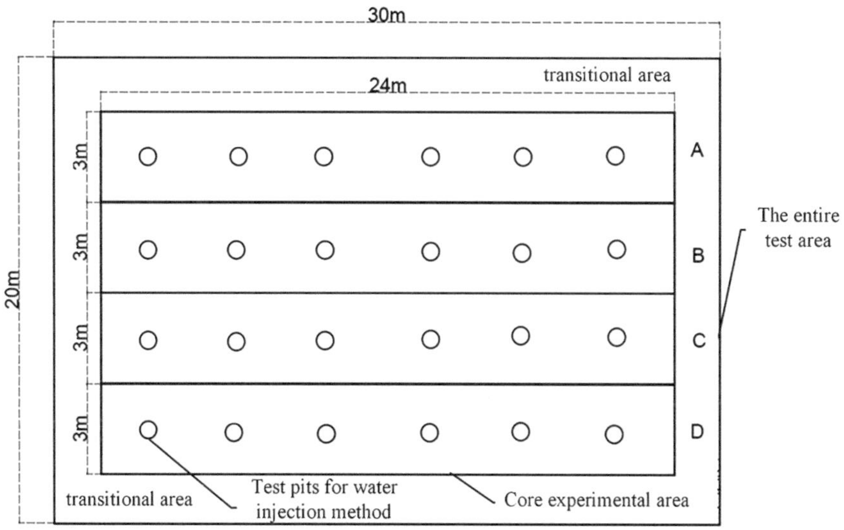

Fig. 6.1 Schematic diagram of the test site (unit/m)

Correlation analysis between different particle sizes and dry density reveals that the correlation coefficients for other particle sizes are lower than that of the P5. Therefore, the content of particles with a diameter smaller than 5 mm (P5) is selected as the indicator for correlation analysis with dry density. The relationship between dry density, *SCV*, P5, and moisture content for rockfill materials is illustrated in Fig. 6.2.

From the figure, it is evident that there is a linear correlation between the dry density of rockfill and *SCV* $\left(R^2 = 0.8387\right)$; a linear correlation with P5 $\left(R^2 = 0.5119\right)$; and a lack of correlation with moisture content $\left(R^2 = 0.2631\right)$. This indicates a strong linear relationship between dry density and *SCV*, while the correlation with moisture content is comparatively weak. This can be attributed to the fact that the material source is rockfill, and all filling materials used in the experiments were taken from the same location in the upstream river channel, resulting in minimal variations in moisture content.

(3) Establishment of regression model for filling materials

Based on the aforementioned correlation analysis results, a simple regression model (Model I) is established using *SCV* as the independent variable and dry density as the dependent variable, as expressed in Eq. (6.1). Furthermore, a multiple linear regression model (Model II) is formulated with *SCV* and P5 content as independent variables and dry density as the dependent variable, as detailed in Eq. (6.2). Lastly, a multiple nonlinear regression model (Model III) is constructed using *SCV* and P5 content as independent variables, with dry density as the dependent variable, as outlined in Eq. (6.3).

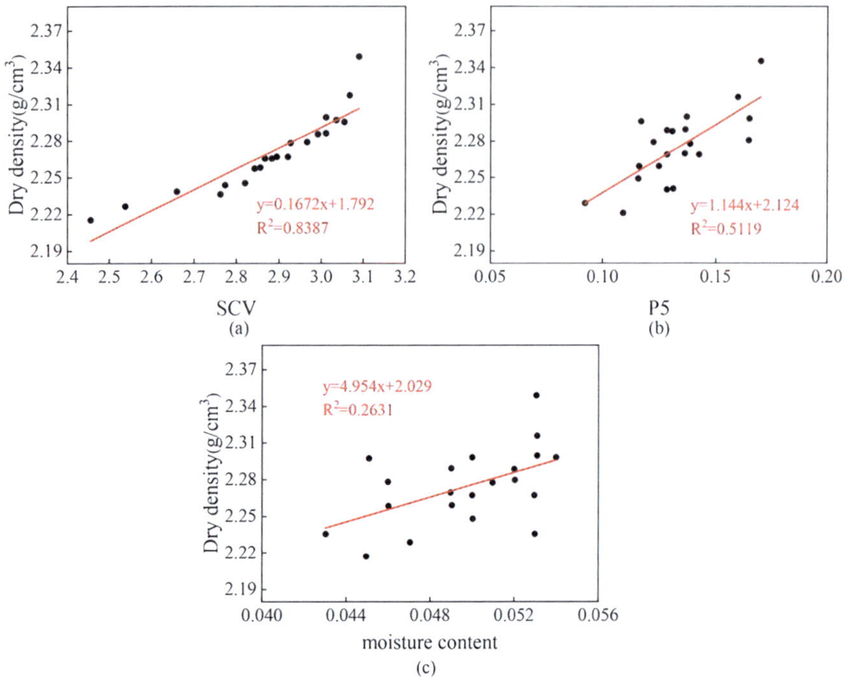

Fig. 6.2 Correlation analysis of rockfill materials **a** dry density and *SCV* **b** dry density and P5 **c** dry density and moisture content

$$\text{Model I}: \overline{\rho_d} = 0.818\overline{SCV} - 0.133 \quad (6.1)$$

$$\text{Model II}: \overline{\rho_d} = 0.711\overline{SCV} + 0.168\overline{p} - 0.148 \quad (6.2)$$

$$\text{Model III}: \overline{\rho_d} = 0.797\overline{SCV} + 0.674\overline{p}^2 - 0.634\overline{p} - 0.009 \quad (6.3)$$

Here, $\overline{\rho_d}$, $\overline{p}$, and $\overline{SCV}$ represent the normalized values of dry density (g/cm^3), P5 (%), and SCV, respectively, where $\rho_d = 0.13\overline{\rho_d} + 2.22$, $p = 0.077\overline{p} + 0.092$, and $SCV = 0.636\overline{SCV} + 2.454$. The correlation coefficients for Models I, II, and III are $R^2 = 0.839$, $R^2 = 0.856$, and $R^2 = 0.903$, respectively. Figure 6.3 illustrates the comparison between the dry density values calculated from the three regression models and the actual dry density values, with detailed error results displayed in Table 6.1.

From Fig. 6.3, it is evident that the relative errors of all three regression models are below 2%, indicating that the predictive outcomes of the established models are quite precise and meet the demands of practical engineering applications. Figure 6.4 compares the relative errors between the predicted dry density values from the various

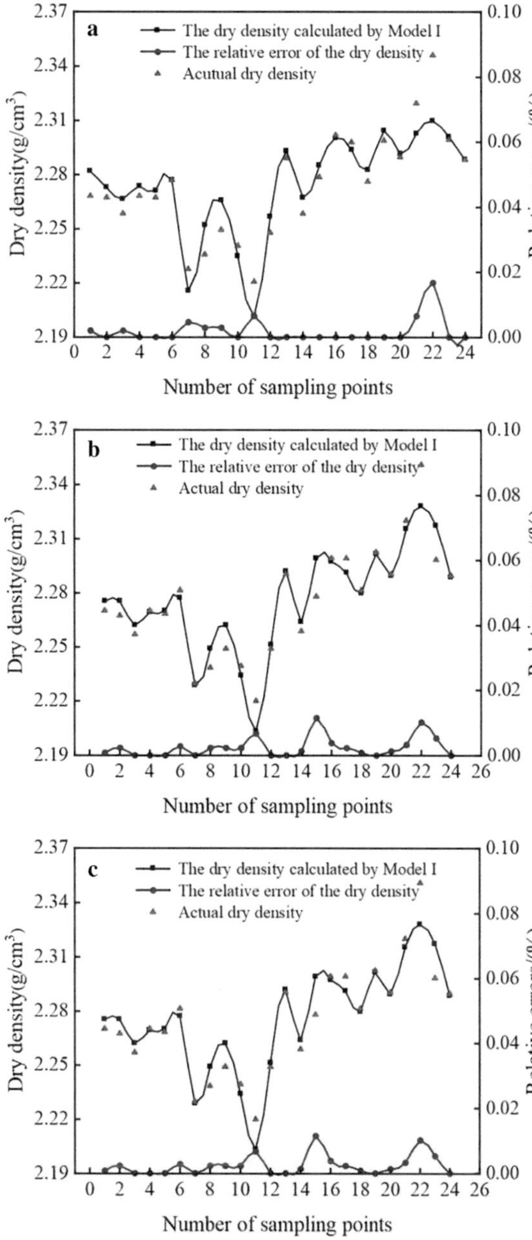

Fig. 6.3 Comparative analysis between actual dry density and dry density predicted by regression model: **a** Model I **b** Model II **c** Model III

Table 6.1 Error table of the regression model of dry density of rockfill materials

Model	Maximum error (g/cm^3)	Average error (g/cm^3)	Maximum relative error (%)	Average relative error (%)
I	0.041	0.0079	1.74	0.35
II	0.035	0.0082	1.48	0.36
III	0.022	0.0068	0.95	0.30

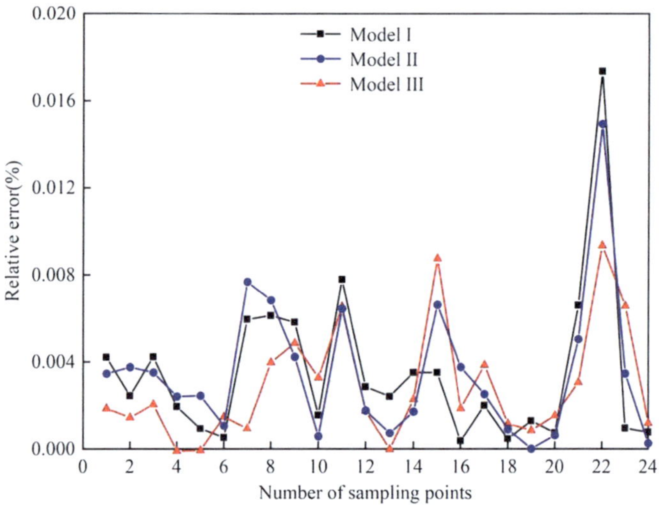

Fig. 6.4 Comparative analysis of the relative errors between the dry density predicted by the three regression models and the actual dry density

regression models and the actual dry density values. In terms of average error, all three regression models exhibit average errors below 0.00811 g/cm^3.

For Models I and II, the relative errors are largely consistent. Although including P5 (%) in the regression model enhances its statistical significance, there is no marked improvement in the correlation between *SCV* and dry density. The analysis suggests that while incorporating P5 into the regression model results in a certain degree of improvement in the correlation and predictive accuracy concerning *SCV* and dry density, the enhancement is not substantial. This may be attributed to the fact that for Models II and III, the fill materials used for the dam construction are sourced from natural gravel in the riverbed, leading to a relatively concentrated material source and a limited range of variation in dry density.

6.1.2 Kriging Interpolation Method

Kriging interpolation is a spatial interpolation method based on the theory of variogram, covariance function and regionalized variable theory [9]. It performs unbiased estimation and interpolation of unknown location data points from the perspective of spatial autocorrelation of data. Kriging interpolation is currently widely used in the interpolation of soil properties [9]. It has the advantages of high approximation, strong extrapolation ability and the ability to consider spatial correlation of data. It can not only improve the scientificity and accuracy of interpolation results, but also is suitable for scenarios that require error assessment of the interpolation result surface. Based on the above advantages, in the application scenario of establishing a full working surface, Kriging interpolation has advantages that other methods do not have.

The variogram reflects the spatial variability of the data, and it is meaningful only at half the maximum distance between the data points. Therefore, let the semivariogram be defined as $r(h)$, where:

$$r(h) = \frac{1}{2N(h)} \sum_{i=1}^{N(h)} [Z(x_i) - Z(x_i + h)]^2 \tag{6.4}$$

In the equation, h represents the distance between two sample points; $N(h)$ denotes the total number of sample point pairs at a distance h; $Z(x)$ is the spatial data function, where $Z(x_i)$ represents the data value at spatial location x_i, and $Z(x_i + h)$ indicates the data value at spatial location $x_i + h$. Specifically applied to the evaluation of pavement subgrade compaction quality, the spatial data refers to the compaction quality indicator data. A typical semivariogram curve is illustrated in Fig. 6.5.

Regionalized variables are those that exhibit spatial distribution, with their values dependent on their specific spatial locations. The observed regionalized variables transition from random variables to deterministic values at specific spatial points, expressed as follows:

$$Z(x_u, x_v, x_w) = Z(x) \tag{6.5}$$

The purpose of the covariance function is to evaluate the differences between regionalized random variables. Its calculation formula is as follows:

$$c(h) = \frac{1}{N(h)} \sum_{i=1}^{N(h)} \left[Z(x_i) - \overline{Z}(x_i)\right]\left[Z(x_i + h) - \overline{Z}(x_i + h)\right] \tag{6.6}$$

In the equation, $\overline{Z}(x_i)$ and $\overline{Z}(x_i + h)$ represent the sample means of $Z(x_i)$ and $Z(x_i + h)$. A typical covariance function curve is illustrated in Fig. 6.6.

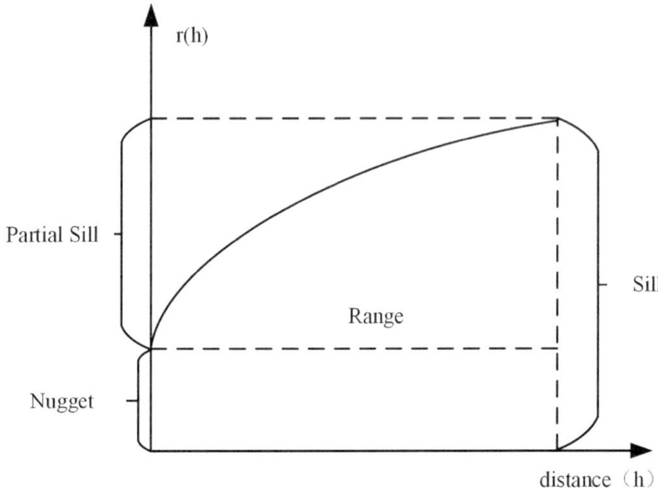

Fig. 6.5 Typical semivariogram plot

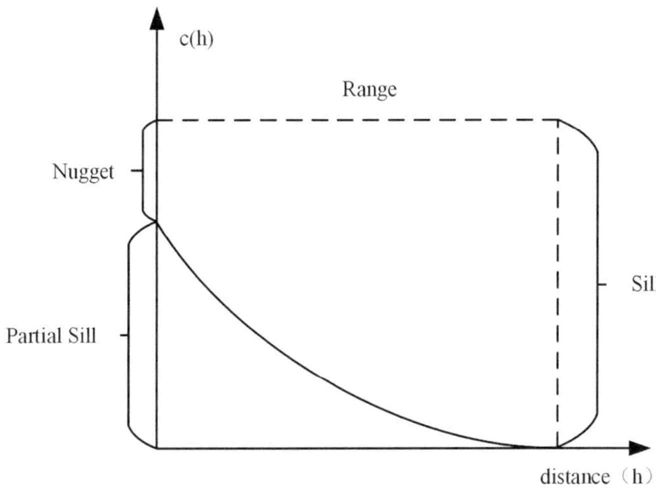

Fig. 6.6 Typical covariance function graph

In Fig. 6.5, the measurement errors and spatial variability determine whether the nugget effect is zero. As the distance between sampling points increases, the semivariogram gradually stabilizes at a constant value known as the sill. If the distance between spatial point pairs exceeds the sill, their spatial correlation ceases to exist. By subtracting the nugget effect from the sill, one can derive the partial sill. The range is defined as the distance over which the semivariogram increases from the nugget effect to the sill. When the distance between sample points falls within the range, the degree of compaction at any location on the compacted surface can be interpolated or extrapolated using Kriging. Conversely, when the distance between sample points exceeds the range, the degree of compaction at any location on the compacted surface cannot be interpolated or extrapolated using Kriging, as spatial correlation between the sample points no longer exists.

When the distance between sample points falls within the range, the compactness at any location on the compacted surface can be estimated using several effective sample values surrounding the point of interest, considering the influence of spatial location. The calculation formula is as follows:

$$Z(x_0) = \sum_{i=1}^{n} \lambda_i Z(x_i) \tag{6.7}$$

In the equation, λ_i represents the weight coefficient associated with the estimated value, which can be determined using the following two formulas:

$$\sum_{i=1}^{n} \lambda_i c\left(x_j - x_i\right) = c(x_0 - x_i) - \mu \tag{6.8}$$

$$\sum_{i=1}^{n} \lambda_i = 1 \tag{6.9}$$

In the equation, $c\left(x_j - x_i\right)$ denotes the semivariance between the sampling point x_j and x_i; $c(x_0 - x_i)$ represents the semivariance between the sampling point x_0 and x_i; and μ is the Lagrange multiplier associated with the minimum variance.

6.1.3 A Rapid Compaction Quality Assessment Method for the Entire Working Surface

Factors influencing compaction quality can be categorized into internal and external factors. Internal factors primarily include the properties or categories of materials (gradation), moisture content, temperature during compaction, and the strength of the foundation or underlying layers. External factors mainly encompass compaction effort (such as the weight of compaction tools, driving speed, and vibration frequency), lift thickness, types of compaction machinery (including sheep's foot rollers, smooth rollers, tire rollers; static or dynamic rollers; continuous contact, partial uplift, double jump, oscillating motion, and chaotic motion), and compaction methods (compaction strategies).

It is evident that the compaction quality of earth and stone filling materials is influenced not only by construction parameters but also by the properties of the filling materials (such as gradation and moisture content). Therefore, evaluating the compaction quality of filling materials solely based on indicators derived from the performance analysis of roller movements, without considering the effects of compaction process parameters and the properties of earth-rock filling materials, presents a one-sided perspective.

To address this, the contact-type roller-integrated acoustic wave detection technology described in Sect. 4.2 will be employed to collect the acoustic signals emitted by the soil materials during compaction. The onboard computer will perform spectral analysis on these signals to obtain continuous *SCV*, which characterizes the compaction quality of soil materials. The specific calculation formula is referenced in Sect. 4.3. Subsequently, a correlation analysis will be conducted between *SCV* and the compaction parameters, as well as the compaction quality measured by conventional testing methods, using statistical regression analysis. This will facilitate the establishment and selection of an appropriate relationship model between the density of earth-rock materials and the CCI (*SCV*).

By integrating this with the Kriging interpolation method discussed in this section, a comprehensive assessment of compaction quality across the entire working surface during the filling process can be achieved. This approach effectively resolves the challenges present in current evaluation techniques and methods for filling quality, particularly in the context of earth-rock dam filling quality assessments, which often suffer from low accuracy, inefficiency, lack of user-friendliness, complex interpretations of compaction states, notable dispersion, significant errors in results, susceptibility to filling material influences, and a neglect of the effects of compaction parameters and material properties.

The specific evaluation steps (Fig. 6.7) are as follows:

(1) By utilizing the earth and stone compaction quality monitoring system installed on the roller, relevant compaction construction parameters—including driving speed v, lift thickness h, vibration frequency f, and number of compaction passes n—are collected in real-time at specified intervals (e.g., every 2 s)

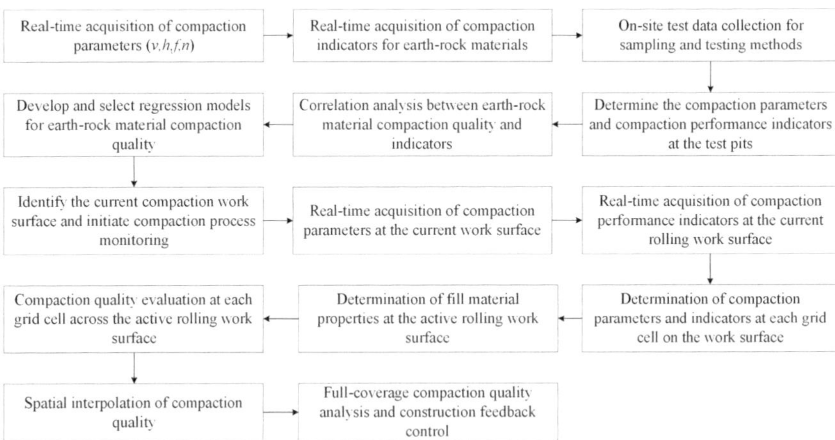

Fig. 6.7 Compaction quality rapid assessment flow chart

during compaction. This data, along with the corresponding spatial location information, is transmitted to a remote monitoring platform and simultaneously stored in the database server of the platform.

(2) By employing the on-board acoustic wave detection system for compaction quality of soil and stone materials installed on the roller (for specific information, refer to Sect. 4.2), the continuous *SCV* and its corresponding spatial location information are collected in real-time at specified intervals (e.g., every 2 s) during compaction. The results are then transmitted to a remote monitoring platform and simultaneously stored in the database server of the platform. The calculation of *SCV* is as follows:

$$SCV = k \times A_2 \tag{6.10}$$

In the equation, k denotes the calibration coefficient of the compacted material, and A_2 represents the second harmonic component of the effective acoustic spectrum after the acoustic field time-domain signal undergoes filtering, spectral analysis, and logarithmic processing. The acoustic signals are measured in real-time by a microphone positioned below the roller's frame. A larger *SCV* indicates a superior compaction quality of the soil and stone materials.

(3) On the compaction construction site, conventional point sampling methods (such as the water replacement method) are employed to collect field and laboratory test results from the test pit locations, including the names of soil and stone materials, moisture content ω, gradation p, and compaction quality (compactness K or dry density ρ). The results are then transmitted to the remote monitoring platform and simultaneously stored in the database server of the platform, while the spatial location information of the test pit points is obtained using RTK-GPS technology.

(4) Utilizing the compaction parameters, the continuous *SCV*, and their corresponding spatial location information collected in real-time from steps (1) and (2), the compaction construction parameters (v, h, f, n) and the *SCV* for each test pit location are determined based on the coordinates of each test pit. Additionally, the moisture content ω, gradation p and compaction quality (compactness K or dry density ρ) of the soil and stone materials at each test pit are established, thereby forming a dataset of sample data for the soil and stone materials at various locations.

(5) Based on the collected sample data from the test pit locations, the correlation between the compaction quality (compactness K or dry density ρ) of different soil and stone materials and the compaction parameters, as well as the compaction characterization indexes, is analyzed. The strength of these correlations is then used to identify the variables for the compaction quality regression model.

(6) Utilizing the variables determined from the correlation analysis of the regression model, along with the sample data from each soil and stone material test pit, distinct regression models are established for the compaction quality (compactness K or dry density ρ) of different sampling points in relation to

the compaction construction parameters (v, h, f, n), the compaction characterization index SCV, and the material properties of the soil and stone materials (moisture content ω, gradation p). This results in an assessment model for the compaction quality of soil and stone materials. The established model generally takes the following form:

$$Z_1 = F(SCV) \tag{6.11}$$

$$Z_2 = F(v, h, f, n, SCV, w, p) \tag{6.12}$$

$$Z_3 = F\left(v^2, h^2, f^2, n^2, SCV, w^2, p^2\right) \tag{6.13}$$

In this context, formulas (6.11), (6.12), and (6.13) represent univariate, multivariate linear, or nonlinear compaction quality assessment models, where Z_1, Z_2, and Z_3 denote the compaction quality of the soil and stone materials (compactness K or dry density ρ), and $F(\cdot)$ serves as the regression function, which can be determined through regression analysis methods. Furthermore, the model must undergo significance testing before it can be utilized; otherwise, additional sample data will be required to establish a regression model (compaction quality assessment model) that meets the criteria for significance testing.

Taking the shell material (gravel) of a clay core dam in a certain reservoir project as an example, its compaction quality can be reflected by dry density ρ. Considering that during the actual compaction process, the lift thickness h, driving speed v, and vibration frequency f essentially remain constant according to design requirements, and that the correlation analysis between moisture content ω and compaction quality leads to a conclusion of non-correlation, this variable can be excluded from the regression model. In this project, when the number of passes during compaction reaches the design requirement of 8 passes, conventional sampling methods (such as the water replacement method) are employed to assess the compaction quality of the filling material (dry density ρ) on-site. Consequently, the compaction quality assessment model for the filling material is established as follows:

$$\overline{Z_1} = F_0\left(\overline{SCV}\right) \tag{6.14}$$

$$\overline{Z_2} = F_0\left(\overline{SCV}, \overline{p}\right) \tag{6.15}$$

$$\overline{Z_3} = F_0\left(\overline{SCV}, \overline{p}^2, \overline{p}\right) \tag{6.16}$$

In this context, $F_0(\cdot)$ denotes the regression function for the compaction quality (dry density ρ) of the filling material (gravel), which can be determined through regression analysis methods; this model must also undergo significance testing. $\overline{\rho}$, $\overline{SCV}$, and $\overline{p}$ represent the normalized values of the dry density ρ, the compaction characterization index SCV and the gradation p of the fill material, respectively; $\rho = a_0\overline{\rho}+b_0, SCV = a_1\overline{SCV}+b_1, p = a_2\overline{p}+b_2$. By analyzing the accuracy of various regression models and considering the need for efficient and convenient application, an appropriate regression model is selected to serve as the compaction quality assessment model.

(7) Based on the construction progress of the earth-rock engineering, the next construction site is identified, along with the roller equipped with the earth-rock compaction monitoring system and the on-board acoustic wave detection system for compaction quality. Subsequently, the compaction operation commences, accompanied by the initiation of monitoring throughout the compaction process.

(8) Utilizing the earth-rock compaction monitoring system installed on the roller, spatial location information of the roller, along with the compaction parameters (v, h, f, n) during compaction, are collected in real-time at specified intervals (e.g., every 2 s). The results are transmitted to a remote monitoring platform and simultaneously stored in the database server of the platform.

(9) Utilizing the on-board acoustic wave detection system for compaction quality installed on the roller, the compaction effect index SCV and its corresponding spatial location information are collected in real-time during the compaction process at specified intervals (e.g., every 2 s). The results are transmitted to a remote monitoring platform and simultaneously stored in the database server of the platform.

(10) The longer sampling interval (e.g., every 2 s) from steps (8) and (9) is taken as the sampling accuracy to determine the distance traveled by the roller during this time interval. The current compaction construction site is then divided into a grid (for instance, subdivided into 1 m × 1 m cells). Utilizing the on-site compaction data collected in steps (8) and (9), values are assigned to each grid based on their spatial location information, thereby determining the compaction parameters (v, h, f, n) and the compaction characterization index SCV for each grid.

In this context, the lift thickness h, the number of compaction passes n, and the compaction characterization index are all values collected during the final pass at the grid according to design requirements. The driving speed v is taken as the average of the roller's multiple passes through that grid. If a grid contains multiple sampling values, the average of these collected values is used as the assigned value for that grid.

(11) Following the completion of the compaction process, on-site point sampling tests (such as the water replacement method) are conducted to collect the material properties of the filling material at the current compaction construction site, including moisture content ω and gradation p. Typically, the point sampling detection method for the compaction construction site involves multiple test pit locations; hence, the moisture content ω and gradation p of the filling material are taken as the averages of the test results from these various pit locations. It

is also assumed that the moisture content ω and gradation p are uniform across all grids within the compaction construction site.

(12) Using the compaction quality assessment model established in step (6), along with the compaction parameters and the values of the compaction characterization index SCV determined in steps (10) and (11) for each grid, the compaction quality (compactness K or dry density ρ) for each grid within the current compaction construction site is calculated.

(13) Considering that the grids are discrete across the entire compaction construction site (for instance, each grid measuring 1 m × 1 m), it is necessary to employ kriging interpolation to assess the compaction quality at any location within the worksite. This will yield the compaction quality values for any position on the current compaction construction site. Subsequently, the results are visualized using ArcGIS and GS+ software to produce a compaction quality cloud map for the entire worksite. The findings are then transmitted to the remote monitoring platform and simultaneously stored in the database server of the platform.

(14) Based on the compaction quality obtained at any given location in step (13), and in conjunction with the established compaction quality control standards (requirements for engineering quality control at the construction site), such as the condition that the dry density ρ of the gravel must be greater than or equal to the acceptable standard value ρ_d, which can analyze the regions of acceptable and unacceptable compaction quality. Moreover, the areas that do not meet the standards can be highlighted using a cloud map representation. The overall compaction quality acceptance rate r for the entire worksite can be calculated using the following formula:

$$r = \frac{S_D}{S} \times 100\% \tag{6.17}$$

Here, r represents the acceptance rate of the compaction quality across the entire compaction construction site; S_D denotes the area of the regions in the compaction quality cloud map where the compaction quality characterization indicator SCV or the compaction quality (compactness K or dry density ρ) exceeds the acceptable value, while S signifies the total area of the entire compaction construction zone. A higher value of r indicates that the control of compaction quality during the filling process has been more effectively managed. Based on the results of the acceptance rate, personnel on site can promptly implement remedial measures for any identified unacceptable areas.

6.2 Compaction Quality Assessment Method Based on Green's Spline Interpolation

Practical applications have revealed several deficiencies associated with Kriging interpolation: firstly, a sufficient number of data samples is required; otherwise, the accuracy of the results may be significantly compromised. Additionally, the calculation process is cumbersome and involves a considerable computational load. Furthermore, the selection of the variogram must be performed manually based on experience, and both the choice of the variogram and the fitting of its model substantially impact the accuracy of the Kriging interpolation results. Consequently, this section will introduce the Green spline interpolation method as an alternative.

6.2.1 *The Green Spline Interpolation Method*

The Green spline interpolation method, proposed in the 1980s [10], is also referred to as biharmonic spline interpolation due to the utilization of the biharmonic operator ∇^4. Today, Green's function has been applied across one-dimensional, two-dimensional, and even multi-dimensional contexts [11]. Considering this field test site as a two-dimensional plane, the two-dimensional Green spline interpolation function $w(x)$ under tensile force P for N data points within the rolling area is expressed by the following Eqs. (6.18) and (6.19):

$$\nabla^2\left[\nabla^2 - P^2\right]\omega(x) = \sum_{j=1}^{N} \alpha_j \delta\left(x - x_j\right) \tag{6.18}$$

$$\omega\left(x_j\right) = \omega_i \quad i = 1, 2, \cdots, N \tag{6.19}$$

wherein, ∇^2 represents the Laplacian operator; $x = (x_1, x_2)$ denotes a position coordinate within the two-dimensional plane; and $(\delta(x))$ is the delta function. When the tensile force $P = 0$, the left-hand side of Eq. (6.18) reduces to $\nabla^4 w(x)$, where ∇^4 is the biharmonic operator, and its specific expression is intricately linked to the choice of coordinates employed [12, 13]. The solutions to Eqs. (6.18) and (6.19) are as follows:

$$\omega(x) = \sum_{j=1}^{N} \alpha_j g_2\left(x - x_j\right) \tag{6.20}$$

wherein, $g_2(x - x_j)$ represents the two-dimensional Green spline function under tensile force, which is derived from the non-homogeneous Eq. (6.21). The coefficient α_j on the right-hand side of the aforementioned equations is computed using the linear simultaneous Eq. (6.22).

$$\nabla^2\left[\nabla^2 - P^2\right]g_2(x, x') = \delta(x - x') \tag{6.21}$$

$$\omega_i = \sum_{j=1}^{N} \alpha_j g_2(x_i - x_j) \quad i = 1, 2, \cdots, N \tag{6.22}$$

The Green spline interpolation method is classified as a global interpolation technique and offers several advantages:

(1) it is user-friendly and simple to operate, allowing for the direct input of existing data samples into established computational programs, irrespective of whether the grids are regular.
(2) the computational process is straightforward, requiring minimal computational effort, resulting in high efficiency and precision. Current computational programs are based on analytical functions, eliminating the need for optimization of numerical models; thus, the results obtained are analytical solutions rather than vague numerical approximations.
(3) this method can suppress spurious fluctuations in regions with sparse data, effectively stabilizing the display of localized anomalous changes. During the computation, the gradient control of the Green function is incorporated to constrain the height and slope of the curves (or surfaces).

6.2.2 Experimental Test

(1) Test site and materials

The Hengyong Highway constitutes a vital segment of the expressway network in Hunan Province, China, spanning a total length of 106.227 km and featuring a subgrade width of 26.5 m. The entire route adheres to the standards of a two-way, four-lane highway, with a design driving speed of 120 km/h and a design vehicle load classification corresponding to Highway I level. Specifically, the main line of the fifth bidding section of the Hengyong Highway commences at K50+000 and concludes at K59+860, covering a total distance of 9.86 km. The subgrade filling materials primarily consist of high and low liquid limit silt. Given the variations in material properties among different filling materials, particularly the distinct characteristics of high and low liquid limit silts, the applicability of the RICM index to these materials necessitates verification through targeted research. This field test is conducted in two test sections, K52+800 to K53+030 and K55+900 to K56+150, respectively, with the subgrades of the two test sections filled with low and high liquid limit silt, respectively.

(2) Analysis of physical and mechanical characteristics of materials

① Indoor compaction test

Moisture content is a significant factor influencing the compaction quality of silt subgrades. The optimal moisture content and maximum dry density for both high and low liquid limit silts were determined through indoor compaction tests, aimed at elucidating the compaction characteristics of these two types of silt and providing a reference for field compaction tests and the engineering applications of RICM technology. Prior to conducting these indoor compaction tests, the liquid and plastic limits of the soil samples were measured using an optoelectronic liquid-plastic limit combined tester. For the high liquid limit silt samples, the moisture contents corresponding to cone depths of 2.5 mm and 20 mm were designated as the plastic limit (ω_P) and the liquid limit (ω_L), respectively, resulting in calculated plasticity indexes of 22.7 and 22.5. Similarly, for the low liquid limit silt samples, the moisture contents corresponding to the same cone depths were regarded as the plastic limit (ω_P) and the liquid limit (ω_L), yielding calculated plasticity indexes of 16.5 and 16.3.

In this indoor compaction test, silt samples were obtained swiftly and efficiently through heavy compaction, after which the optimal moisture content and maximum dry density of the silt samples were determined. The compaction apparatus utilized in this test is depicted in Fig. 6.8, detailing several parameters: the drop hammer has a diameter of 50 mm and a weight of 4.5 kg, while the test cylinder features an inner diameter of 100 mm and a height of 127 mm, accompanied by a lantern ring height of 50 mm. Additional testing equipment included electronic scales, metal plates, and round hole sieves, among others.

The compaction test method involved compacting the soil samples in five layers, with twenty-seven compaction cycles performed per layer. The resulting compaction

Fig. 6.8 Compaction device

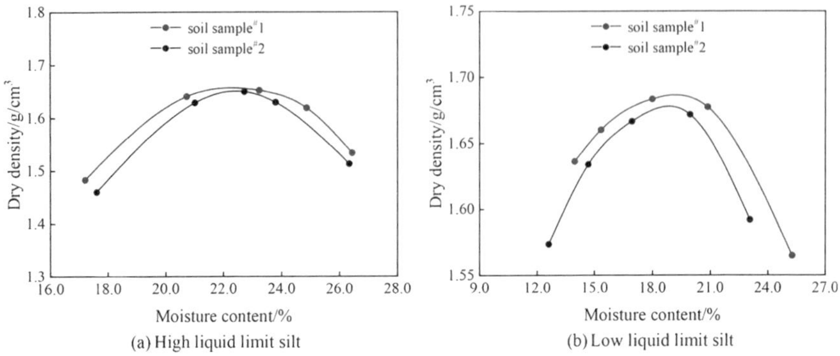

Fig. 6.9 Compaction curve of high and low liquid limit silt

curves for high and low liquid limit silts are illustrated in Fig. 6.9. From Fig. 6.9, it is evident that the maximum dry density and optimal moisture content for the high liquid limit silt sample one were 1.64 g/cm^3 and 22.7%, respectively, while for sample two, these values were 1.66 g/cm^3 and 22.5%. For the low liquid limit silt, sample one exhibited a maximum dry density of 1.69 g/cm^3 and an optimal moisture content of 19.1%, whereas sample two had values of 1.68 g/cm^3 and 18.8%. A comparison of the curves reveals that the optimal moisture content of high liquid limit silt was marginally higher than that of low liquid limit silt overall. In summary, the liquid and plastic limits, maximum dry density, and optimal moisture content for the two types of silt samples have been compiled, as presented in Table 6.2.

② Particle analysis test

Silt comprises sand and clay particles, the proportions of which significantly influence its physical and mechanical properties. Therefore, investigating the particle composition and size distribution of high and low liquid limit silt through particle analysis testing holds considerable importance for understanding its engineering characteristics. To this end, high and low liquid limit silt samples of a specific mass were collected from the field test site and subsequently divided into four equal portions, each weighing 300 g (with a maximum error not exceeding 0.1 g). The distribution of different particle sizes within the high and low liquid limit silt was determined by sieving eight samples through various mesh sizes and weighing them according to the

Table 6.2 Basic material properties of high and low liquid limit silt

Subgrade materials	Pile number	Maximum dry density ρ_{max} (g/cm^3)	Natural moisture content ω_n (%)	Optimum moisture content ω_{max} (%)	Liquid limit ω_L (%)	Plastic limit ω_P (%)	Plasticity index I_P
LLL silt	K52+800–K53+030	1.69	27.0	18.9	49.3	32.6	16.7
HLL silt	K55+900–K56+150	1.65	27.1	22.6	58.3	35.6	22.7

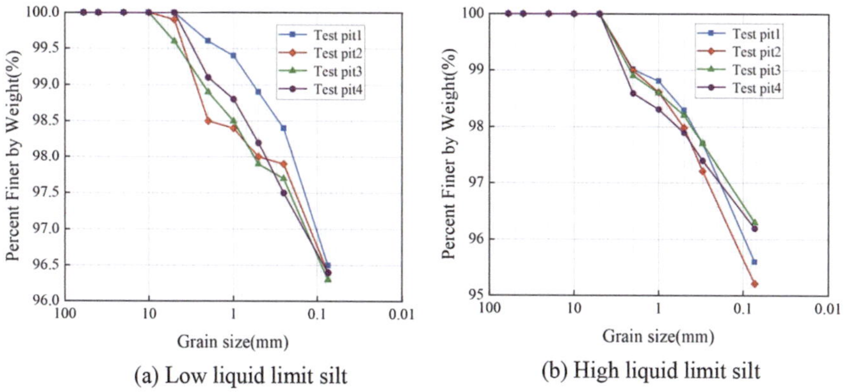

Fig. 6.10 Grain size analysis results for field samples taken from in situ test locations

sieve pore sizes, thereby facilitating a comprehensive understanding of their characteristics. The particle grading curves for high and low liquid limit silt are illustrated in Fig. 6.10. The maximum particle size for low liquid limit silt is approximately 2–5 mm, with particles smaller than 0.075 mm comprising 96.4 to 96.5% of the total. In contrast, the maximum particle size for high liquid limit silt ranges from 5 to 10 mm, with 95.2–96.3% of the particles being smaller than 0.075 mm.

③ Dynamic triaxial test

As illustrated in Fig. 6.11a, the geotechnical triaxial test compaction instrument was employed to create remolded samples with a moisture content of 22.6% and dry densities of 1.3, 1.4, 1.5, and 1.6 g/cm^3. The back pressure saturation method was utilized to saturate the samples, with the degree of saturation assessed by measuring the pore water pressure coefficient (B); all samples exhibited (B) values exceeding 0.95. The samples were then mounted on a dynamic triaxial apparatus, as shown in Fig. 6.11c. With a consolidation stress ratio of 1.0, the consolidation stresses applied were 100 kPa, 200 kPa, and 300 kPa, respectively. A vibration frequency of 1.0 Hz was employed to conduct dynamic cyclic loading under constant strain conditions, allowing for an examination of how consolidation stress and dry density influence the relationship between dynamic stress and dynamic strain. This investigation serves as a reference for selecting an appropriate elastoplastic model for the soil in question. The test was conducted under consolidation and undrained conditions, with all tests consolidated at equal pressure. The stability criterion for consolidation was controlled such that the change in consolidation drainage did not exceed 0.1 cm^3 within a one-hour period.

Under the condition of a dry density of 1.6 g/cm^3 and a strain amplitude of 1%, the typical dynamic stress-dynamic strain time-history curve for compacted high liquid limit silt subjected to a confining pressure of 30 kPa is presented in Fig. 6.12b. The evolution of dynamic stress can be categorized into three distinct stages. The initial stage of vibration marks the acceleration phase of dynamic stress attenuation, during

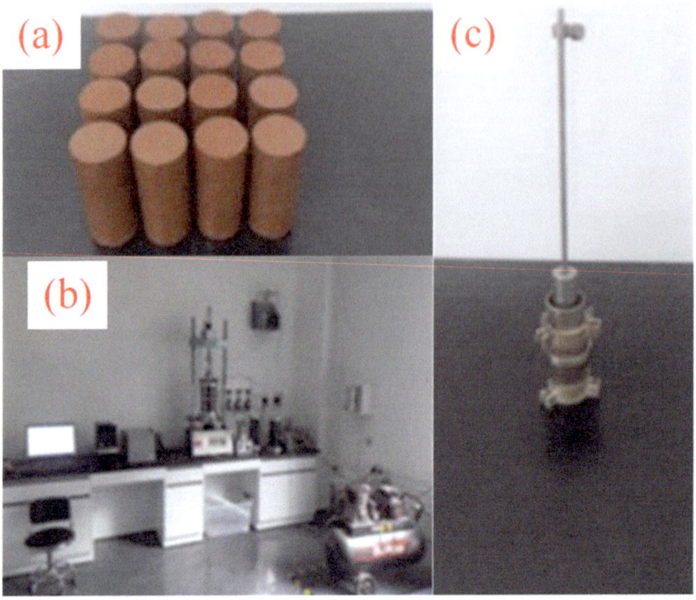

Fig. 6.11 Soil sample preparation and test equipment: **a** remodeled soil samples; **b** dynamic triaxial apparatus; **c** geotechnical triaxial test compaction instrument

which the soil sample experiences elastic deformation, and the rate of dynamic stress reduction is notably high. The intermediate stage represents a deceleration phase of dynamic stress decay, while the final stage is characterized by a constant rate of dynamic stress decay. After the 600th loading cycle, dynamic stress diminishes very slowly or even stabilizes, indicating a state of negligible decay.

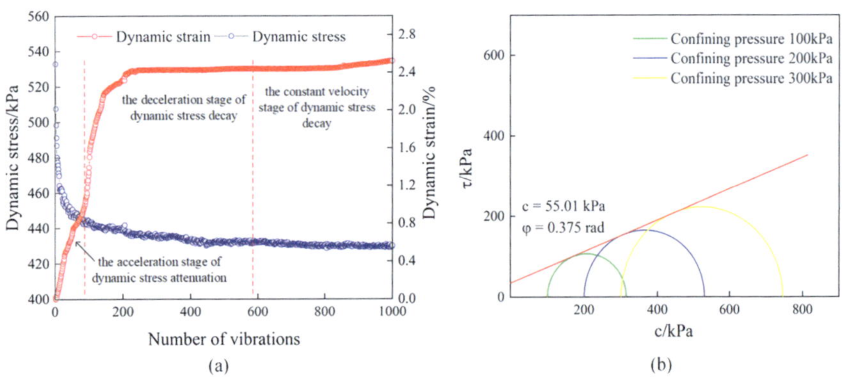

Fig. 6.12 Result curve of dynamic triaxial test: **a** relationship curve of dynamic stress, dynamic strain, and vibration times; **b** Mohr circle of principal stress

By processing the data, the Mohr circle of principal stress under various working conditions can be derived. The Mohr curve of principal stress at a dry density of 1.6 g/cm^3 is illustrated in Fig. 6.12a. With the same dry density, the shear strength index of graded soil under specific working conditions can be determined by drawing the common tangent line of the Mohr circle for different confining pressure states, as summarized in Table 6.3. The combination of Fig. 6.12 and Table 6.3 reveals that the constitutive relationship of graded soil approximates linearity under the consolidation stress state.

(3) Field test program

The experimental test scheme is detailed in Fig. 6.13 and Table 6.4. Each test strip measures 3 m in width and 10 m in length, with an ample adjustment area for the roller surrounding the perimeter of the effective test strip. In the low liquid limit silt test strip, five test pit points for the sand replacement method were evenly spaced along the length, located at 1 m, 3 m, 5 m, 7 m, and 9 m, respectively. Conversely, in the high liquid limit silt test strip, three test pits were evenly distributed along the length, positioned at 2, 5, and 8 m. A 26-ton single-drum vibratory roller compactor (XS265JS) was employed for the compaction of high liquid limit silt, with a vibration frequency of 27/32 Hz and an excitation force of 405/290 kN. For the corresponding low liquid limit silt, a 22-ton single-drum vibratory roller (SSR220AC) was utilized, featuring a vibration frequency of 29/35 Hz and an excitation force of 410/300 kN.

As illustrated in Fig. 6.14, the field rolling test for high and low liquid limit silt can be categorized into three phases: before rolling, during rolling, and after rolling. Figure 6.14a depicts the paved silt subgrade prior to rolling, featuring a lift thickness of 40 cm, with the test strips marked on the surfaced subgrade. Figure 6.14b captures

Table 6.3 Basic parameters of graded silt (moisture content $\omega = 22.6\%$)

Number	Dry density (g cm^{-3})	Poisson's ratio	Cohesion (kPa)	Friction angle (rad)	Elastic modulus (MPa)
No. 1	1.30	0.20	5.45	0.35	16.99
No. 3	1.40	0.20	9.76	0.36	21.04
No. 5	1.50	0.22	16.56	0.37	27.16
No. 7	1.60	0.25	31.26	0.37	32.62

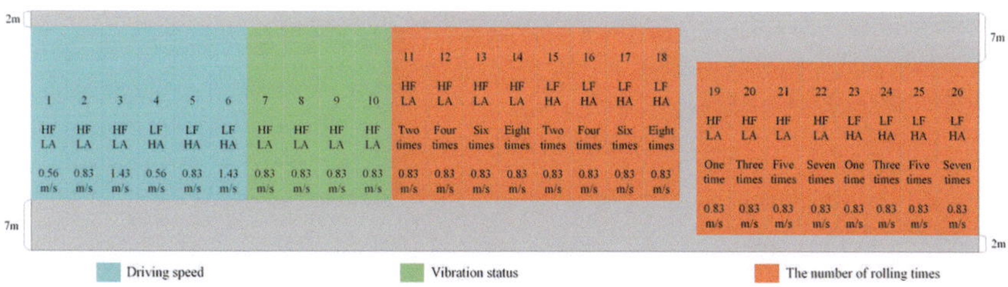

Fig. 6.13 Site layout of field rolling test

Table 6.4 Pre-designed experimental test scheme

Working condition	Vibration status	Driving speed/(m/s)	Passes (static)	Passes (dynamic)
1	LFHA	0.56	2	1–8
2		0.83	2	1–8
3		1.43	2	1–8
4	HFLA	0.56	2	1–8
5		0.83	2	1–8
6		1.43	2	1–8

Fig. 6.14 Photos of the field test at the subgrade area filled with high and low liquid limit silt

the rolling operation, during which the corresponding compaction parameters were implemented according to the test plan; data on RICM measure values, compaction parameters, and additional information (including the real-time spatial location of the roller) were collected continuously. Figure 6.14c presents the pre-arranged test pits, which are utilized to assess the compactness of high and low liquid limit silt through the sand replacement method after rolling, while simultaneously conducting particle size analysis on the high and low liquid limit silt extracted from the test pits.

6.2.3 *Result Analysis*

(1) Correlations between compaction parameters and RICM index

The number of compaction passes serves as an index of the magnitude of compaction energy and constitutes one of the critical control parameters during the construction process, significantly influencing the compaction of subgrade materials. To minimize disparities in gradation and moisture content that could skew the analysis results, strip 1 was filled with high liquid limit silt, while strip 2 contained low liquid limit silt. A test pit was established every 3 m in strip 1, and measurement points were positioned every 2 m in strip 2 (as shown in Fig. 6.14). Initially, static rolling was conducted twice by the roller to achieve a smooth surface on the soil. Subsequently, both strip 1 and strip 2 underwent vibration at a constant speed for eight cycles at a velocity of 0.83 m/s (working conditions 2 and 5), with the corresponding RICM measure values recorded for each test pit following each vibration rolling.

Figure 6.15 illustrates the correlation between various RICM indexes and the number of compaction passes. At a given measuring point, both the *CMV* and *E* values for high and low liquid limit silt exhibit an increasing trend with the number of compaction passes. Under a specific number of compaction passes, the *CMV* and *E* for high liquid limit silt are consistently lower than those for low liquid limit silt, indicating a disparity in the stiffness of the two materials. Due to the influences of material source characteristics and construction techniques, the *CMV* and *E* values at different measuring points, despite the same number of compaction passes, display slight variations. Consequently, it is essential to establish a linear regression equation relating the number of compaction passes to the average *CMV* and average *E*, and to compute the coefficient of determination (R^2) for the relationship between the RICM indexes and the number of compaction passes. The trend diagram depicting the variations of different RICM indexes in relation to the number of compaction passes for high and low liquid limit silt subgrades is presented in Fig. 6.15.

Upon comparing Fig. 6.15, it is evident that the RICM measure values for both high and low liquid limit silts increase with the number of compaction passes. In Fig. 6.16, the linear correlation between the *E* value and the number of compaction passes for high liquid limit silt is the strongest, with a coefficient of determination $\left(R^2\right)$ reaching 0.994; conversely, the linear correlation between the *CMV* and the number of compaction passes is comparatively weak, yielding a coefficient of determination $\left(R^2\right)$ of 0.798. Similarly, for low liquid limit silt, the linear correlation between the *E*

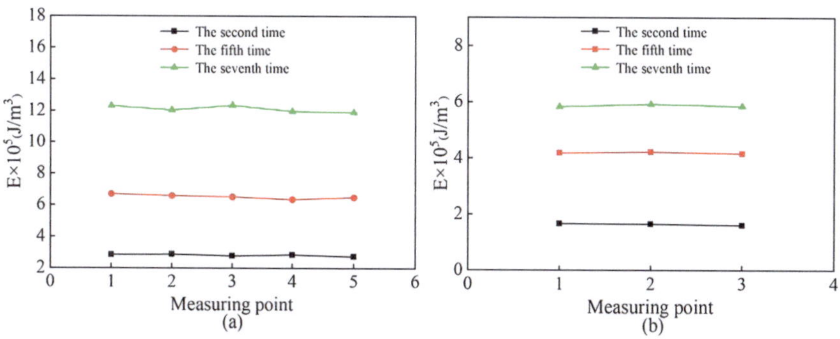

Fig. 6.15 Relationships between the number of compaction passes and the RICM measure values: **a** Low liquid limit silt; **b** High liquid limit silt

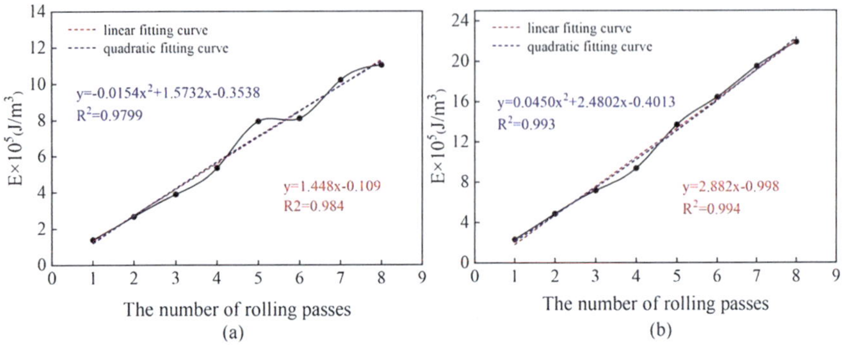

Fig. 6.16 Variation trend of RICM index with the number of compaction passes: **a** Low liquid limit silt; **b** High liquid limit silt

value and the number of compaction passes remains the strongest, with $\left(R^2\right)$ reaching 0.984, while the linear correlation between the CMV and the number of compaction passes is weak as well, with $\left(R^2\right)$ at 0.827.

Figure 6.16 reveals that at the same number of compaction passes, the CMV values for high and low liquid limit silts exhibit significant differences. This disparity can be attributed to two primary factors: first, considerable variations exist in the self-weight, vibration state, and construction technology of the roller, leading to differences in the compaction energy imparted by various types of rollers on the subgrade materials; second, there are inherent differences in the material properties of the two subgrade types. Furthermore, Fig. 6.16c illustrates that the CMV for high liquid limit silt exhibits noticeable fluctuations during the eighth pass, suggesting that as soil compaction increases, the vibration rolling may prematurely enter the "over-rolling stage." This phenomenon can cause shearing damage to the soil's particle structure, resulting in the expansion of internal pores and a subsequent decrease in the compactness of the high liquid limit silt.

To validate the correlation between the two RICM indexes and the number of compaction passes, as well as the stability of the data, an F-test was conducted using Eq. (6.23). The significance values p for all four models under the specified confidence level are below 0.05, signifying that the experimental results are statistically significant. Calculations reveal that the standard deviation coefficient V_σ for each model is less than 6.70%, indicating a high degree of fit and minimal data variability. Table 6.5 demonstrates that both the CMV and E exhibit a clear linear positive correlation with the number of compaction passes.

$$F = \frac{ESS/k}{RSS/(m-k-1)} \tag{6.23}$$

where k denotes the number of independent variables; m represents the sample size, and the expression follows the F distribution with degrees of freedom $(k, m-k-1)$.

Table 6.5 Regression analysis results between the number of compaction passes and the RICM measure values in different types of silt subgrades

Subgrade material	RICM	Regression model	R^2	V_σ (%)
Low liquid limit silt	*CMV*	y = 1.098x + 3.684	0.83	2.54
	E	y = 1.448x − 0.109	0.98	3.35
High liquid limit silt	*CMV*	y = 0.564x − 0.049	0.80	1.30
	E	y = 2.882x − 0.998	0.99	6.67

(2) Correlations between the compactness and RICM index

As an index of soil compaction quality, a stronger correlation between the RICM index and compaction quality enhances the credibility of the evaluation results. To assess whether the RICM index accurately and effectively reflects the compaction quality of high and low liquid limit silt, and to determine which index is more suitable, a preliminary correlation analysis of the data obtained from field tests is necessary. Figure 6.17 illustrates the linear regression relationship between the two RICM indexes and in situ measurements for the two types of subgrade materials.

As shown in Table 6.6, for high liquid limit silt, the linear correlation between *CMV* and compactness *K* is the strongest, with a coefficient of determination (R^2) reaching 0.715 and a standard deviation coefficient V_σ of 1.765%; conversely, the linear correlation between *E* and compactness *K* is weaker, with (R^2) at 0.587 and a standard deviation coefficient V_σ of 3.786%. For low liquid limit silt, the linear correlation between *CMV* and compactness *K* remains the strongest, with (R^2) at 0.757 and a standard deviation coefficient (V_σ) of 2.174%; meanwhile, the correlation for *E* is weak, with (R^2) at 0.539 and a standard deviation coefficient V_σ of 1.077%. Evidently, whether for high or low liquid limit silt, the correlation between *CMV* and compactness *K* is superior to that of *E*, thereby indicating that *CMV* serves as a more effective characterization index.

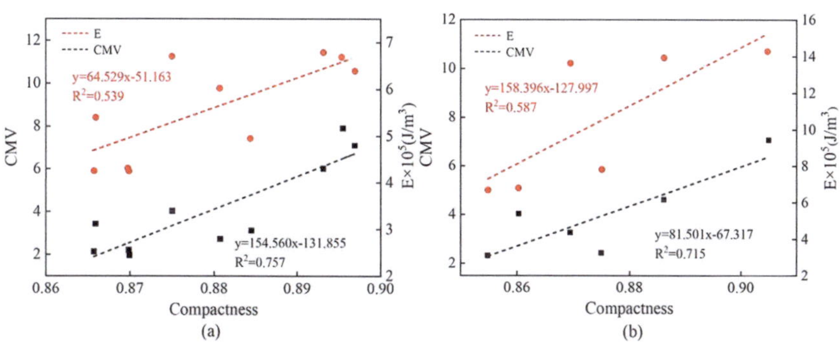

Fig. 6.17 Correlation between RICM index and compactness: **a** low liquid limit silt; **b** high liquid limit silt

Table 6.6 Regression analysis results between the compactness and the RICM values indifferent types of silt

Subgrade material	RICM	Regression model	R^2	$V_\sigma(\%)$
Low liquid limit silt	*CMV*	y = 154.560x − 131.855	0.78	2.17
	E	y = 64.529x − 51.163	0.54	1.08
High liquid limit silt	*CMV*	y = 81.501x − 67.317	0.72	1.76
	E	y = 158.396x − 127.997	0.59	3.79

The preceding analysis reveals that all RICM measure values exhibit consistent correlations with in-situ measurements. Consequently, the two RICM indexes demonstrate varying applicability to different subgrade materials within highway projects. In summary, the *CMV* shows a superior correlation with compactness, accompanied by a relatively low standard deviation coefficient. This indicates that the *CMV* possesses higher data stability and reduced dispersion, thereby enabling it to more accurately characterize the compaction quality of both high and low liquid limit silt. Based on the established quantitative relationships, it is evident that the *CMV* can, to a certain extent, reflect the compaction characteristics of high and low liquid limit silt under comparable conditions. Therefore, the *CMV* is recommended as the compaction quality control index for silt subgrades on the Hengyong highway.

6.2.4 Compaction Quality Assessment Models

Based on Sect. 6.2.3, it is evident that the *CMV* demonstrates a superior linear relationship with compactness compared to the *E* index. Given the complexity of the dynamic coupling problem, there may exist an uncertain nonlinear relationship between the RICM index and compactness during the compaction process. Therefore, to derive the most suitable assessment model, we can pursue a high-order nonlinear extension of simple linear regression. By examining the uncertainty inherent in the model, we can investigate the validity of the nonlinear assumptions relating the RICM index to compactness, as well as the explanatory power of the *CMV* and *E* index regarding compactness. Ultimately, building upon the linear regression analysis results presented in Sect. 6.2.3, we can identify a compaction quality assessment model that exhibits superior performance for high and low liquid limit silt subgrades.

Based on the preceding analysis, a multiple regression equation relating the RICM index to compactness has been established (refer to Eq. (6.24)), with the root mean square error (RMSE) chosen as the performance evaluation criterion for the model.

$$y = a + \sum_{i=1}^{m} b_i x^i \tag{6.24}$$

where y represents the compactness of silt test pit points with high and low liquid limits; x^i denotes the RICM index (either CMV or E); a and b_i are the coefficients of the regression equation; and m is the order of the model. In this section, the maximum order m for the model's argument variable is set to 3. Utilizing the sampled data (RICM index and compactness) from 12 groups of high liquid limit silt subgrades as training samples, univariate regression models were established with CMV as the independent variable and E as the independent variable, respectively. Similarly, 15 groups of low liquid limit silt samples were employed to construct the equations, using the same methodology as that applied to the high liquid limit silt. As the order of the independent variables increased, the change curves of the coefficient of determination $\left(R^2\right)$ for the two index models of high and low liquid limit silt were plotted, as illustrated in Fig. 6.18. Figure 6.19 depicts the variation trend of the RMSE between the predicted compactness and the measured compactness across different orders.

From Figs. 6.18 and 6.19, it is evident that for high liquid limit silt, as the order of the model's independent variable increases, the coefficient of determination $\left(R^2\right)$ for both indexes gradually declines, while the RMSE rises. A similar trend is observed in the low liquid limit silt model. Furthermore, at the same order, the R^2 values for CMV consistently exceed those of E indicating a stronger correlation between CMV and compactness compared to that between E and compactness. Additionally, the mean absolute percentage error (MAPE) corresponding to CMV is slightly lower than that of E demonstrating that under identical orders, the nonlinear predictive performance of CMV is superior.

In summary, whether for high or low liquid limit silt, the fitting relationship between CMV and compactness is significantly more pronounced. As the order of the independent variable increases, the degree of fit progressively improves, whereas the correlation between E and the nonlinear model of compactness remains weak. This suggests that CMV possesses a greater degree of nonlinear explanatory

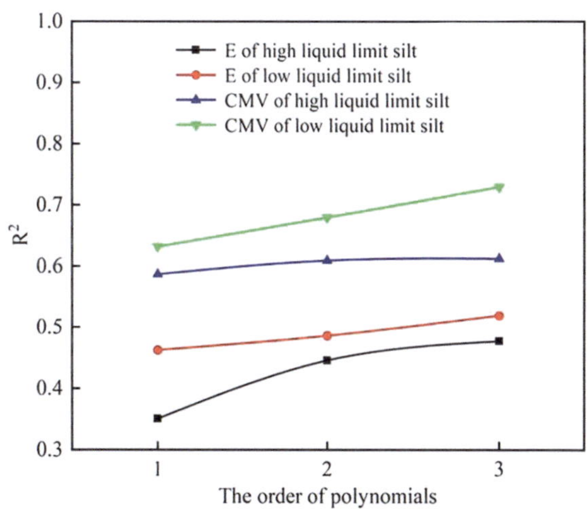

Fig. 6.18 Changing trend of coefficient of determination R^2 for high and low liquid limit silt models under different order of independent variables

Fig. 6.19 Variation trend of the RMSE for high and low liquid limit silt models under different order of independent variables

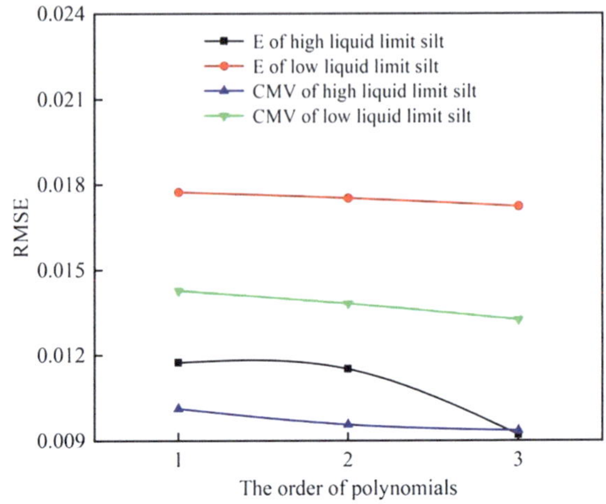

power regarding compactness than E. Considering the fitness results of the sample data for the linear correlations between the two indexes, and taking into account factors such as uncertainty, accuracy, convenience, and real-time applicability, the six models established were validated using test data. Ultimately, Model I and Model II were selected as the assessment models for the compaction quality of high and low liquid limit silt, respectively, to fulfill actual engineering requirements. Their linear expressions are as follows:

$$\text{Model I: } K_1 = 5.53 \times 10^{-3} CMV_2 + 0.8482 \tag{6.25}$$

$$\text{Model II: } K_1 = 6.653 \times 10^{-3} CMV_2 + 0.8534 \tag{6.26}$$

where, K_1 and CMV_1 are the compactness and RICM measurements of high liquid limit silt. K_2 and CMV_2 are the compactness and RICM measurements of low liquid limit silt.

Model I, serving as a compaction quality assessment model for high liquid limit silt, exhibits a coefficient of determination $\left(R^2\right)$ of 0.589 (with an adjusted R^2 of 0.548). In contrast, Model II, designed for low liquid limit silt, demonstrates a coefficient of determination $\left(R^2\right)$ of 0.633 (with an adjusted R^2 of 0.605). To evaluate the predictive accuracy of these two models, the prediction results from twelve sets of training sample data for high liquid limit silt and fifteen sets for low liquid limit silt were analyzed. The relationship between the predicted compaction values generated by each regression model and the actual measured compaction values was compared and examined.

Figure 6.20 illustrates the comparison between the predicted compactness from the six aforementioned regression models and the actual compactness measured using the sand replacement method. Table 6.7 presents the errors associated with the compaction quality models for high and low liquid limit silt. From the figure and

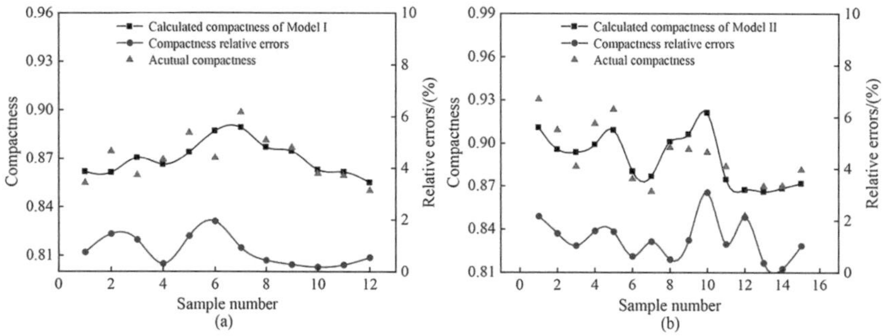

Fig. 6.20 Comparison between the actual compactness and the calculated compactness: **a** Model I; **b** Model II

Table 6.7 Error analysis of compaction quality models of high and low liquid limit silt

Model	Maximum error (g/cm^3)	Average error (g/cm^3)	Maximum relative error (%)	Average relative error (%)
Model I	0.017	0.007	1.94	0.82
Model II	0.028	0.012	3.08	1.29

table, it is evident that the maximum absolute error across these models is less than or equal to 0.028 g/cm^3, while the average relative error for all models is less than or equal to 1.29%. This indicates that the constructed compaction assessment models for both high and low liquid limit silt are capable of meeting actual engineering requirements.

6.2.5 Compaction Quality Assessment Using Green Spline Interpolation Algorithm

Given that the stability of the subgrade is contingent upon the compaction quality of the filling materials, the adoption of an effective compaction quality assessment method is of paramount importance. Existing research [14, 15] indicates that the factors influencing the compaction quality and effectiveness of subgrade include material properties (such as gradation and moisture), the strength of the foundation or underlying layer, compaction parameters, roller type and vibration mode, as well as construction methods and strategies. Generally, control over compaction quality can be achieved by managing result control indexes (compactness), material source control indexes (moisture and gradation), and compaction parameter control indexes (such as the number of compaction passes).

From the preceding analysis, it is evident that RICM indexes can directly represent the stiffness (or compaction status) of subgrade materials in highway projects.

However, the *CMV* and other RICM indexes (such as *E*) derived from the same strip may reflect differing compaction statuses of the subgrade materials when accounting for the spatial variability of soil properties and measurement variability [16]. To address this issue, a regression model and a green spline interpolation algorithm have been introduced to assess compaction quality comprehensively across 100% of the work area. The detailed assessment procedure for the compaction quality assessment method is outlined as follows: (1) conduct an analysis of the compaction characteristics of silt subgrade; (2) implement field rolling tests and collect data; (3) perform correlation analysis; (4) establish and validate the regression model while estimating compaction quality; (5) conduct global interpolation statistical analysis; (6) assess compaction quality across 100% of the work area. The overall compaction quality of the entire rolling compaction area can be estimated by executing steps (1) through (5).

Based on Sect. 6.2.4, to evaluate the effectiveness of the acceptance assessment model, five test pits were randomly excavated from test strips filled with high and low liquid limit silt, and their true compaction values were measured using the sand replacement method as validation data. The correlation between the estimated compaction values derived from the compaction quality assessment method and the measured compaction values was analyzed, with the results depicted in Fig. 6.21.

The findings indicate that for the low liquid limit silt, the coefficient of determination between the compactness estimated by the assessment method and the in situ measurements is 0.888, with a maximum absolute error of 0.0181 g/cm^3 and an average relative error of 1.162%. In contrast, for the high liquid limit silt, the coefficient of determination between the estimated compactness and the in situ measurements is 0.862, with a maximum absolute error of 0.0089 g/cm^3 and an average relative error of 1.205%.

The small absolute and relative errors in the compaction quality of both types of silt subgrades, coupled with the high correlation between the measured and predicted values, substantiate the efficacy of the compaction quality assessment method across the entirety of the work area.

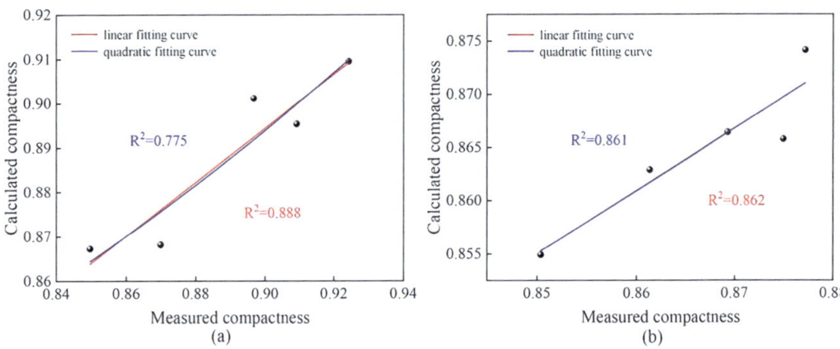

Fig. 6.21 Correlation between actual compactness and predicted compactness: **a** low liquid limit silt; **b** high liquid limit silt

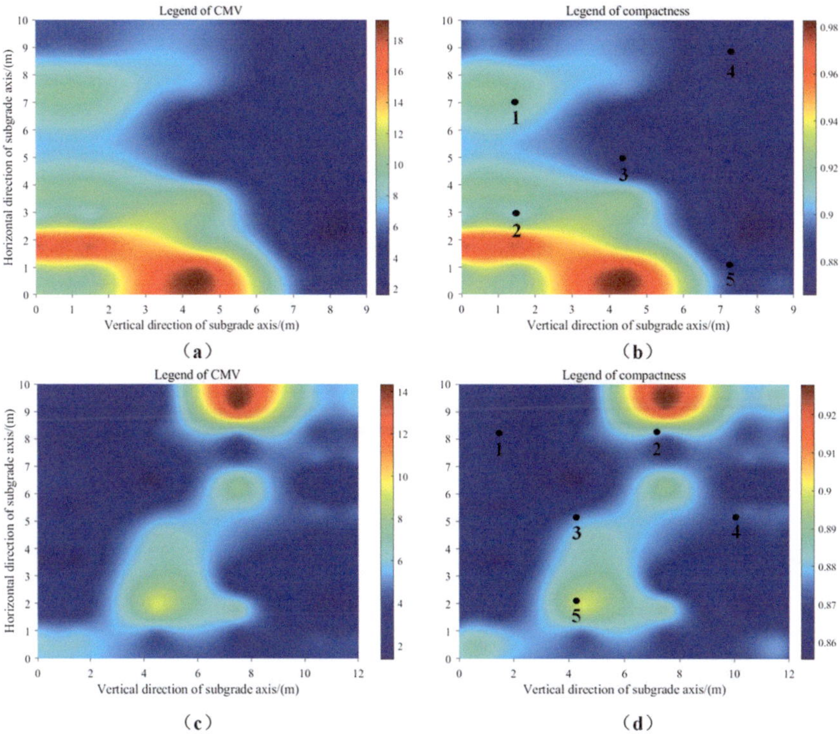

Fig. 6.22 Compaction color-coded maps of low and high liquid limit silt subgrade using the Green spline interpolation method: **a**, **b** low liquid limit silt; **c**, **d** high liquid limit silt

Figure 6.22 presents a color-coded map of the field test areas situated within the high and low liquid limit silt subgrade, employing the Green spline interpolation method, which utilizes *CMV* and compactness to illustrate the compaction quality of the test area. The five randomly selected test pits are depicted in Fig. 6.22b, d, which were utilized to validate the effectiveness of the assessment method. Table 6.8 enumerates the errors and relative errors between the calculated compactness and the actual compactness at the five verification sampling points.

Utilizing the compaction quality assessment method and the color-coded map produced through user-friendly drawing technology, the compaction quality of the work areas within the high and low liquid limit silt subgrade can be effectively represented and leveraged for quality control purposes.

Table 6.8 (a) Error between the predicted compactness and the actual compactness. (b) Error analysis between the predicted compactness and the actual compactness

(a)

Subgrade material	Pit point	Error (g/cm^3)	Relative error (%)
Low liquid limit silt	1	0.014	1.51
	2	0.014	1.56
	3	0.004	0.46
	4	0.018	2.14
	5	0.001	0.15
High liquid limit silt	1	0.013	1.47
	2	0.003	0.32
	3	0.017	1.95
	4	0.002	0.20
	5	0.005	0.56

(b)

Subgrade material	Maximum error (g/cm^3)	Average error (g/cm^3)	Maximum relative Error (%)	Average relative error (%)
Low liquid limit silt	0.018	0.010	2.14	1.16
High liquid limit silt	0.0170	0.008	1.95	0.89

References

1. Liu DH, Gao L (2018) Analysis and improvement of roller vibration behavior-based indexes for monitoring compaction quality of earth-rock dams. J Hydroelectr Eng 37(4):111–120
2. Liu DH, Liu ZL, Feng YW (2019) Analysis on applicability and influence factors of real-time monitoring compaction indexes for rock-fill dam materials. J Hydroelectr Eng 38(06):1–10
3. An ZZ, Liu TY, Huangfu ZH, Zhang QL, Li CS, Li QB (2020) Neural network model for evaluating compaction quality of rockfill materials by compaction meter value. J Hydroelectr Eng 39:110–120
4. Wang F, Zhong D-H, Yan Y-L, Ren B-Y, Wu B-P (2018) Rockfill dam compaction quality evaluation based on cloud-fuzzy model. J Zhejiang Univ A 19:289–303
5. Lin WW, Zhong DH, Hu W et al (2018) Research on the dynamic evaluation of the compaction quality of earth-rock dams based on the random forest algorithm. J Hydraul Eng 49(08):945–955
6. Pan HY, Zhang WH, Liang TT et al (2023) Inducement analysis of taxi drivers' traffic accidents based on MIMIC and machine learning. J Jilin Univ (Eng Technol Ed) 53(02):457–467
7. Wang JJ, Zhong DH, Guan T et al (2019) Real-time evaluation of compaction quality of earth-rock dams using fuzzy logic based on KM and AC-BFA. J Hydroelectr Eng 38(03):165–178
8. Liu B, Zhao YF, Chen ZY et al (2022) Research on real-time monitoring indexes of the compaction quality of rockfill dams based on the rolling wave velocity. J China Inst Water Resour Hydropower Res 20(01):20–29
9. Li BG, Gao J, Liu WX et al (2006) Geostatistical analysis and kriging of hexachorocyclohexane residues in topsoil from Tianjin, China. Environ Pollut 142(03):567–575

10. Sandwell DT (1987) Biharmonic spline interpolation of GEOS-3 and SEASAT altimeter data. Geophys Res Lett 14(2):139–142
11. Vennapusa RKP, White JD, Morris DM (2010) Geostatistical analysis for spatially referenced roller-integrated compaction measurements. J Geotech Geoenviron Eng 136(6):813–822
12. Liu D, Li Z, Lian Z (2014) Compaction quality assessment of earth-rock dam materials using roller-integrated compaction monitoring technology. Autom Constr 44:234–246
13. Xiao CZ, Li XF, Zhang JJ (2017) Effect of compaction degree and water content on performance of sandy silt. J Shenzhen Univ Sci Eng 34(5):501–508
14. Zhu XY, Bai SJ, Xue GP, Yang J, Cai YS, Hu W, Jia XY, Huang BS (2018) Assessment of compaction quality of multi-layer pavement structure based on intelligent compaction technology. Constr Build Mater 161:316–329
15. Qin W, Sha AM (2010) Research on the relationship between fractal features of particle size distribution and compaction effect of silts. J Wuhan Univ Technol 32(14):58–61
16. Thompson MJ, White DJ (2007) Field calibration and spatial analysis of compaction monitoring technology measurements. Transp Res Rec 2004(1):69–79

Chapter 7
Compaction Quality Intelligent Assessment Methods and Models

Since its inception, Artificial Neural Networks (ANN) have garnered widespread attention from scholars both domestically and internationally due to their formidable capabilities in data processing. They have yielded varying degrees of research outcomes across numerous fields, primarily categorized into Convolutional Neural Networks (CNN) and Recurrent Neural Networks (RNN). A typical neural network comprises an input layer, multiple hidden layers, and an output layer. The input layer serves as the data entry point, accommodating one-dimensional to multi-dimensional data sets; the hidden layers function as units for data computation and optimization, with the option to increase the number of hidden layers to extract additional data features; the output layer represents the target structure output. The architecture of the neural network is illustrated in Fig. 7.1.

In addressing the prediction of high-dimensional nonlinear data, researchers have integrated neural networks and made continuous advancements, resulting in the emergence of models such as the Back Propagation (BP) model, Long Short-Term Memory (LSTM) model, Extreme Learning Machine (ELM) model, RBF model, and various corresponding derivative models. However, when applied to the field of compaction quality assessment, traditional neural network models reveal a significant dependency on high-quality data samples and challenges in rapidly training critical model parameters.

To effectively tackle these issues, improvements must be made on both data samples and model constructs. Nevertheless, due to the constraints of project timelines, researchers often find it challenging to conduct extensive long-term experiments on-site, making the acquisition of a sufficient number of high-quality sample data quite difficult. Consequently, researchers are compelled to enhance models to improve the extraction efficiency of features from small data samples.

Moreover, while the evaluation of compaction quality indexes for filling materials encompasses source parameters and compaction parameters, the source parameter data necessitates extensive manual testing to achieve optimal results, which is time-consuming and labor-intensive. Additionally, the diversity of filling materials

Q. Zhang, *Compaction Quality Nondestructive Testing and Intelligent Control Technology for Infrastructure*, https://doi.org/10.1007/978-981-95-1264-5_7

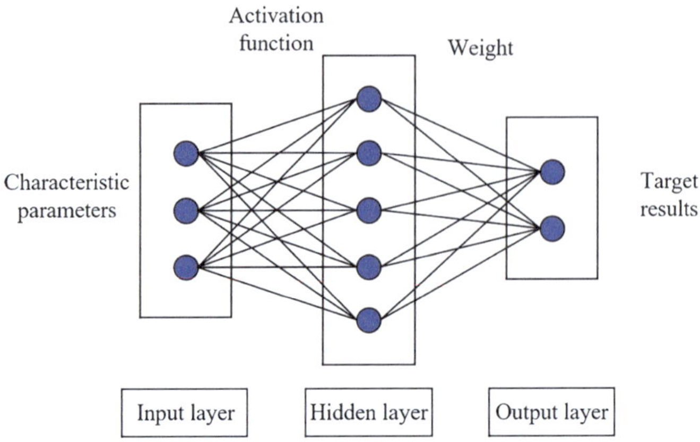

Fig. 7.1 Neural network structure diagram

leads to variability in source parameters, thereby diminishing the applicability of the assessment models.

To mitigate the impact of source parameters on the assessment models, the compaction parameters and roller-integrated compaction monitoring indexes will be utilized as inputs for the assessment models. Thus, this chapter will introduce compaction quality intelligent assessment models based on the BP model, LSTM model, and RBF model, subsequently incorporating an intelligent assessment model that accounts for small sample uncertainties. The effectiveness of the assessment models will be validated through specific case analyses.

7.1 Compaction Quality Intelligent Assessment Method Based on BP

7.1.1 BP Model

The BP neural network, introduced in the 1980s, has evolved through continuous development and application to become a relatively mature algorithm. It is one of the most widely used models in convolutional neural networks and is classified as a feedforward multilayer neural network model. This model is capable of inductively summarizing the inherent characteristics of sample data, thereby forming specific data chain rules internally, which are utilized for model training and testing. The structure of the model is illustrated in Fig. 7.2.

The training process of the BP model occurs along two pathways: the forward path, where data samples enter from the input layer to the hidden layers (comprising convolutional and pooling layers) for layer-by-layer analysis and computation, subsequently reaching the output layer, where nonlinear transformations yield the target

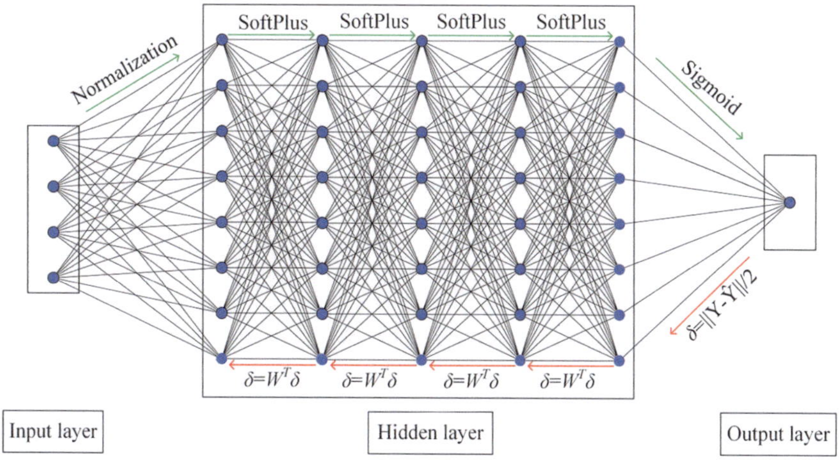

Fig. 7.2 BP model structure diagram

results. Concurrently, the backward path involves the iterative correction of the weight coefficients W and δ connecting the input layer to the hidden layers, among the hidden layers, and from the hidden layers to the output layer. This process continues until the errors across all layers are minimized, yielding the final adjusted results and achieving the optimal target outcome. Consequently, the BP model is characterized by its ability to facilitate forward information propagation and backward error adjustment.

The BP model for compaction quality assessment, incorporating the compaction quality assessment index *KECV* proposed in Sect. 7.2.1, is specifically represented by the formula (7.1):

$$X_{c,p} = f(n, v, V, f, KECV) \tag{7.1}$$

In the formula, $X_{c.p}$ represents the compaction quality index value, while n, v, V, and f denote the number of compaction passes, driving speed, drive direction, vibration frequency, respectively. According to the model, the input consists of five-dimensional sample data, posing significant challenges in capturing high-dimensional data features. Therefore, the number of hidden layers in the BP model is set to five, facilitating the search for a fitting plane and accelerating convergence. Additionally, the five-dimensional sample data undergo normalization prior to training to eliminate errors arising from disparities in magnitude among the dimensions. The output layer is one-dimensional, representing the index values of compaction quality, such as compactness and relative density.

Based on relevant documentation, the maximum training iteration is set to 1000, with the model's learning rate and desired fitting error established at 0.1 and 0.01, respectively. Furthermore, the BP model encounters the issue of local convergence, which can lead to suboptimal precision in the target results. This problem arises from the commonly used sigmoid function, which serves as the activation function during model training. The first derivative of this function is most effective near zero; however, as it moves towards positive values, a sudden decline in the gradient occurs, resulting in a rapid decrease in the error curve during training, ultimately reaching zero and prematurely concluding the training process, leading to larger discrepancies in the target results.

To address this issue, a near-linear function, SoftPlus, can be selected as the activation function for the training process when constructing the BP model. This function maintains a certain level of activation near zero while ensuring that the derivative value continues to increase as it moves towards positive values. This adjustment mitigates the accumulation of errors during multi-layer iterations, enhances the stability of the backward propagation, and significantly reduces the likelihood of premature local convergence when training on small sample data. The pseudocode for the BP model, based on Matlab software, is presented in Table 7.1.

7.1.2 Compaction Quality Intelligent Assessment Based on BP

(1) Compaction quality assessment of water stabilization layer of pavement

Based on the field experiments outlined in Sect. 2.3, a total of 36 sets of field compaction test data and compaction quality data were obtained for the base layer, along with 42 sets for the sub-base layer. Of this data, 80% was designated as the training set, while the remaining 20% (6 sets and 7 sets, respectively) served as the test set [1, 2]. For the pavement's water-stabilized layer, the constraints of on-site construction standards and progress requirements resulted in a singularity of test conditions and data. Consequently, the input layer dimension of the BP neural network model was set to 1, allowing for a "one-to-one" intelligent assessment and prediction. The compaction quality assessment index selected for the filling material was the *KECV* index, with the assessment results depicted in Figs. 7.3, 7.4, 7.5 and 7.6 The specific results are presented in Table 7.2.

From the combination of the figures and tables, it is evident that the coefficient of determination for the BP model's predicted values compared to the actual values was 0.403 for the base layer and 0.350 for the sub-base layer, indicating a relatively poor predictive performance. This shortcoming primarily arises from the fact that the BP neural network evaluation model is better suited for predicting high-dimensional nonlinear compaction data, while its applicability for low-dimensional or "one-to-one" predictions is limited. Furthermore, the scarcity of sample data also contributes to the lack of precision in the predictions.

Table 7.1 BP model pseudo code

Input: training set D; learning rate η; maximum number of iterations Z; test set T

Output: trained BP model and prediction results

Data normalization: [training set] = mapminmax(training set, 0, 1)

Create network: net

Train the network

  Initialize the weights and biases of the network.

 While the number of iterations is less than the maximum number of iterations Z

  repeat for D

  ##Forward propagation

  for each unit j in the hidden layer or output layer.

    $I_j = \sum i w_{ij} O_j + \delta_j$; Calculate the net output of unit j

    $I_j = 1/(1 + e^{-Ij})$; Calculate the net output of unit j

  end

  ##Backpropagate the error

  for Each unit j in the output layer

    $Err_j = O_j(1 - O_j)(T_j - O_j)$; Calculate the error

  end

  for From the last hidden layer to the first hidden layer, each unit j in the hidden layer

    $Err_j = O_j(1 - O_j)\sum_k Err_k w_{kj}$; k is a neuron in the subsequent layer

  end

  for each weight w_{ij} in the network

    $w_{ij} = (\eta) Err_j O_i$; Weight increment

    $w_{ij} = w_{ij} + \Delta w_{ij}$; Weight adjustment

  end

  for each bias δ_j in the network

    $\Delta\delta_j = (\eta) Err_j$; Bias increment

    $\delta_j = \delta_j + \Delta\delta_j$; Bias adjustment

  end

 end

Test the network using T

[prediction value] = [net, test set]

Reverse normalization of prediction results

[prediction value] = mapminmax(prediction value, 0, 1)

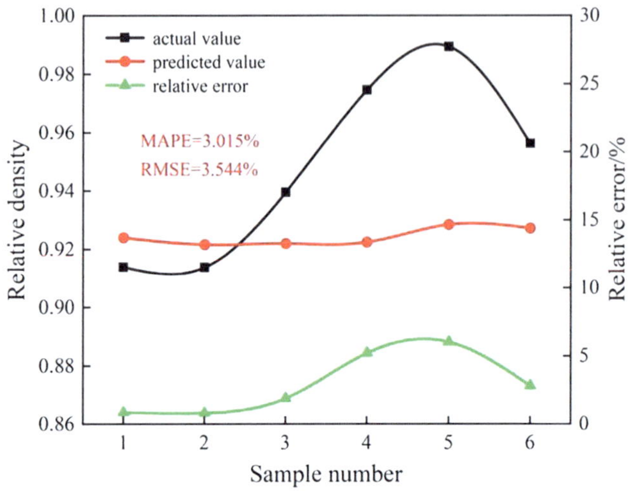

Fig. 7.3 Analysis of fitting error between predicted relative density and measured relative density of base filling materials based on BP

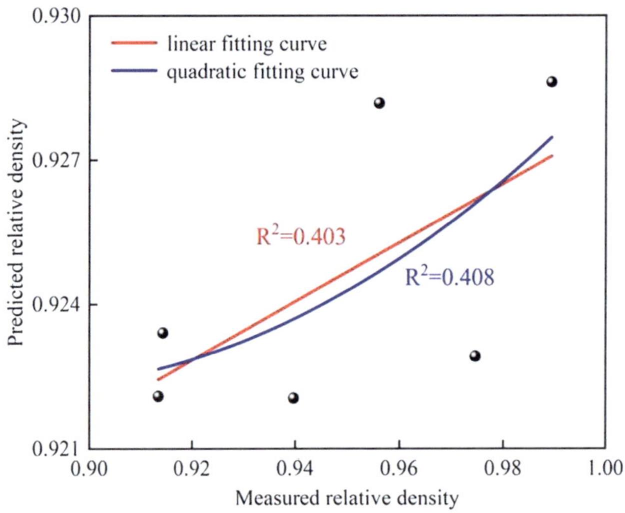

Fig. 7.4 Correlation analysis between predicted relative density and measured relative density of base filling materials based on BP

(2) Assessment of the compaction quality of the base layer

According to the field experiments detailed in Sect. 7.2.3, a total of 115 sets of field compaction test data and compaction quality data were obtained for the base layer. Of this data, 80% was designated as the training set, while the remaining 20% (23 sets) served as the test set [3, 4]. For the base layer, the input layer dimension of

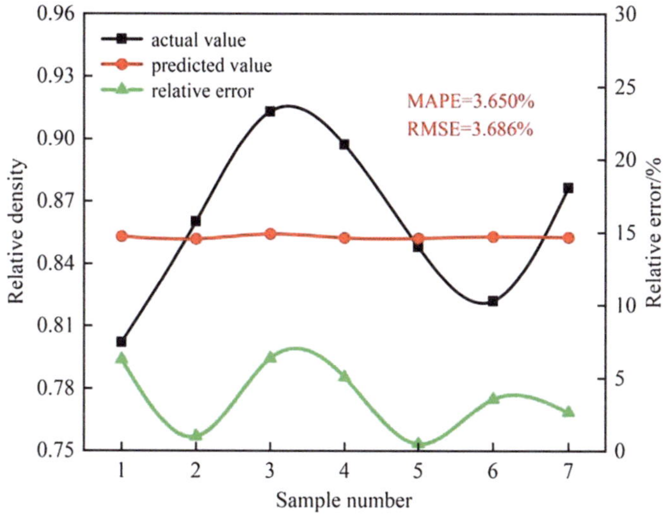

Fig. 7.5 Analysis of the fitting error between predicted relative density and measured relative density for the subbase filling materials

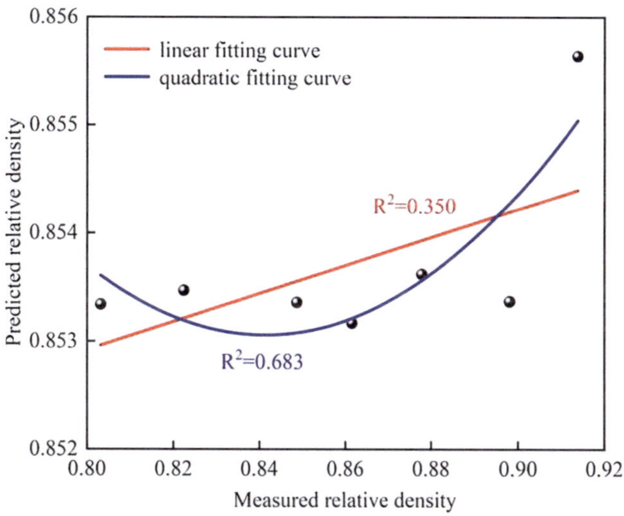

Fig. 7.6 Correlation analysis between predicted relative density and measured relative density for the subbase filling materials

Table 7.2 Assessment results of the BP model for pavement layer

Filling layer	R^2	MAPE/%	RMSE/%
Pavement base layer	0.40	3.02	3.54
Pavement subbase layer	0.35	3.65	3.69

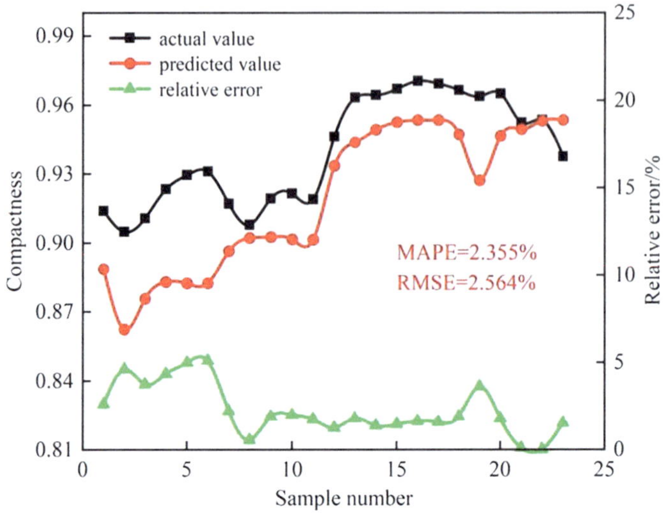

Fig. 7.7 Analysis of the fitting error between predicted compactness and measured compactness for the subgrade filling material

the BP neural network model was set to 5, comprising the *KECV* index, number of compaction passes, drive direction, exciting force, and vibration frequency. Here, forward drive was recorded as 2, while reverse drive was noted as 1. The output layer represented the compactness, enabling a "five-to-one" nonlinear assessment and prediction of compaction quality, with the assessment results illustrated in Figs. 7.7 and 7.8. The specific results are presented in Table 7.3.

From the analysis of the figures and table, it is evident that for the compaction quality assessment of the base layer, the coefficient of determination between the BP model's predicted values and the actual values was 0.775. This indicates a higher predictive accuracy compared to the assessment of the compaction quality of the water-stabilized layer. This also demonstrates that the BP neural network assessment model exhibits superior adaptability for predicting high-dimensional nonlinear compaction data.

7.2 Compaction Quality Intelligent Assessment Method Based on LSTM

7.2.1 LSTM Model

LSTM networks are a type of model that belongs to the family of RNNs. They introduce a gating mechanism, consisting of the "forget gate," "input gate," and "output gate," which allows a cell to regulate the passage or blockage of input data. This mechanism effectively addresses the issues of gradient explosion or vanishing

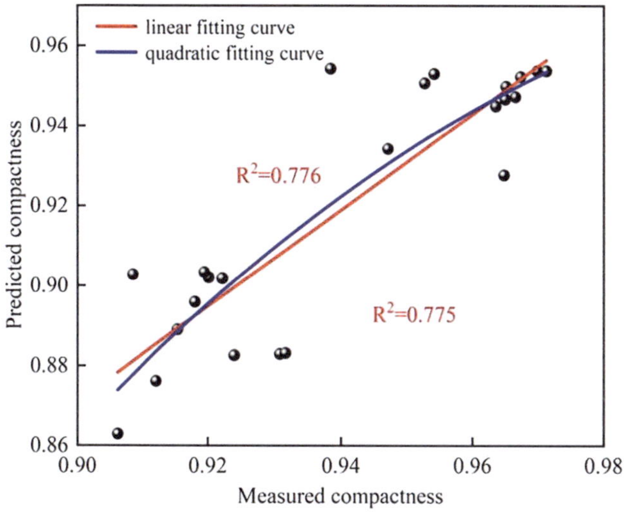

Fig. 7.8 Correlation analysis between predicted compactness and measured compactness for the subgrade filling material

Table 7.3 Assessment results of the BP model for subgrade layer

Filling layer	R^2	MAPE/%	RMSE/%
Subgrade	0.78	2.36	2.56

that can occur during training [5], thereby enhancing the capability to process time series data. The structure of the LSTM is illustrated in Fig. 7.9.

Let t represent the time step t, and the input time series be ($Y = \{y_1, y_2, y_3, \cdots, y_T,\}$). The LSTM cell state is computed using formulas (7.2), (7.3), (7.4), (7.5), (7.6), and (7.7):

$$f_i = \sigma\left(W_f \cdot \left[h_{t-1}, x_t\right] + b_f\right) \tag{7.2}$$

$$i_t = \sigma\left(W_i \cdot \left[h_{t-1}, x_t\right] + b_i\right) \tag{7.3}$$

$$C_t = \tan h\left(W_c \cdot \left[h_{t-1}, x_t\right] + b_c\right) \tag{7.4}$$

$$C_t = f_t \cdot C_{t-1} + i_t \cdot \widetilde{C_t} \tag{7.5}$$

$$o_t = \sigma\left(W_o \cdot \left[h_{t-1}, x_t\right] + b_o\right) \tag{7.6}$$

$$h_t = o_t \tan h(C_t) \tag{7.7}$$

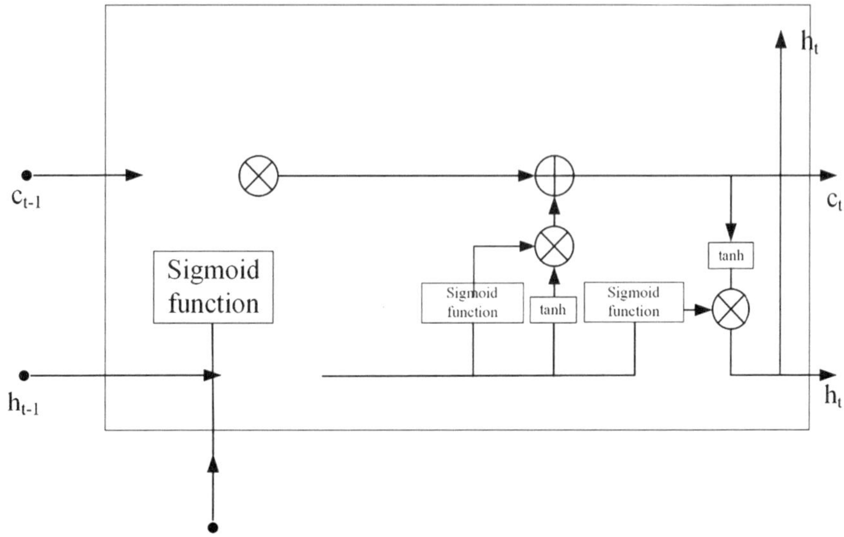

Fig. 7.9 Structure diagram of the LSTM model

In the equations, f represents the activation value of the forget gate; h_t denotes the output of the model at time t; h_{t-1} is the state at time $t-1$; i signifies the activation value of the input gate; o indicates the activation value of the output gate; $\widetilde{C}_t$, C_t, C_{t-1} represent the LSTM memory cells, with the amount of information transmitted internally dependent on the forget gate's passage; x_t is the input at time t; W stands for the weight matrix; b refers to the bias; and σ is the Sigmoid function, which serves to activate the transmission.

First, the activation value of the forget gate f_t at time t is obtained using (x_t, h_{t-1}) and the bias matrix b_f, followed by normalization of f_t using the Sigmoid function. Next, the activation value of the input gate i_t is computed based on the steps for calculating the forget gate's activation value, and, in conjunction with the intermediate state value $\widetilde{C}_t$ at time t, the memory cell C_t is determined. Finally, the model's output result h_t at time t is derived from the computed activation value o_t of the output gate and the memory cell C_t. The pseudocode for the LSTM model is presented in the Table 7.4.

7.2.2 *Compaction Quality Assessment Based on the LSTM Model*

(1) Compaction quality assessment for the water-stabilized layer of the pavement

Based on the field tests described in Sect. 7.2.3, a total of 36 sets and 42 sets of field compaction test data and compaction quality data were collected for the base

Table 7.4 Pseudocode for the LSTM model

Input: Training set D; network parameters; number of training iterations N; error tolerance; test set T
Output: Trained LSTM model and predicted results
1 Data normalization: ([Training Set] = {mapminmax}('reverse', Training Set, outputns))
2 Network parameter initialization: Set the number of nodes and biases for the gates in the network
3 Network weight initialization: Initialize weights and network states
4 Network training and learning
5 for 1:Number of training iterations N
6 repeat for D
7 for 1:Hidden layer nodes
8 Compute the activation values f、i、o
9 end
10 Calculate the memory cell values C_t
11 Calculate the output h_t and the error
12 if Error is less than the specified tolerance
13 End training
14 else
15 Update weights, cell states, and hidden states
16 end
17 end
18 Test using the network T
19 [Predicted values] = Output at the last time step of the test set
20 Reverse normalization of the predicted results
21 [Predicted values] = mapminmax('reverse', Training set, outputns)

and sub-base layers, respectively. Eighty percent of these were used as the training set, while the remaining 20% (6 sets and 7 sets) served as the test set [2, 4]. For the water-stabilized layer of the pavement, the singularity of test conditions and data arose due to construction regulations and scheduling requirements, leading to an input layer dimension of 1 for the LSTM neural network model, enabling a "one-to-one"

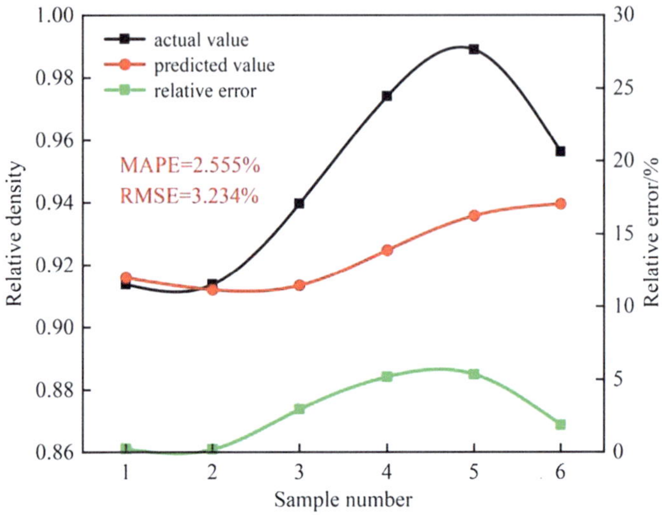

Fig. 7.10 Analysis of fitting error between predicted relative density and measured relative density of base filling materials based on LSTM

intelligent assessment prediction. The compaction quality assessment index selected for the fill material was the *KECV* index, with the assessment results illustrated in Figs. 7.10, 7.11, 7.12 and 7.13 The specific results are detailed in Table 7.5. Comparatively, as noted in Sect. 7.2.1, it is evident that for both types of filling layers, the coefficient of determination (R^2) of the LSTM model surpasses that of the BP model, while the former also exhibits lower MAPE and Root Mean Square Error (RMSE) than the latter.

(2) Compaction quality assessment for the subgrade layer

Based on the field tests described in Sect. 7.2.3, a total of 115 sets of field compaction test data and compaction quality data were collected for the pavement base layer, with 80% allocated as the training set and the remaining 20% (23 sets) serving as the test set [2, 4]. For the pavement base layer, the input layer dimension of the LSTM neural network model is set to 5, which includes the *KECV* index, number of compaction passes, drive direction, exciting force, and vibration frequency. In this context, forward drive is recorded as 2 and reverse drive as 1. The output layer represents the compactness, facilitating a "five-to-one" nonlinear compaction quality assessment prediction. The assessment results are illustrated in Figs. 7.14 and 7.15, with specific results detailed in Table 7.6. Similarly, as compared to Sect. 7.2.1, it is evident that for the pavement base layer, the coefficient of determination (R^2) of the LSTM model outperforms that of the BP model, while the former also demonstrates lower MAPE and RMSE than the latter.

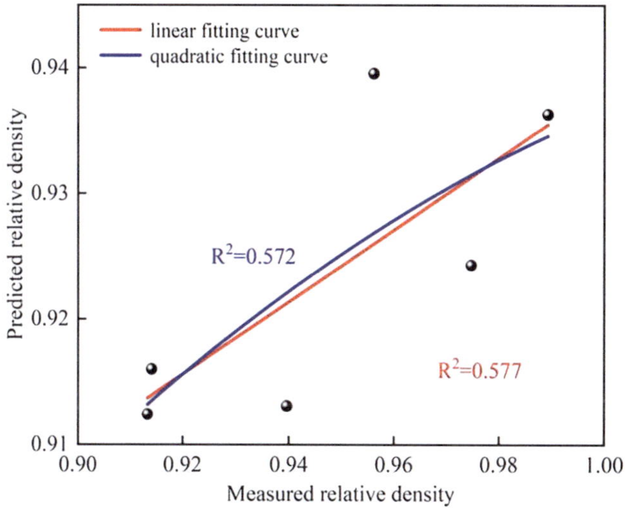

Fig. 7.11 Correlation analysis between predicted relative density and measured relative density of base filling materials based on LSTM

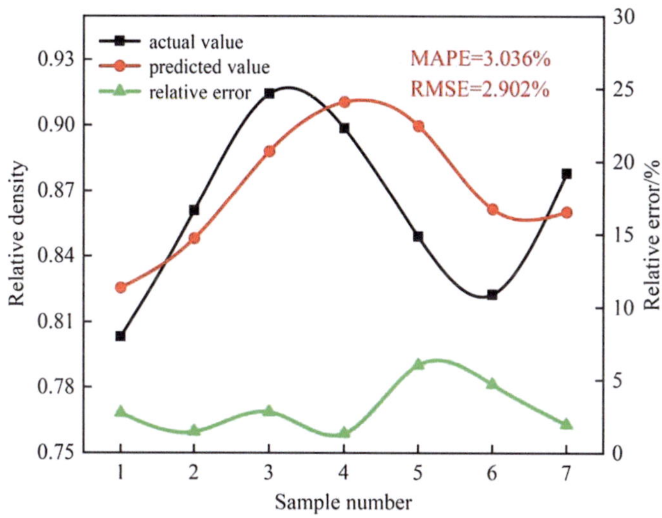

Fig. 7.12 Analysis of fitting error between predicted relative density and measured relative density of subbase filling materials

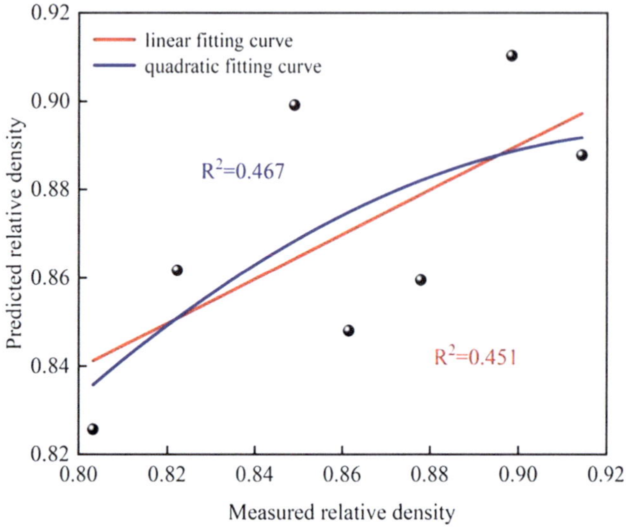

Fig. 7.13 Correlation analysis between predicted relative density and measured relative density of subbase filling materials

Table 7.5 Assessment results of LSTM model for pavement layer

Filling layer	R^2	MAPE/%	RMSE/%
Pavement base layer	0.58	2.56	3.23
Pavement subbase layer	0.45	3.04	2.90

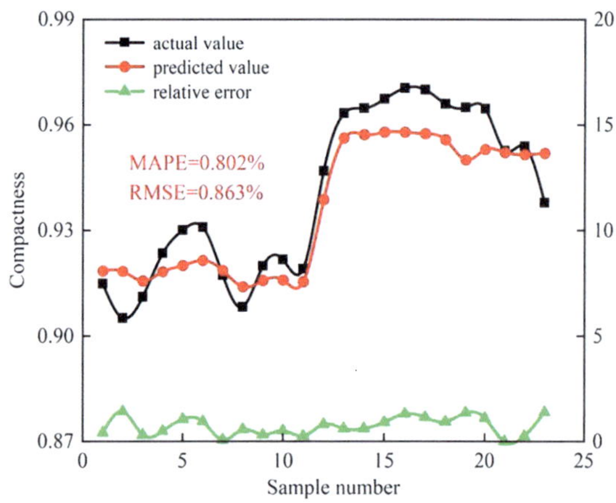

Fig. 7.14 Error analysis between predicted compactness and measured compactness of subgrade filling materials

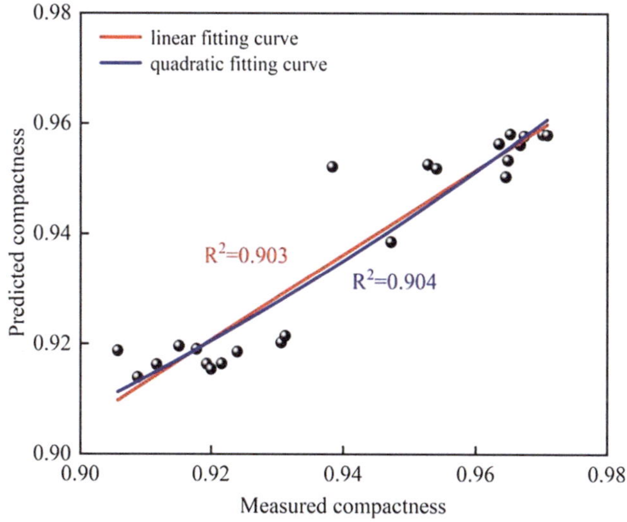

Fig. 7.15 Correlation analysis between predicted compactness and measured compactness of subgrade filling materials

Table 7.6 Assessment results of LSTM model for subgrade layer

Filling layer	R^2	MAPE/%	RMSE/%
Subgrade	0.90	0.80	0.86

7.3 Compaction Quality Intelligent Assessment Method Based RBF

7.3.1 RBF Model

The Radial Basis Function (RBF) neural network model, introduced in the late 1980s, is a type of artificial neural network that employs radial basis functions as activation functions. The fundamental RBF model is a three-layer feedforward network that realizes the mapping from the input layer space to the output layer space through a nonlinear transformation from the input layer to the hidden layer, followed by a linear transformation from the hidden layer to the output layer. The activation function of the hidden layer nodes is the radial basis function, which utilizes the distance between the input vector and the node centers as the independent variable, directly mapping the input vector to the hidden layer. The activation function of the output layer nodes is linear, which linearly weights the output information from the hidden layer before producing the final output [6]. The model structure is illustrated in Fig. 7.16. The RBF neural network boasts several advantages, including high computational efficiency, a simple structure, straightforward training, and rapid convergence in learning. It is capable of approximating continuous functions to any desired precision

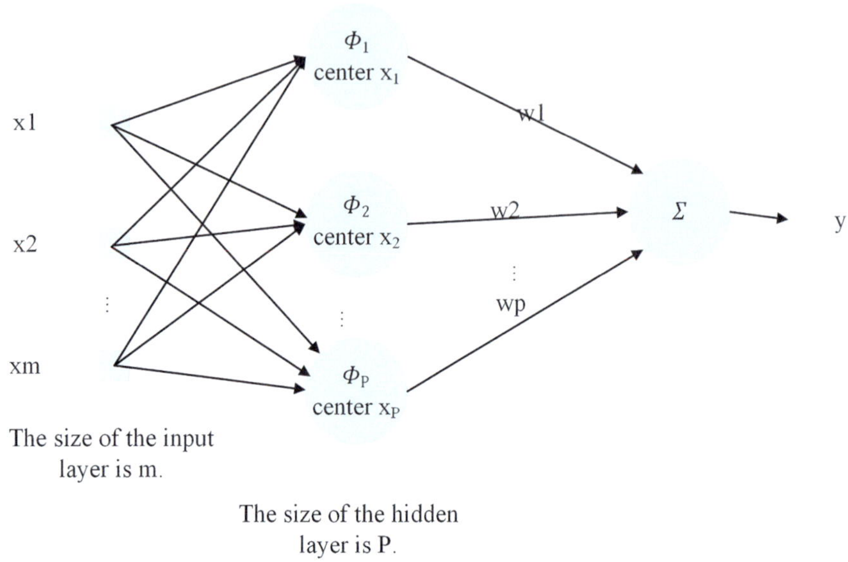

Fig. 7.16 Structure diagram of RBF model

and effectively addresses issues related to local minima. Overall, it outperforms the BP neural network model in terms of approximation capability, classification ability, and learning speed.

The radial basis function is a type of scalar function that is symmetric about a radial axis, typically defined as a monotonically decreasing function of the Euclidean distance $\sum$ between a point x in space and a certain center c_j. The most commonly used radial basis function is the Gaussian function, expressed as follows:

$$\varphi_j(\|x - c_j\|) = \exp\left(-\frac{\|x - c_j\|}{2\sigma_j^2}\right) \tag{7.8}$$

In this expression, c_j represents the center vector of the function, while σ_j denotes the width vector.

7.3.2 *Field Tests*

(1) Testing equipment

The testing equipment used is the SANY SSR260-5H 26-ton vibratory roller, as depicted in Fig. 7.17. To facilitate intelligent compaction of earth-rock dams, the roller has been modified to include a GPS receiver, wireless communication antenna, accelerometer, variable pump, and onboard controller. An automatic

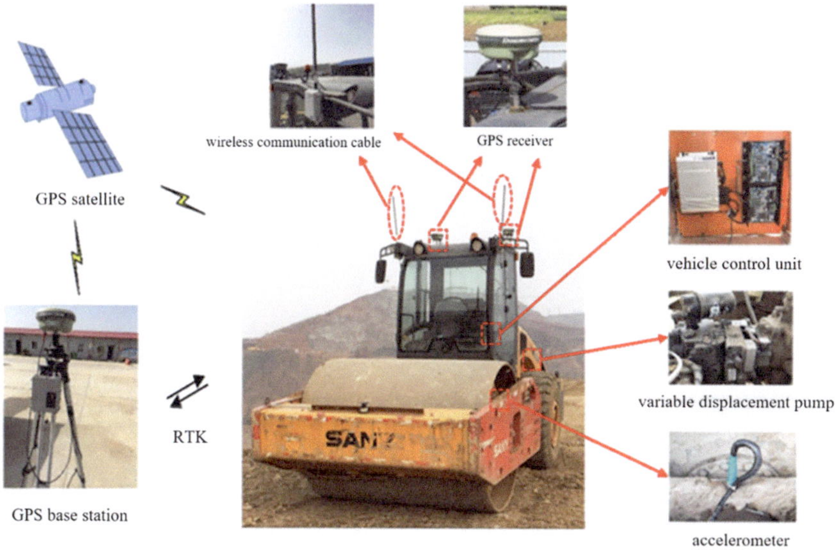

Fig. 7.17 Roller and intelligent compaction equipment used in the test

compaction system based on RTK-GPS and closed-loop feedback control enables precise measurement and automatic control of the roller's position and driving speed. The variable pump regulates the hydraulic oil flow to the vibrating motor, allowing for stepless adjustment of the vibration frequency within the range of 0–34 Hz. The accelerometer is mounted on the vibratory wheel to collect real-time vertical acceleration signals, while the GPS and acceleration signals are stored and processed by the on-board controller.

(2) Testing site and materials

The field compaction tests were conducted at the Qianping Reservoir in Ruya County, Luoyang City, Henan Province. The main dam of the reservoir is an earth-rock dam with a clay core, standing at a height of 90.3 m, and the dam shell material consists of rockfill materials sourced from the upstream stockpile. The particle size distribution curve for the onsite rockfill material, obtained through sieve analysis, is illustrated in Fig. 7.18. As shown in the figure, the maximum particle size of the tested rockfill material reaches 400 mm, with the content of particles smaller than 5 mm ranging from 15 to 24%, and the moisture content of the rockfill material is between 3 and 4%.

(3) Testing plan

To validate the effectiveness of the *CMV* index in characterizing the compaction quality of rockfill material and to investigate the relationship between the *CMV* index and compaction parameters (vibration frequency, driving speed, and number of compaction passes), a series of field compaction tests were designed, as detailed in

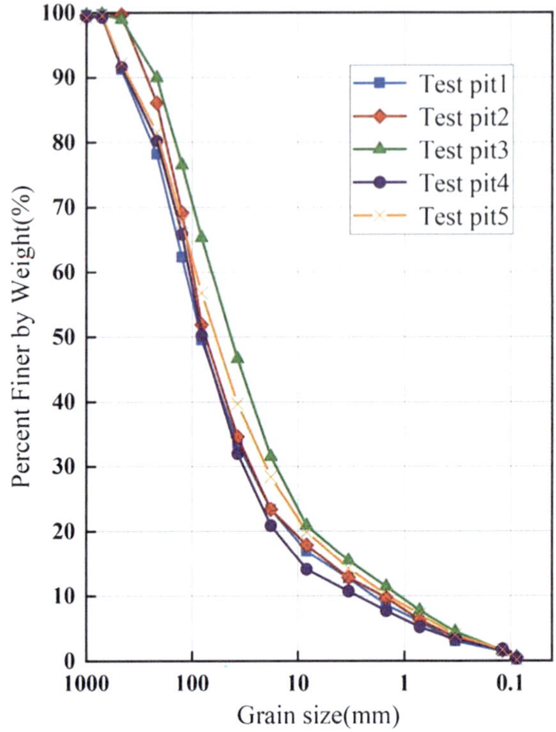

Fig. 7.18 Grading curves of rockfill materials

Table 7.7. The protocol involved two static passes and one to eight vibratory passes. The testing area was divided into individual test strips, each with an actual width of 3 m, suitable for single-pass compaction by the roller, with an effective length of 12 m. Additional lengths were provided at both ends to eliminate the effects of vibration initiation and cessation on the vibration signal waveform, ensuring the acquisition of stable acceleration signals. Within each strip, three test pits for in-situ testing were sequentially arranged. Furthermore, sufficiently sized adjustment areas were established between the strips to facilitate the positioning of the compaction machine. The testing was conducted using the SANY SSR260-5H model 26-ton single-drum vibratory roller (as shown in Fig. 7.17). Given the suboptimal compaction results at vibration frequencies below 20 Hz, the employed frequency range was set between 20 and 32 Hz.

The specific steps for the field compaction tests of rockfill material are divided into three phases: pre-vibration compaction, during compaction, and post-compaction, as illustrated in Fig. 7.19. Figure 7.19a depicts the preparation of the rockfill material prior to vibration compaction, where the thickness of the test rockfill material is set at 80 cm, and the initial delineation of test strips and pit points is conducted on

Table 7.7 On-site rolling test plan

Working condition No	Vibration frequency/(Hz)	Driving speed/(m/s)	Number of compaction passes (Static)	Number of compaction passes (Vibration)
1	20	0.60	2	1–8
2	24	0.69	2	1–8
3	28	0.60	2	1–8
4	32	0.69	2	1–8

the leveled surface of the rockfill. Figure 7.19b illustrates the compaction process, during which appropriate compaction parameters are selected based on the testing protocol, and real-time data on the *CMV* index, compaction parameters, and other relevant information (such as the real-time spatial location of the roller) are collected. Figure 7.19c shows the measurement of in-situ test pit data, such as dry density, following vibration compaction. The dry density of the rockfill is measured using the water displacement method, and relative density values are calculated by combining these measurements with the maximum and minimum dry densities obtained from laboratory sieve analyses, thereby characterizing the compaction quality of the rockfill. The specific formula for calculating relative density is as follows:

$$D_r = \frac{\rho_{\max}(\rho_d - \rho_{\min})}{\rho_d(\rho_{\max} - \rho_{\min})} \tag{7.9}$$

In the formula: D_r represents the relative density in percentage; $\rho_{\max}$ denotes the maximum dry density of the test pit in g/cm^3; $\rho_{\min}$ indicates the minimum dry density of the test pit in g/cm^3; and ρ_d signifies the dry density of the test pit in g/cm^3.

7.3.3 *Test Results Analysis*

(1) Experimental data processing

① Vibration acceleration signal processing

The accelerometer selected for this experiment has a sampling frequency of 1000 Hz. Due to the presence of high-frequency noise in the raw signal, a low-pass filter using a FIR filter was first applied to the acceleration signal. Given that the maximum vibration frequency of the roller is 32 Hz, the cutoff frequency for the low-pass filter was set to 100 Hz to encompass the third harmonic component of the vibrational acceleration. The filtered acceleration time-domain signal underwent a Fast Fourier

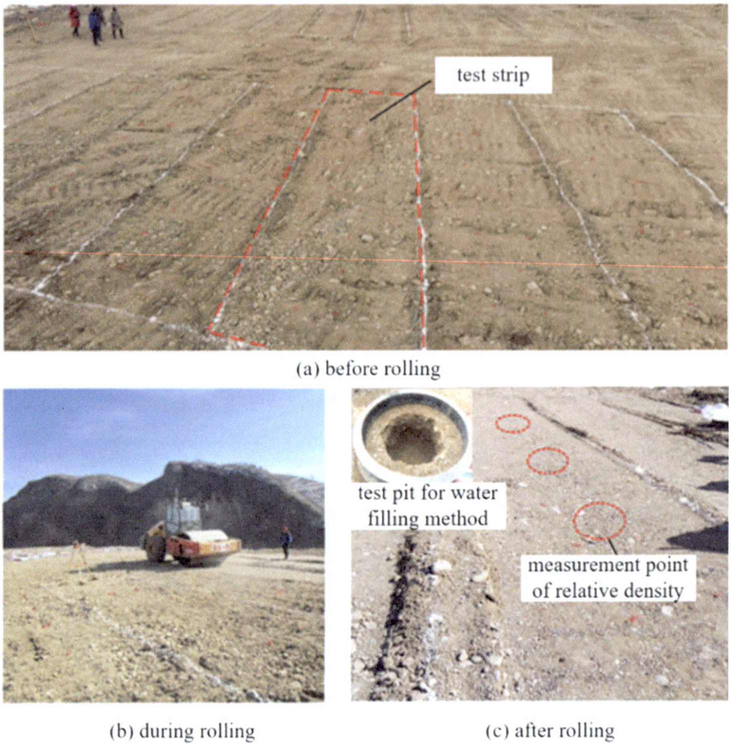

Fig. 7.19 Procedures of field compaction tests of rockfill materials

Transform (FFT), yielding the frequency spectrum of the acceleration signal. The frequency spectra for the first and eighth passes of a particular test strip are depicted in Fig. 7.20. The figure clearly reveals the fundamental frequency and the second harmonic, while the third harmonic is less pronounced. In the figure, A_0 and A_2 represent the amplitudes of the fundamental frequency and the second harmonic, respectively. It is evident from the figure that the amplitude of the second harmonic in the eighth pass has significantly increased compared to the first pass.

② Index calculation

Based on the acceleration frequency spectrum, the calculation formula for the *CMV* value is as follows:

$$CMV = C\frac{A_2}{A_0} \tag{7.10}$$

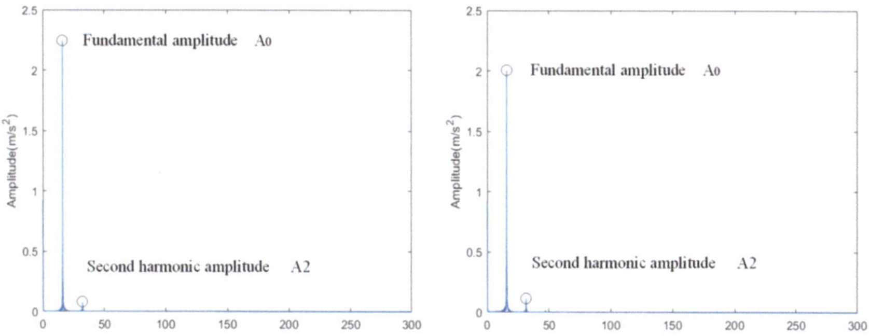

Fig. 7.20 Spectra of vibration acceleration signals

In the formula: C is a constant, taken as 300 for this experiment; A_0 represents the amplitude corresponding to the fundamental frequency in the acceleration spectrum; and A_2 denotes the amplitude corresponding to the second harmonic in the acceleration spectrum.

(2) Correlation analysis between *CMV* and relative density of rockfill materials

Following the testing protocol outlined in Sect. 7.3.2, in-situ measurements of the wet density at the test pit points were conducted using the water displacement method after the completion of the field compaction tests on the rockfill material. This data was combined with laboratory test results to derive the relative density values of the rockfill. Based on the experimental data, the relationship between the *CMV* index and the relative density of the filling material was established, as illustrated in Fig. 7.21. The figure indicates that, under four different working conditions, the *CMV* consistently increases with the rise in the relative density of the rockfill, with correlation coefficients of 0.780, 0.746, 0.829, and 0.822, respectively. This demonstrates a strong correlation between the *CMV* and the relative density of the rockfill, suggesting that it can serve as an effective monitoring index for the compaction quality of the rockfill.

(3) Correlation analysis between *CMV* and compaction parameters

① Correlation analysis between the number of compaction passes and *CMV*

By processing and analyzing the experimental data obtained from the field tests in Sect. 7.3.2, the *CMV* values for two test strips, each subjected to eight passes under two different working conditions, were recorded for the 2nd, 4th, 6th, and 8th passes, as illustrated in Fig. 7.22. The figure indicates that, under both working conditions, the *CMV* significantly increases with the number of compaction passes.

② The influence of driving speed on *CMV* value

When analyzing the impact of driving speed on the *CMV* value, it is essential to control for vibration frequency and maintain the same drive direction. During the

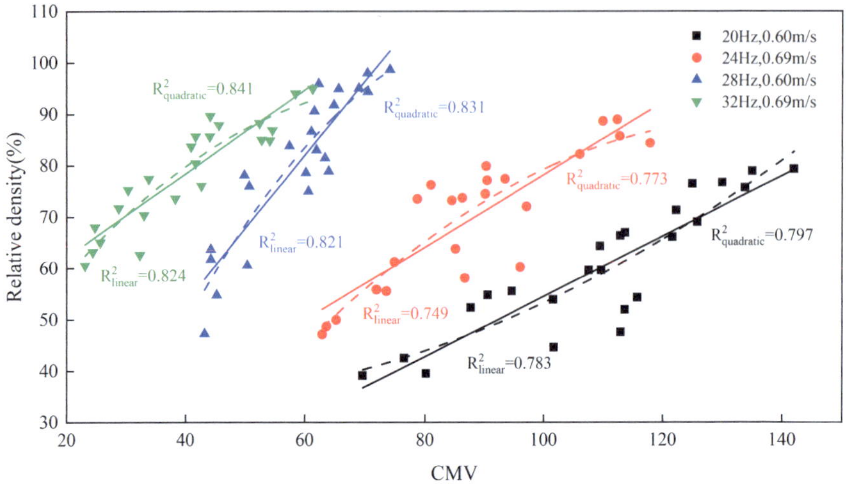

Fig. 7.21 Relationship between *CMV* and relative density of rockfill material

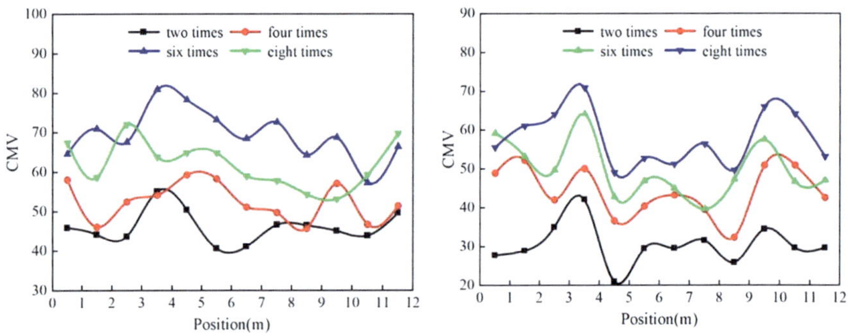

(a) frequency 28Hz, driving speed 0.60m/s (b) frequency 32Hz, driving speed 0.69m/s

Fig. 7.22 Relationship between *CMV* and the number of compaction passes

experiment, three different driving speeds were employed: 0.40 m/s, 0.69 m/s, and 0.83 m/s, all with a vibration frequency of 28 Hz, and the roller operated in reverse. Figure 7.23 illustrates the trend of *CMV* values for the rockfill material after the eighth pass at varying travel speeds. It is evident from the figure that driving speed has a significant influence on the *CMV* value, with average *CMV* values of 75.18, 69.61, and 62.20 at the three different speeds, respectively. The *CMV* increases as driving speed rises. This phenomenon can primarily be attributed to the effect of driving speed on the contact area between the vibratory drum and the dam material. As indicated in the following formula, the soil stiffness in the vibratory roller-soil dynamic system is not only related to soil properties but also to the contact area between the vibratory drum and the soil. Under the same soil conditions, a larger contact area results in greater soil stiffness:

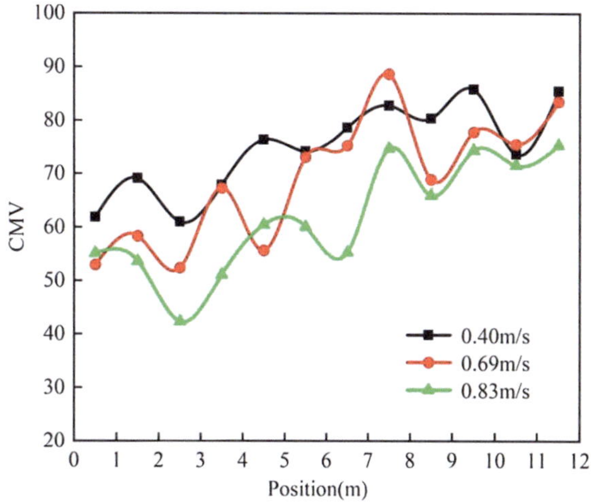

Fig. 7.23 Influence of driving speed on *CMV*

$$k_s = \frac{4Gr_0}{1 - v} \tag{7.11}$$

In the formula: G represents the shear modulus in Pascals (Pa); v is the Poisson's ratio; and r_0 denotes the equivalent radius of the contact area, measured in meters, which is determined by the contact area between the vibrating drum and the soil.

During the compaction process, a slower driving speed results in greater compaction energy being transmitted from the vibratory drum to the dam material, leading to increased settlement and deformation of the material. Consequently, this enhances the contact area between the vibratory drum and the dam material. Therefore, a reduced driving speed increases soil stiffness and the reactive force of the soil, ultimately resulting in an elevation of the *CMV*.

③ Effect of vibration frequency on *CMV* value

To analyze the impact of vibration frequency on the *CMV* value, different vibration frequencies were employed during vibratory compaction, while maintaining a driving speed of 0.69 m/s with the roller operating in reverse. Figure 7.24 illustrates the trend of *CMV* values following the eighth pass at varying vibration frequencies. The figure reveals that the *CMV* value decreases as the vibration frequency increases. According to the vibratory roller-soil dynamics theory, the soil reactive force is maximized at the system's resonance frequency, where the nonlinear vibrations of the vibratory drum are also most pronounced. Consequently, the *CMV* value is highest within this resonance frequency range, while moving further from the resonance frequency results in diminished soil reactive forces and lower *CMV* values. In this experiment, the resonance frequency of the vibratory roller-soil system is approximately 22 Hz; thus, within the vibration frequency range of 20 to 32 Hz, the *CMV* value decreases as the vibration frequency increases.

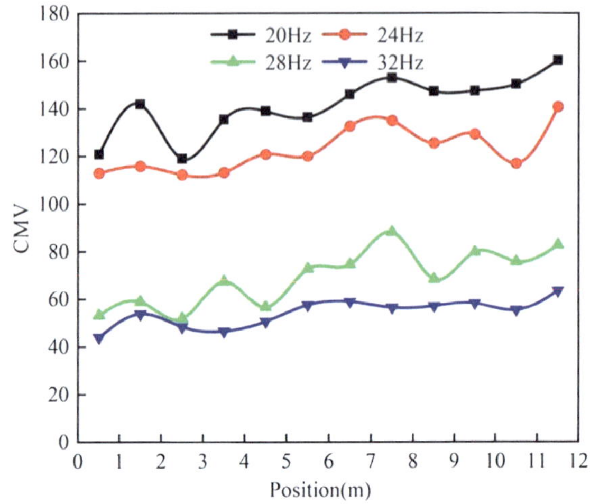

Fig. 7.24 Influence of vibration frequency on *CMV*

④ The influence of drive direction on *CMV* value

When the roller operates in the forward and reverse directions, the rotational direction of the eccentric weights differs from the rolling direction of the vibratory drum, resulting in unequal forces acting on the drum under these two conditions. Consequently, the drive direction of the roller also influences the *CMV* value. In the experiment, compaction was conducted in both forward and reverse directions while maintaining a vibration frequency of 28 Hz and a driving speed of 0.69 m/s. Figure 7.25 presents the *CMV* values after the eighth pass for both drive directions. The average *CMV* values for forward and reverse operation were 73.62 and 69.61, respectively, indicating that the drive direction does have a noticeable impact on the *CMV* value; however, this effect is comparatively minor when assessed against the influences of driving speed and vibration frequency.

7.3.4 Compaction Quality Assessment Based on RBF Model

According to the analysis in Sect. 7.3.3, the *CMV* can serve as a monitoring indicator for the compaction quality of rockfill material; however, the compaction parameters significantly influence the *CMV*. Therefore, it is essential to consider the effects of compaction parameters in the assessment model for compaction quality based on the RBF model. In light of this analysis, a rockfill compaction quality assessment model has been established, as depicted in Fig. 7.26. This model consists of three layers, with the input layer encompassing *CMV*, vibration frequency, driving speed, and the direction of roller operation, while the output layer represents the relative density of the rockfill.

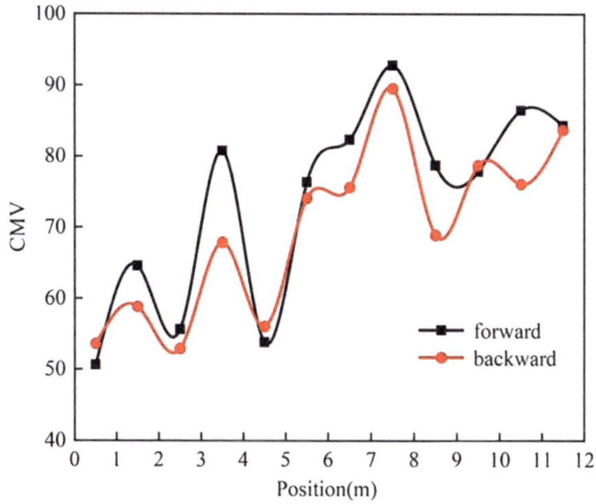

Fig. 7.25 Influence of drive direction on *CMV*

$$D_r = f(CMV, f, v, d) \tag{7.12}$$

In the formula, D_r denotes the relative density of the rockfill material, expressed as a percentage; f represents the vibration frequency in Hertz (Hz); v indicates the driving speed in meters per second (m/s); and d signifies the direction of roller operation, with a value of 0 for forward and 1 for reverse.

Through a series of field compaction tests, a total of 129 sets of valid data were obtained, with 103 sets designated as the training group and 26 sets as the testing group. To eliminate the influence of dimensional differences among the various parameters, the data should first undergo linear normalization using the following

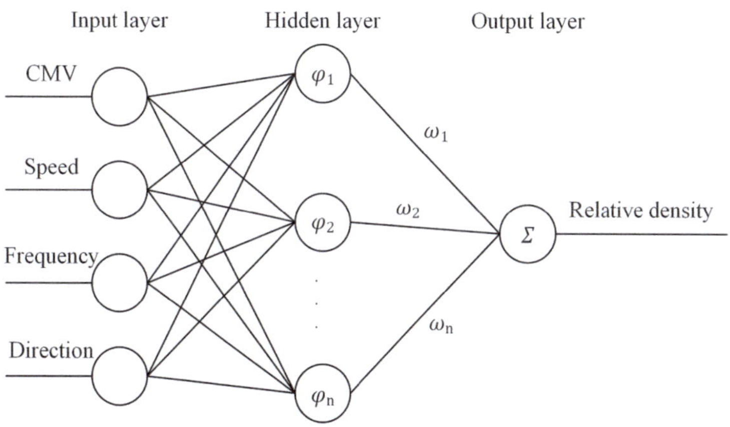

Fig. 7.26 Compaction quality assessment model based on RBF neural network

formula:

$$x^* = \frac{x - x_{\min}}{x_{\max} - x_{\min}} \tag{7.13}$$

In the formula, x^* represents the normalized value; x denotes the data value; $x_{\max}$ is the maximum data value; and $x_{\min}$ is the minimum data value.

The compaction quality assessment model was trained using the training group data, and then the test group data was substituted into the trained model for relative density prediction. Figure 7.27 shows the comparison between the predicted relative density of the test group and the measured relative density. As can be seen from the figure, the correlation coefficient between the model-predicted relative density and the measured relative density is 0.859, which has a strong correlation. Figure 7.28 shows the error between the relative density predicted by the test group model and the measured relative density. The average relative error of the model is 3.68%, and the maximum relative error is 8.98%, which shows that the model has sufficient accuracy and can be used for the compaction quality assessment of rockfill materials.

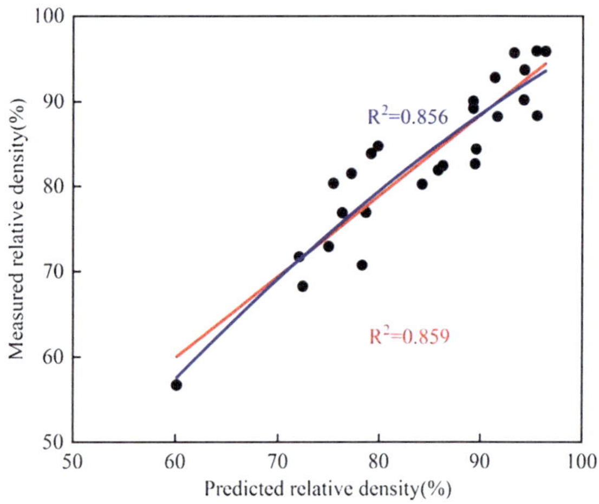

Fig. 7.27 Comparison between estimated and measured relative densities

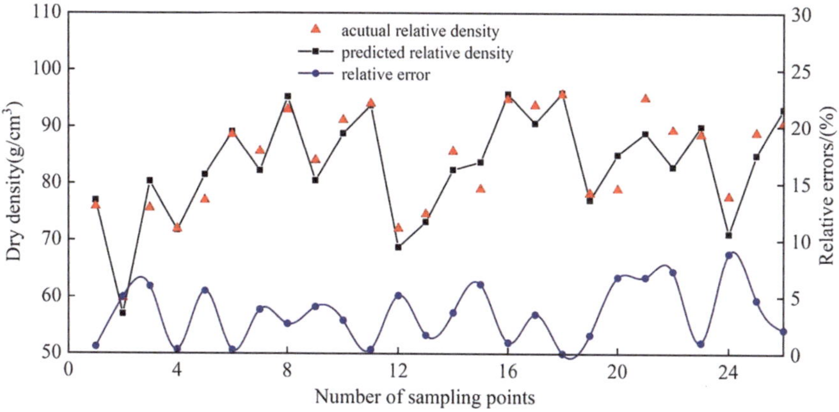

Fig. 7.28 Errors in estimated relative densities against measurements

7.4 Compaction Quality Intelligent Assessment Considering Small Sample Uncertainty

7.4.1 UB-MPGA-BP Model

The relative density or compactness of the filling materials serves as a compaction quality index influenced by multiple factors, with the relationship between in-situ measurements and CCC/RICM/IC/IRC data being nonlinear and highly intricate. The prerequisite for accurately predicting and assessing compaction quality is to elucidate the high-dimensional mapping relationship between in situ measurements and CCC/RICM/IC/IRC data. Traditional linear regression models fall short in capturing these nonlinear characteristics; therefore, this section introduces a BP neural network combined with an improved Multi-Population Genetic Algorithm (MPGA) and uncertainty analysis for compaction quality assessment, referred to as UB-MPGA-BP.

(1) BP NN

This section presents a BP-based compaction quality assessment model (refer to Fig. 7.29), expressed by the following Eq. (7.14):

$$Q = f(n, V, F_{ac}, f, CMV) \tag{7.14}$$

In this context, n, V, F_{ac}, f, and CMV represent the number of compaction passes, driving speed, actual exciting force, vibration frequency, and compaction meter value, respectively. Q denotes the compactness, relative density, or other indexes of compaction quality. Due to the high dimensionality and complexity of the input parameters in the compaction quality assessment model, feature extraction from high-dimensional data often necessitates additional hidden layers.

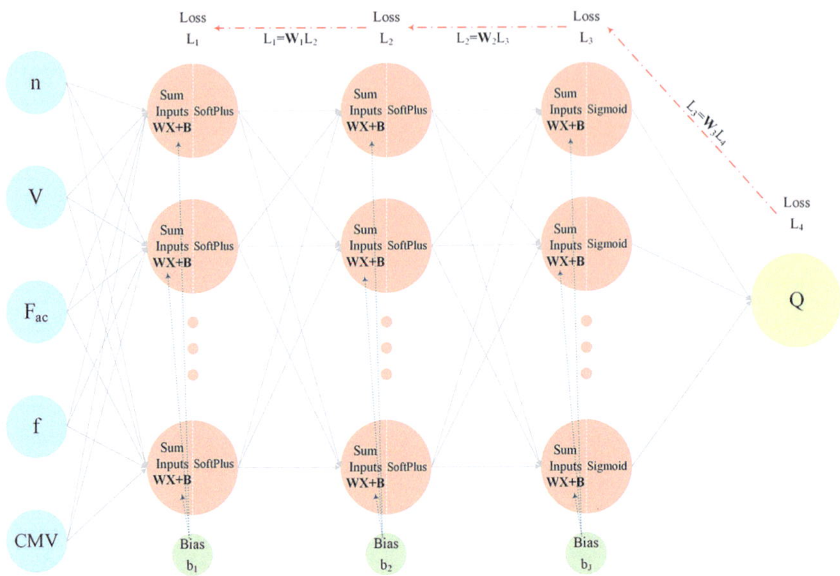

Fig. 7.29 Structure of BP NN

To develop a lightweight and easily deployable compaction quality assessment model for swift on-site implementation, a balance between time cost and regression performance was established. Consequently, the hidden layers (HLs) of the BP network were configured to three, each containing six neurons (NRs). A learning rate (LR) of 0.1, an expected training error (EE) of ($10e^{-5}$), and a maximum training epoch of 1000 were selected.

To eliminate discrepancies within the input data, a normalized mapping process must be applied to the input data:

$$y = y_{\min} + (y_{\max} - y_{\min}) \cdot \frac{x - x_{\min}}{x_{\max} - x_{\min}} \tag{7.15}$$

Here, $x_{\max}$ and $x_{\min}$ represent the maximum and minimum values of the input data, respectively. x signifies the input data, while $y_{\max}$ and $y_{\min}$ denote the upper and lower bounds for normalization, respectively, with y being constrained within the interval ($y \in [0, 1]$).

Furthermore, this model employs the Sigmoid function as the output activation function, which is defined as:

$$S = \frac{1}{1 + e^{-x}} \tag{7.16}$$

Here, x represents the input variables in the output layer, while S denotes the activation variables of the output layer. The value range of S is (0, 1), allowing for a more effective fitting of the normalized variables.

BP neural networks are prone to becoming trapped in local optima when fitting high-dimensional data. This vulnerability arises from the necessity of calculating the derivative of the transfer activation function for each layer, particularly as the number of hidden layers increases during the backward error propagation process. Traditional transfer activation functions, typically represented by a sigmoid series function such as the hyperbolic tangent function, have a derivative value range of less than one. Consequently, only those derivatives approaching zero exhibit effective activation, while the gradients of values exceeding zero diminish sharply. This phenomenon results in a continuous decay of error throughout the backward error propagation process, leading to the derivative of the transfer activation function gradually approaching zero, thus resulting in the gradient vanishing problem. Ultimately, this situation culminates in premature training, confining the model to a local optimal solution.

In evaluating the transfer activation functions, the performances of Sigmoid, SoftPlus, ReLU, Leaky ReLU, ELU, and other activation functions were compared, with the specific results presented in Table 7.8. As indicated in the table, the SoftPlus, ReLU, Leaky ReLU, and ELU functions outperformed the Sigmoid function due to the latter's inadequate gain on both sides of the signal. However, the SoftPlus, ReLU, Leaky ReLU, and ELU functions exhibited no significant advantages or disadvantages; this is attributable to the fact that the inputs in this experiment were exclusively positive, resulting in slightly varied distributions among the different inputs. Activation functions such as SoftPlus, ReLU, Leaky ReLU, and ELU demonstrated favorable responses within the central region and on both sides.

Furthermore, SoftPlus was chosen as the transfer activation function. Nevertheless, in theory, ReLU, Leaky ReLU, and ELU could also be viable alternatives. SoftPlus is defined as follows:

$$SoftPlus = \ln\left(1 + e^{x}\right) \tag{7.17}$$

Table 7.8 Performance comparison of different transfer activation function (average results in 10 times of testing)

Activation function	R^2	MAPE (%)	RMSE (%)
Sigmoid	0.853 ± 0.011	0.510 ± 0.007	0.0052 ± 0.0001
SoftPlus	0.894 ± 0.011	0.417 ± 0.011	0.0048 ± 0.0001
Relu	0.889 ± 0.010	0.420 ± 0.011	0.0049 ± 0.0001
LeakyRelu	0.894 ± 0.010	0.419 ± 0.009	0.0047 ± 0.0001
Elu	0.893 ± 0.010	0.418 ± 0.010	0.0048 ± 0.0001

Here, x represents the input or transfer variable between the hidden layers. The SoftPlus function serves as a transfer activation function, exhibiting an increasing derivative for values greater than zero, thereby mitigating the issue of gradient vanishing. Meanwhile, the activation performance of the derivative near zero closely resembles that of the traditional sigmoid function, enhancing the reliability of feature transfer and the backward propagation of error.

The typical loss function of BP neural networks can be articulated as follows:

$$\begin{aligned} L &= \min \frac{1}{N} \sum_{i=1}^{N} \left\| y_i - \widehat{y_i} \right\| \\ &= \min \frac{1}{N} \sum_{i=1}^{N} \sum_{k=1}^{K} \left(y_{i,k} - \widehat{y_{i,k}} \right)^2 \end{aligned} \tag{7.18}$$

Here, y_i and $\hat{y}_i$ represent the outputs of the BP neural network and the expected output (ground truth value), respectively. N indicates the size of the training sample, while K denotes the dimension of the output, which is assumed to be 1.

(2) Uncertainty analysis

The weight and bias parameters of the BP neural network play a crucial role in determining the accuracy of the compaction quality assessment model, while the generalizability of these model parameters significantly influences both the accuracy and generalizability of the assessment. The training loss function of traditional neural networks is typically represented as shown in Eq. (7.18), which fails to fully address the model's poor generalization due to the divergence in model parameter distribution and the presence of data noise.

Consequently, the epistemic and aleatoric uncertainties of the compaction quality assessment model have been examined from the perspectives of model parameter optimization and data noise tolerance. A combined loss function, which accounts for the coupling effects of both epistemic and aleatoric uncertainties, has been established to replace Eq. (7.18) in order to enhance prediction accuracy and generalization capability.

① Epistemic uncertainty analysis

Epistemic uncertainty arises from the insufficient training of the model, primarily due to an inadequate number of training samples. When test samples do not align with the distribution of the training samples, the model may encounter difficulties in accurately discerning the characteristics of these outlier samples, as the sample space was not comprehensively covered during training. This issue is particularly pronounced in studies assessing the compaction quality of filling materials, especially when the acquisition of in situ measurements (such as relative density and compactness) relies on field sampling pit tests, which are challenging to obtain in large quantities and efficiently. Consequently, data sources are often insufficient for compaction operations. In this context, investigating epistemic uncertainty within the compaction quality assessment model becomes particularly crucial.

Epistemic uncertainty is associated with the distribution of model parameters, and an insufficient sample size can result in a biased estimation of the parameters learned by the model. Therefore, Bayesian inference can be employed to ascertain the true distribution of the model parameters. Assuming that the true model parameter distribution follows a Gaussian prior distribution Z ~ N(0, Δ), and considering the input dataset X, the posterior distribution of the model parameters can be computed using Bayesian inference as follows:

$$P(Z|X) = \frac{P(X|Z)P(Z)}{P(X)} \tag{7.19}$$

Here, $P(X|Z)$ represents the likelihood estimate of X based on the model parameter distribution Z. $P(Z)$ corresponds to the prior information regarding the model parameters, while $P(X)$ denotes the true distribution of the input dataset. Due to the high dimensionality of the input data, it is nearly impossible to experimentally obtain the true distribution $P(X)$; thus, directly calculating Eq. (7.19) is impractical. Consequently, the Markov Chain Monte Carlo (MCMC) method was employed to approximate the posterior distribution through sampling from the prior distribution [7].

In practice, the MCMC method often encounters slow convergence when the data dimensions are excessively high. To address this challenge, multiple dropout operations were applied to the hidden layers of the BP network (refer to Fig. 3 for a specific dropout scheme), wherein the weight values and biases between any hidden layers were randomly discarded during the training process, achieving an effect analogous to that of MCMC. By iterating this method a predetermined number of times, the model output results obtained through sampling under the true parameter distribution can be directly acquired [8]. Therefore, the epistemic uncertainty evaluation metric for the compaction quality assessment model can be articulated as the variance in the input results following the dropout operation. That is:

$$L_e = \frac{1}{S}\sum_{S=1}^{S}\left(f^Z(x_s) - E(y)\right)^2 \tag{7.20}$$

Here, L_e denotes epistemic uncertainty, reflecting the model's capacity to generalize across different datasets; a value of L_e closer to zero indicates superior generalization of the model. $f^Z(x_s)$ represents the model output derived from the sampling distribution under the model parameter distribution Z, while $E(y)$ signifies the mean of the model's sampling output. S denotes the number of samples taken Fig. 7.30.

② Aleatoric uncertainty analysis

Aleatoric uncertainty is typically characterized as the variance in model training attributable to noise present in the input data. Given that the input data were gathered from sensors and manual pit tests, the most straightforward approach to evaluate the aleatoric uncertainty of the data is to utilize the variance of the input data.

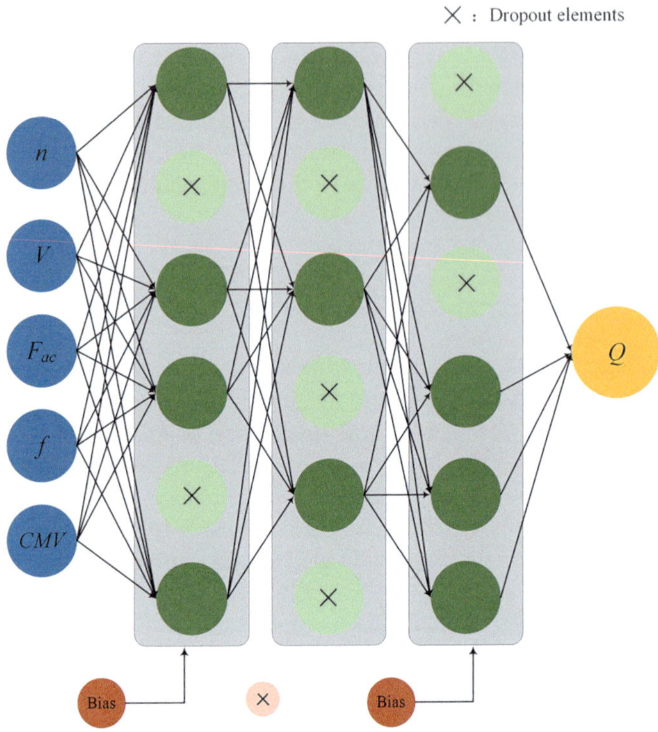

Fig. 7.30 Schematic diagram of a dropout operation in BP NN

$$L_a = \sigma^2 \tag{7.21}$$

L_a represents the aleatoric uncertainty associated with observation noise. In the homoscedastic regression model, the observation noise σ can be assumed to be constant for every input x; however, the inputs in Model (7.14) may vary and are heteroscedastic. Consequently, the observation noise σ is data-dependent and can be learned during the training process. Previous research has posited that the observed inputs can be considered as data samples drawn from a Gaussian distribution (heteroscedastic). Thus, the observation noise (aleatoric uncertainty) can be estimated using the following minimization function [9]:

$$L_{NN}(\theta) = \frac{1}{N}\sum_{i=1}^{N}\frac{1}{2\sigma^2(x_i)}\left\|y_i - \widehat{y_i}\right\|^2 + \frac{1}{2}\log\sigma^2(x_i) \tag{7.22}$$

Here, θ denotes the parameters of the neural network model. x_i represents the input, while y_i signifies the actual compactness or relative density, and $\hat{y}_i$ is the model output. Aleatoric uncertainty can, to some degree, indicate the model's tolerance to data quality; that is, when the performance of the models is comparable, a higher level of aleatoric uncertainty reflects superior data tolerance.

③ Combinatorial uncertainty analysis

To more accurately assess the performance, generalization capability, and data tolerance of the compaction quality model, Eq. (7.23) is employed as a combined uncertainty assessment index, derived from existing literature [10].

$$L_c = \frac{1}{S}\sum_{i=1}^{S}\widehat{y_i}^2 - \left(\frac{1}{S}\sum_{i=1}^{S}\widehat{y_i}\right)^2 + \frac{1}{S}\sum_{i=1}^{S}\widehat{\sigma_i}^2 \tag{7.23}$$

Here, $\hat{\sigma}_i$ represents the data variance in each training epoch. The first two terms in Eq. (7.23) correspond to epistemic uncertainty, whereas the final term signifies aleatoric uncertainty.

Therefore, to incorporate both types of uncertainties into the model training, the output of the model should take the form of $\{f^Z(x), \sigma^2\}$. Here, $f^Z(x)$ represents the prediction result of the model, which can be substituted into Eq. (7.20) to derive the epistemic uncertainty, while σ^2 denotes the aleatoric uncertainty. Consequently, the training loss function of the compaction quality assessment model can be defined as follows:

$$L = \frac{1}{2N}\sum_{i=1}^{N} exp(-log(L_a))\|y_i - \widehat{y_i}\| + \frac{1}{2}\log(L_a) \tag{7.24}$$

The first term in Eq. (7.24) represents the conventional loss function detailed in Eq. (7.18). Traditional BP networks strive to fit each data point as closely as possible, which often results in overfitting and consequently undermines the model's generalization capabilities. By incorporating the regularization term $\log(L_a)$, the first and second terms of Eq. (7.24) create a balance, whereby excessive or insufficient noise will cause a significant increase in the loss function. As a result, the model endeavors to identify the optimal combination of outputs to enhance the precision of its predictions while also providing a degree of generalization ability. In practice, $\log(L_a)$ acts as an adaptive weight that escalates when the data exhibit high intrinsic noise and diminishes for samples with low intrinsic noise. This enables the model to distinguish between samples during training, thereby enhancing its generalizability across diverse compaction datasets.

④ Improved MPGA

The essence of BP neural networks lies in error backpropagation, which is primarily influenced by the edge weight matrix and bias vector between the hidden layers. The phenomenon of forgetfulness within BP networks, coupled with the backpropagation mechanism, results in a deficiency of global adjustment for weights and biases. Consequently, it becomes challenging for BP networks to derive optimal edge weight matrices and bias vectors when processing high-dimensional data.

As a global optimization algorithm, the Genetic Algorithm (GA) not only adjusts the fitness of individuals within the population on a global scale but also preserves the fitness of trained individuals from previous generations to a certain extent. Consequently, employing the GA to optimize the edge weight matrix and bias vector can effectively mitigate the inherent shortcomings of the BP neural network. Building upon this foundation, the Modified Genetic Algorithm (MPGA) can further enhance the diversity of evolutionary directions and improve global search capabilities.

Additionally, to augment the feature selection performance and efficiency of the crossover and mutation operations within the algorithm, binary coding was utilized to encode the population of individuals. In the context of the BP network for compaction quality assessment, let us assume that the edge weight matrix of the hidden layer is denoted by W, where each component $\omega_{i,j}$ represents the edge weight of the i-th dimensional input variable transitioning from the j-th to the $(j+1)$-th hidden layer. The edge bias vector of the hidden layer is represented by B, with each component b_j signifying the edge bias of the j-th layer. Thus, the binary code of the v-th individual within the u-th population can be expressed as:

$$\begin{aligned} S^{u,v} &= (W; B)_u^v \\ &= \left[\omega_{1,1}, \cdots, \omega_{1,q}; \cdots, \omega_{p,1}, \cdots, \omega_{p,q}; b_1, \cdots, b_q\right]_u^v \end{aligned} \tag{7.25}$$

The migration operator of the traditional MPGA directly replaces the worst individual in the subsequent population with an optimal individual from the preceding generation. This design of the migration operator overlooks the possibility that the worst individual may, to some extent, be closer to the global optimum, while the optimal individual could merely represent a locally optimal solution. Consequently, in light of the update characteristics of the MPGA, a more robust migration operator has been devised to preserve the diversity characteristics of the population and to prevent convergence to local optimal solutions. The fitness of the MPGA is defined by Eq. (7.24). Let us assume that the worst and optimal individuals in the u-th population are denoted as $S^{u,worst}$ and $S^{u,optimal}$, respectively; thus, the improved migration operation can be articulated as:

$$\begin{gathered} \forall u \in U \cap u \neq u_0, N_\Theta = \sum_{u \neq u_0}^{U} N_u \\ where\ N_u = \begin{cases} 1\ if \left|L^{u_0,worst} - L^{u,optimal}\right| < M, u \neq u_0 \\ 0 \qquad\qquad else \end{cases} \\ then:\ (1) if N_\Theta < U/2, S^{u_0,worst} = S^{u_{-1},optimal}; \\ (2)\ if\ N_\Theta \geq U/2, S^{u_0,worst} = S^{u_0,worst} + F\left(S^{u_{-1},optimal} - S^{u_{-1},optimal}\right), F \in [0, 1] \end{gathered} \tag{7.26}$$

The improved migration operation posits that when the difference in fitness between the worst individual of the current population and the optimal individual from any other population is less than the threshold M, N_u is set to 1; otherwise, it is assigned a value of 0. By summing all N_u, if the total N_Θ is less than half of the population U, the worst individual of the current population is replaced by the optimal individual from its previous generation, aligning with the traditional migration operator. Conversely, if N_Θ is greater than or equal to half of the population, the worst individual in the current population is generated as a combination of itself and the optimal individuals from the preceding and succeeding populations, where F signifies the combination ratio.

This enhanced migration operator can effectively break the "deadlock" of local optimal individuals by amalgamating the genetic traits of the worst and optimal individuals across different populations when the disparity among individuals decreases. This strategy improves the feasibility of escaping the "locally optimal point," thereby bolstering the global search capabilities of the MPGA.

By incorporating the aforementioned migration operator, the pseudocode for UB-MPGA-BP is presented in Table 7.9. The pseudocode indicates that the time complexity does not exceed $O(MAXG \cdot MP \cdot Pf(N))$, where $f(N)$ represents the time complexity of the BP neural network. This integration does not augment the time complexity of the improved migration operator. Therefore, the enhanced MPGA is achieved without incurring additional time costs.

7.4.2 Experimental Test

(1) Test site and materials

The Hengyong Highway constitutes a significant component of the expressway network in Hunan Province, China, spanning a total length of 106.227 km and featuring a width of 26.5 m. The entire route adheres to a two-way, four-lane expressway standard, designed for a driving speed of 120 km/h and classified under highway Level I for load capacity. The fifth section of the Hengyong Highway commences at K50 + 000 and concludes at K59 + 860, encompassing a total length of 9.86 km. The filling materials utilized for the subgrade primarily consist of LLLS, HLLS, gravelly soil, and loess.

Given the substantial differences in the material properties of various filling materials, particularly the distinct characteristics of Low Liquid Limit Silt (LLLS) and High Liquid Limit Silt (HLLS), it is essential to validate the applicability of a single compaction quality assessment model across different filling materials. The tests were conducted in two sections: K52 + 800 to K53 + 030 and K55 + 900 to K56 + 150, respectively. The subgrades in these two test sections were constructed using LLLS and HLLS filling materials (refer to Table 7.10 for their fundamental properties). Both LLLS and HLLS are classified as silty soils, primarily composed of powder and sand, with a minimal presence of clay particles. The particle gradations are illustrated in

Table 7.9 The pseudocode of UB-MPGA-BP

```
Input: P = 50: population size in each population; MP = 10: number of populations;
  MAXG = 5: maximum generation to maintain the optimal fitness; epochs = 1000:
  The training epochs of BP NN
Output: S^optimal: optimal individual which contains the edge weight matrix and bias
  vector; L^optimal: optimal fitness
    initial g = 0; S^{u,v} randomly
  while g≤MAXG
      g = g + 1;
      for i = 1 : MP do
      Selection operation within each population;
      Crossover operation within each population;
      Mutation operation within each population;

      for j = 1 : P

          BP network training operation in epochs times;
          Compute the fitness L = 1/2N · ∑_{i=1}^{N} exp(− log(L_a))‖y_i − ŷ_i‖^2 +
  1/2 · log(L_a)
      end
    end
    for k = 1 : MP do
      Compare S^{u_0,worst} and S^{u_0,optimal}, u ≠ u_0;

      Compute N_Θ;

    if N_Θ < U/2

        S^{u_0,worst} = S^{u_{-1},optimal};
    else
        S^{u_0,worst} = S^{u_0,worst} + F(S^{u_{-1},optimal} − S^{u_1,optimal}), F ∈ [0, 1].
    end
    end
    for l = 1 : MP do
        Find the optimal individual S^{u,optimal} with the smallest fitness L^{u,optimal}
   within each population;
    end
    Find the optimal individual S^optimal with the smallest fitness L^optimal;
  end
```

Fig. 7.31. The maximum particle size for LLLS was approximately 2–5 mm, with a particle content of less than 0.075 mm ranging from 96.4% to 96.5%. In contrast, the maximum particle size for the high liquid-limit silt was around 5–10 mm, with particles smaller than 0.075 mm constituting 93.1% to 94.2%.

(2) Testing program

To investigate the impact of the compaction quality assessment model on the characterization of LLLS and HLLS, two sets of working conditions were employed: LFHA and HFLA exciting forces. Each soil material underwent dynamic rolling for one to eight iterations prior to two static compaction tests, as detailed in Table 7.11. Figure 7.32 illustrates the schematic layout of the 16 strips designated for each soil material in the field. Each test strip measured 3 m in width and 10 m in length, with ample surrounding area to facilitate the roller's ability to reverse, stagger, and maneuver.

Table 7.10 Basic properties of LLLS and HLLS

Subgrade materials	Mileage	Maximum dry density (g/cm^3)	Natural moisture content ω_n (%)	Optimal moisture content ω_O (%)	Liquid limit ω_L(%)	Plastic limit ω_P(%)	Plasticity index I_P
LLLS	K52 + 800 ~ K53 + 030	1.69	27.0	18.9	49.3	32.6	16.7
HLLS	K55 + 900 ~ K56 + 150	1.65	27.1	22.6	58.3	35.6	22.7

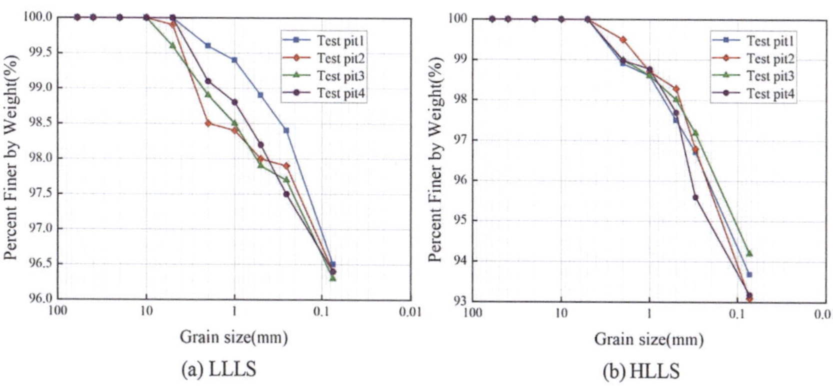

Fig. 7.31 Grain curves of LLLS and HLLS from several test pits

Table 7.11 Operation conditions on-site

Soil type	Operation condition	Vibration status	Vibration frequency (Hz)	Driving speed (m/s)	Passes (Static)	Passes (Dynamic)	Lift thickness (cm)
LLLS	1	LFHA	29	0.83	8	1–8	40
	2	HFLA	35		8	1–8	40
HLLS	1	LFHA	27	0.83	8	1–8	40
	2	HFLA	32		8	1–8	40

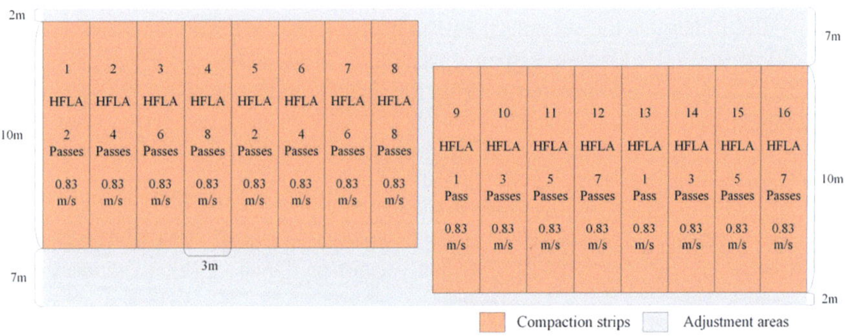

Fig. 7.32 Schematic layout of 16 strips for each soil material

For the LLLS test strips, five measurement points for compactness were evenly distributed along the length of the strip at the 1st, 3rd, 5th, 7th, and 9th meters, respectively. In contrast, the HLLS test strips featured three compactness measurement points, evenly arranged along the length at the 2nd, 5th, and 8th meters.

Figure 7.33 depicts the site work arrangement of the test strips, including the CMV-computed sections within each strip, the onsite pit test points, and the sand replacement method utilized to ascertain the actual compactness following compaction. During the onsite compaction tests, a Sany Heavy Industry SSR220AC single-drum vibratory roller, weighing 22 tons, was employed for the LLLS. Its vibration frequency and excitation force were set at 29/35 Hz and 410/300 kN, respectively. For the HLLS, an XCMG XS265JS single-drum vibratory roller, weighing 26 tons, was utilized, with a vibration frequency and exciting force of 27/32 Hz and 405/290 kN, respectively. Upon completion of the compaction process, the IRC terminal automatically generated the *CMV* for each section, while onsite workers conducted sand cone tests according to the initial arrangement, as illustrated in Fig. 7.33.

Fig. 7.33 Field compaction test

7.4.3 Result Analysis

(1) Data process

By employing the compaction assessment model delineated in Eq. (7.14), the training dataset can be methodically processed and organized into a matrix comprising seven rows and N columns. Each row encapsulates the index of the corresponding sample, along with its relative density or compactness, the number of passes, the vibration frequency expressed in Hertz (Hz), the driving velocity measured in meters per second (m/s), the actual exciting force represented in kilonewtons (kN), and the *CMV*. The number of columns is indicative of the total sample size.

(2) Outliers filtering

Prior to employing the collected data for compaction quality assessment, it is imperative to conduct outlier detection to eliminate overtly abnormal values and preserve the integrity of the overall data distribution. Figure 7.34 illustrates the analysis results for Cook's distance in both LLLS and HLLS. There exist numerous criteria to interpret the results of Cook's distance; for instance, a relatively straightforward method for filtering out outliers is to apply thresholds such as $D_i > 1$ or $D_i > 0.5$, both of which are significantly higher than the Cook's distance derived from our dataset. Another criterion for outlier detection involves the homogeneity of variance test, where D_i surpasses the median of the F distribution, $F(0.5, p, n-p)$. In this analysis, the thresholds for LLLS and HLLS were determined to be 0.8782 and 0.8841, respectively, with results indicating that no influential samples were identified. Consequently, outliers were not removed.

(3) Compaction quality assessment results

Compaction quality assessment tests were conducted using 80 collected data samples from the LLLS and 48 samples from the HLLS. The first 80% of the collected data were designated as the training set for the compaction quality assessment model, while the remaining 20% served as the test set. Consequently, there were 64 training samples and 16 test samples for the LLLS, in addition to 38 training samples and 10 test samples for the HLLS. The model's performance is evaluated by calculating the R^2, the MAPE, and the RMSE of the assessment results. R^2 reflects the overall

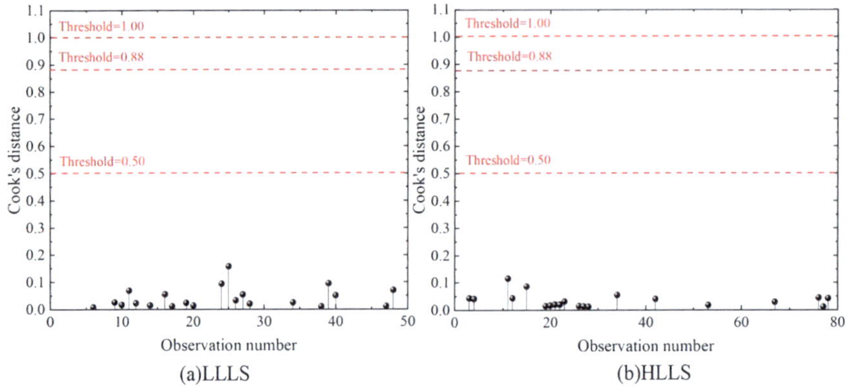

Fig. 7.34 Analysis result of the Cook's distance in each subgrade material

goodness of fit, MAPE indicates the relative estimation error, and RMSE assesses the influence of significant prediction errors.

Furthermore, to illustrate the superiority of the UB-MPGA-BP model proposed in this section, a comparative analysis includes the BP neural network (NN), radial basis neural network (RBF), generalized neural network (GRNN), extreme learning machine (ELM), long short-term memory network (LSTM), particle swarm algorithm-based BP neural network (PSO-BP), difference algorithm-based BP neural network (DE-BP), genetic algorithm-based BP neural network (GA-BP), and multi-population genetic algorithm-based BP neural network (MPGA-BP), along with an analytic hierarchy process (AHP) based model and adaptive simulated annealing particle swarm optimization applied to Elman neural network (ASAPSO-ENN). In this comparative analysis, the initial ten models were specifically constructed for this case study, while the subsequent four models were derived from other studies (refer to Table 7.12).

Among the ten models developed in this study, the hyperparameter settings for the first five neural networks (NNs) were meticulously determined through rigorous trials to ensure their regression performance remained at relatively superior levels. For the latter five BP neural networks, which are enhanced by various metaheuristic algorithms, the hyperparameter configurations were kept as consistent as possible to facilitate a more accurate comparison of the performance improvements afforded by the enhanced algorithms on the BP NNs. The hyperparameters for the four reference models were sourced from existing literature.

A summary of the compaction-quality assessment results of the different methods is presented in Table 7.12. The proposed method (UB-MPGA-BP) outperforms the other models with much higher R^2 and lower errors (MAPE and RMSE) in both LLLS and HLLS, achieving $R^2 = 0.89$, MAPE = 0.42% ~ 0.63%, and RMSE = 0.005 ~ 0.007 for both soil materials. As mentioned earlier, the time complexity of UB-MPGA-BP in the training phase is higher than that of the traditional MPGA-BP owing to the improved mitigation operator and uncertainty-based loss function.

Table 7.12 Comparison of different compaction quality assessment methods

Method	HLLS			LLLS			Parameter setting	Source
	R^2	MAPE (%)	RMSE	R^2	MAPE (%)	RMSE		
BP	0.632	1.458	0.011	0.725	0.761	0.009	HL = 3; NRs = 6; LR = 0.1; EE = 10e-5; epochs = 1000	Constructed in this section
ELM	0.412	1.649	0.016	0.420	1.493	0.015	NRs = 6	
GRNN	0.647	1.462	0.014	0.712	0.999	0.010	Spread = 0.2	
RBF	0.321	1.617	0.017	0.759	0.723	0.008	Spread = 0.2	
LSTM	0.395	1.459	0.016	0.304	1.576	0.016	HL = 3; NRs = 200; initial LR = 0.1; LR drop period = 125; LR drop factor = 0.2 EE = 10e-5; epochs = 1000	
BP	0.631	2.304	0.024	0.690	1.890	0.018	HL = 4; NRs = 9; LR = 0.1; EE = 10e-5; epochs = 1000	[11]
DE-BP	0.855	1.358	0.014	0.677	0.678	0.007	HL = 3; NRs = 6; LR = 0.1; EE = 10e-5; epochs = 1000; P = 500;	
PSO-BP	0.681	1.818	0.018	0.734	0.856	0.009	HL = 3; NRs = 6; LR = 0.1; EE = 10e-5; epochs = 1000; P = 500; c1 = 1.49; c2 = 1.49; w = [0.1,1,1]	
GA-BP	0.736	0.981	0.011	0.824	0.436	0.005	CR = 0.7; MR = 0.1	
MPGA-BP	0.832	0.764	0.009	0.810	0.474	0.005	HL = 3; NRs = 6; LR = 0.1; EE = 10e-5; epochs = 1000; P = 50; MP = 10	
UB-MPGA-BP	**0.888**	**0.631**	**0.007**	**0.894**	**0.417**	**0.005**	*MAXG* = 5; CR = 0.7; MR = 0.1	
AHP based mmodel	0.741	1.273	0.014	0.773	1.035	0.010	–	[12]

(continued)

Table 7.12 (continued)

Method	HLLS			LLLS			Parameter setting	Source
	R^2	MAPE (%)	RMSE	R^2	MAPE (%)	RMSE		
PSO-BP							EE = 10e-5;LR = 1.6135E-3; epochs = 100; P = 20; c1 = 2; c2 = 2	[13]
ASAPSO-ENN							P = 30; LR = 0.01; EE = 10E-4; HL = 8; c1 = 2; c2 = 2	[14]

However, the inference time complexity was the same as that of traditional BP NNs. Therefore, UB-MPGA-BP can satisfy the requirements of real-time on-site IRC assessment.

As shown in Table 7.12, ASAPSO-ENN exhibited the best performance among the assessment models from other studies. However, despite achieving an R^2 value of 0.98 in its respective literature, its overall performance within the context of the small dataset used in this study was still inferior to that of UB-MPGA-BP. This is because existing research typically achieves high performance under conditions of hundreds or even thousands of training samples, rarely considering the compaction quality assessment performance under small-sized datasets. Consequently, this study addresses this issue by constructing the first ten evaluation models presented in Table 7.4, and through a comparative analysis, it demonstrates the superiority of UB-MPGA-BP. As shown in Fig. 7.35, the fitted regression line of the UB-MPGA-BP model almost coincides with the diagonal line, achieving a low prediction error in both the LLLS and HLLS.

(4) Compaction quality assessment on the whole compacted layers

Based on the above results, the compaction quality assessment model proposed in this study has the feasibility and robust reliability of the LLLS and HLLS verified in this section. Therefore, it is worthwhile to assess the compaction quality of all the

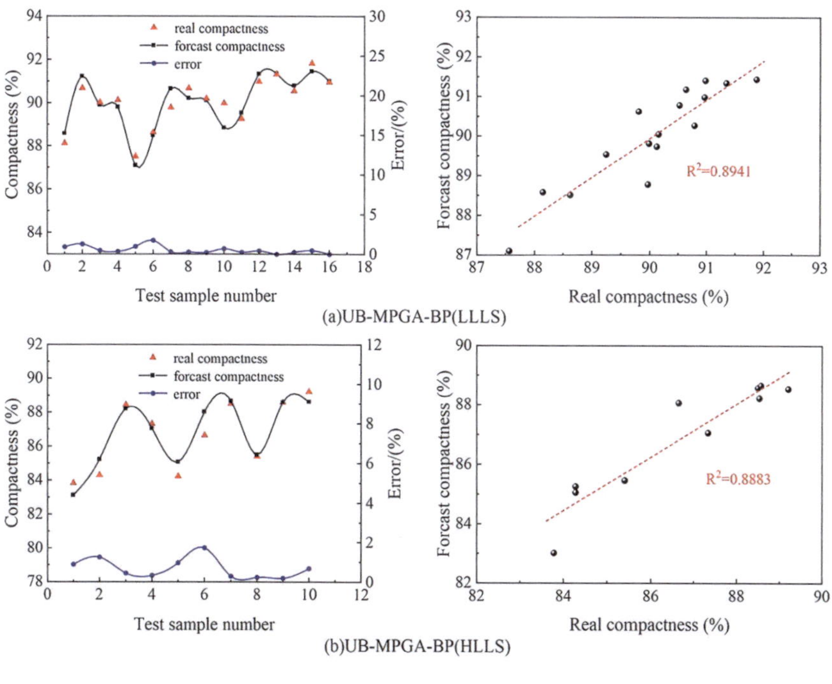

Fig. 7.35 Comparison of different methods on compaction quality assessment with heuristic algorithm optimization

compacted layers. In this study, each test strip of the LLLS and HLLS was divided into 10 sections, and the CMV of each section center was calculated separately. The trained UB-MPGA-BP model was then used to evaluate the compactness of the ten sections on each strip. Finally, a biharmonic spline interpolation [15] was used to interpolate the entire compacted layer and obtain an overall compaction nephogram.

Figure 7.36 presents the compaction quality nephograms of the entire compacted layer in the LLLS and HLLS. The variation in compactness is depicted by different color scales, with red indicating higher compactness and blue indicating lower compactness. It is evident that the overall compactness, as predicted by our method, increases with the number of compaction passes, except for a few abnormally compacted areas highlighted by black dashed circles. These areas of sudden surges in compactness may be attributed to outliers in the CMV. As more sections split a strip, the available vibration frequency information in each section decreased, causing any fluctuation in the vibration frequency to significantly affect the CMV of each section. Furthermore, the compactness nephogram generated by the proposed method can identify areas that are less compact, as indicated by the red dashed circles in Fig. 7.36. This allowed timely recompaction during rolling.

In general, a compact quality assessment model in an IRC system is reliable and practical for guiding practical construction. In a specific compaction project, a limited amount of training data can be initially fed into the UB-MPGA-BP model to acquire the robust compaction features of a specific type of filling material. Subsequently,

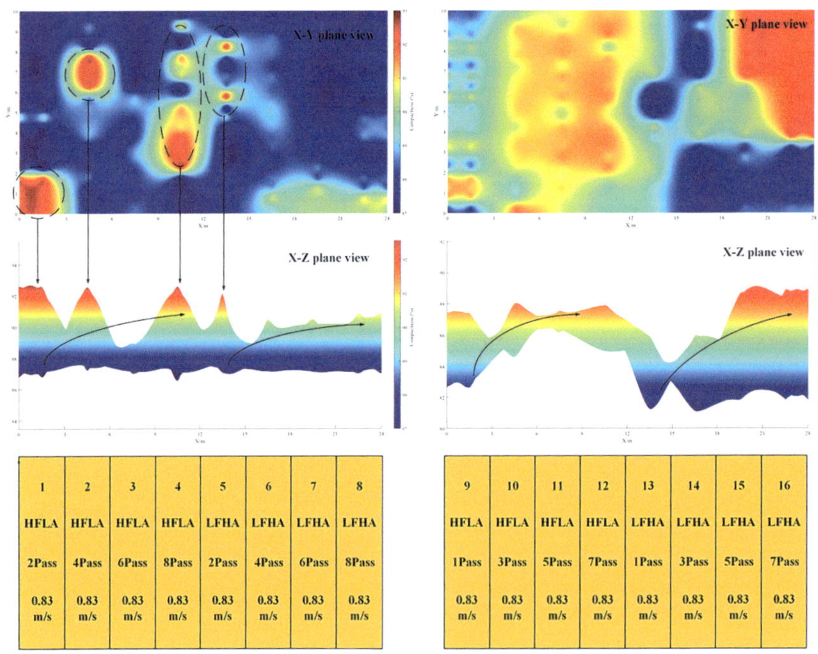

Fig. 7.36 Compaction nephogram of the whole compacted layers (LLLS)

a well-trained model can be integrated into the IRC system, which can be installed on the rollers. These rollers, equipped with an IRC system, can effectively gather compaction parameter data and predict the compactness of the compacted regions. The IRC system demonstrated computational efficiency as a result of the pretraining of the UB-MPGA-BP model, allowing for the rapid generation of compactness outputs and real-time visualization of compaction nephograms at the terminal.

References

1. Yu WF (2023) Research and implementation of aquatic text classification system based on deep learning [D]. Dalian Ocean University
2. Yu XJ (2023) Evaluation of direct economic losses from earthquakes in Yunnan Province based on GA-BP neural network [D]. Yunnan University of Finance and Economics
3. An ZZ, Liu TY, Huangfu ZH, Zhang QL, Li CS, Li QB (2020) Neural network model for evaluating compaction quality of rockfill materials by compaction meter value. J Hydroelectr Eng 39:110–120
4. Dong C, Liu WJ, Li L (2018) Four-parameter prediction model and its FE realization for dynamic resilient modulus of subgrade compacted silty clay [J]. J Vib Shock 37(9):61–68
5. Zhou X (2020) Research on improved TF-IDF feature selection and short text classification algorithm [D]. Anhui University
6. Chai J, Jiang QY, Cao ZK (2002) The function approximation ability of the rbf neural network and its algorithm [J]. Pattern Recogn Artif Intell 15(03):310–316
7. ATTM (2021) Standard practice for measuring ultrasonic velocity in materials by comparative pulse-echo method. E494–20, West Conshohocken, PA
8. Gal Y, Ghahramani Z (2016) Dropout as a bayesian approximation: representing model uncertainty in deep learning [C]. International conference on machine learning. PMLR, pp 1050–1059
9. ASTM (2000) Standard test methods for laboratory compaction characteristics of soils using standard effort [S]. D698, West Conshohocken, PA
10. Lakshminarayanan B, Pritzel A, Blundell C (2017) Simple and scalable predictive uncertainty estimation using deep ensembles [J]. Advances in neural information processing systems, p 30
11. Krebs RD, Waller ED (1971) Highway materials [M]. McGraw-Hill, New York
12. Juan GZ (2011) Relativity analysis on the detection methods of compaction degree for expressway subgrade [J]. Adv Mater Res 413(413–413):252–257
13. ASTM (2007) Standard test method for in-place density and water content of soil and soil-aggregate by nuclear methods [S]. D6938, West Conshohocken, PA
14. Shangguan P, Al-Qadi IL (2015) Calibration of FDTD simulation of GPR signal for asphalt pavement compaction monitoring [J]. Geoscience Remote Sens IEEE Trans 53(03):1538–1548
15. Sandwell DT, Biharmonic spline interpolation of GEOS-3 and SEASAT altimeter data. Geophysi Res Lett 14(2):139–142

Chapter 8
Compaction Quality Intelligent Control Technology

Compaction is the process by which the volume of rockfill materials is diminished to enhance its density. Compaction quality control (QC) serves as a crucial link in ensuring the stability and longevity of earth-rock dams [1, 2]. Inadequate compaction and quality control lead to numerous quality deficiencies, including reduced strength, increased settlement, and heightened permeability [3, 4]. According to existing specifications for earth-rock dam construction, compaction quality predominantly hinges on the manual oversight of compaction parameters—such as the number of compaction passes, compaction trajectory, vibration frequency, lift thickness, and driving speed—during construction, alongside random spot tests conducted post-construction [5].

However, conventional compaction quality control methods present several challenges: (1) The management of compaction parameters is executed by on-site construction personnel, categorizing it as an empirical control method, which results in a higher incidence of control errors and quality defects; (2) Spot tests, due to their destructive nature, are both time-consuming and labor-intensive, rendering them inadequate for the demands of modern mechanized construction; and (3) The compaction quality of the entire rolling operation cannot be recorded in real time, leading to significant deficiencies in data traceability. Consequently, conventional control methods not only fail to effectively manage the compaction quality of the construction area but also fall short of meeting the increasingly stringent quality control requirements for earth-rock dam construction.

Compaction monitoring systems (CMSs) present an innovative solution to the challenges posed by conventional control methods [3, 6, 7]. Research on CMSs [7–9] has primarily concentrated on monitoring compaction parameters [10–14] and the continuous assessment of compaction quality [3, 15]. With advancements in technology and research, several notable CMSs have reached maturity, including continuous compaction control (CCC) [16, 17], roller-integrated compaction monitoring (RICM) [18, 19], and intelligent compaction (IC) [20, 21] systems. The CCC system

Q. Zhang, *Compaction Quality Nondestructive Testing and Intelligent Control Technology for Infrastructure*, https://doi.org/10.1007/978-981-95-1264-5_8

functions as a data acquisition system installed on compaction equipment, continuously gathering real-time information regarding the operation and performance of the compactor [3, 16, 22].

Since the advent of CCC technology, numerous studies have endeavored to correlate roller-measured values with various location-specific metrics, such as density, modulus, penetration resistance, and the California bearing ratio, thereby enhancing the control of soil compaction quality [23–26]. RICM technology incorporates internal sensors installed on the roller to monitor vibrational data, including harmonic frequencies and accelerations, which are utilized to derive the RICM measurement value (RICMMV) for characterizing the real-time compaction status of compacted materials, including subgrade and granular soils, as well as asphalt [18, 27, 28]. Additionally, an on-board computer equipped with a rugged display and a global positioning satellite (GPS) system is integrated into the monitoring technology to map the spatial location of the roller [29].

As a tool for compaction quality control of earthen materials, RICM holds substantial promise for safeguarding investments, enhancing construction quality, and expediting construction progress [18]. Recent initiatives in the United States and Europe have directed attention toward the application of RICM technologies in road construction [22, 30, 31], alongside efforts to relate its parameters to mechanistic pavement design values [18]. Several manufacturers currently offer RICM technologies on smooth drum vibratory roller models to monitor the compaction quality of granular materials and asphalt, as well as non-vibratory configurations for cohesive materials. These manufacturers have established various RICMMVs to characterize the real-time compaction quality of soil, aggregates, and hot mix asphalt (HMA), such as the compaction metre value (*CMV*) utilized by Geodynamic [32, 33], k_s adopted by Ammann [34], machine drive power (*MDP*) employed by Caterpillar [8, 35, 36], and OMEGA implemented by Bomag [37].

To realize these benefits and expedite the adoption of RICM, comprehensive and statistically robust field studies are essential to enhance the understanding of the relationships between machine parameters (i.e., RICMMV) and diverse in situ compaction measurements [32]. Consequently, White et al. [32] conducted a field study involving 30-m test strips utilizing five granular materials to evaluate CMV and MDP roller-integrated compaction technologies. Furthermore, Zhong et al. [11] harnessed RICM technology for precise automatic online monitoring of compaction parameters throughout the entire process, including the number of compaction passes, compaction trajectory, driving speed, vibration status, and lift thickness. This proposed methodology offers a novel approach to quality control in the construction of high-core rockfill dams. Building upon Zhong's research and RICM technology, Liu et al. [1] collected real-time field operation data on driving speed, compaction passes, vibration status, and lift thickness of gravel-mixed cohesive soil in the core wall zone of an earth-rock dam, alongside the corresponding moisture content and gradation of dam materials, to ensure effective control of compaction quality during the construction of earth-rock dams.

When the CCC/RICM system provides automatic feedback control for vibration amplitude, vibration frequency, or driving speed, it is designated as an Intelligent

Compaction (IC) system [18]. IC technology has been successfully implemented in Europe and Japan for several years and made its debut in the United States in the late 2000s [34, 38, 39]. Currently, a variety of CCC/RICM/IC systems from different manufacturers are available, including AccuGrade developed by Caterpillar and Trimble [36], ACE-Plus developed by Ammann [34], DCA by Dynapac [3], Aithon MT by Sakai [19], and BCM05 by BOMAG [40].

Historically, intelligent construction systems based on IC technology have been explored to enhance efficiency and safety, encompassing intelligent navigation strategies [41], autonomous systems employing artificial intelligence methodologies [42], and instrumented roller compactor systems that monitor vibrational behavior [43]. Other intelligent systems studied within the asphalt paving industry include those utilizing neural network modeling [44, 45] and on-board density reading systems [44, 46]. Presently, IC technology is widely adopted in earthworks, particularly for quality control in road construction [19, 21, 34, 37, 47].

In the realm of dam construction, some researchers have engaged in applied studies of CMSs. Zhong et al. [7, 11, 48] and Liu et al. [1, 28] developed real-time monitoring systems to comprehensively assess compaction parameters—such as the number of compaction passes, vibration states, lift thickness, and driving speed—in the construction of earth-core rock-fill dams. Huang et al. [14] created a supervisory system leveraging global navigation satellite systems, wireless data communication, the Internet of Things, and computer technologies to oversee real-time roller compaction parameters of the working surface, including compaction trajectory, number of compaction passes, driving speed, lift thickness, and surface smoothness. Additionally, Karray et al. [49] employed "modal analysis of surface waves" technology to evaluate the deep compaction of the Péribonka Dam foundation.

Despite the widespread application of the aforementioned CMSs in recent real projects, their use for monitoring the compaction quality of filling materials presents several challenges. These include low detection accuracy, considerable variability in detection results, complexities in interpreting compaction status, increased errors in outcomes, susceptibility to material-related factors, misjudgments and human errors, as well as construction quality issues stemming from manual operation.

Consequently, this section will introduce two systems designed for the quality control of earth-rock dams during rolling construction: the unmanned rolling compaction system and the intelligent rolling compaction system.

8.1 Unmanned Rolling Compaction System

In recent years, researchers have increasingly embraced real-time monitoring systems for rolling compaction construction parameters. The accuracy of GPS measurements, which serve as the foundation for these systems, is directly linked to the effectiveness of monitoring compaction parameters, which in turn are closely associated with the compaction quality of filling materials [5]. Consequently, GPS technology has found widespread application in the monitoring of compaction parameters and overall construction quality.

Liu et al. [28] utilized RICM technology based on Real-Time Kinematic (RTK) GPS to monitor the number of compaction passes, compaction trajectory, driving speed, and vibration status during earth-rock dam construction. Xu and Chang [50] assessed the efficacy of IC technology founded on RTK GPS for asphalt compaction, revealing that the IC roller equipped with the RTK GPS system could maintain a continuous record of the number of compaction passes, roller locations, driving speed, and vibration frequency. Kassem et al. [7] developed a compaction monitoring system leveraging cutting-edge GPS technologies to oversee asphalt pavement compaction, employing color-coded maps derived from GPS data to illustrate the number of compaction passes, the roller's location, and other relevant metrics. Additionally, Anderegg et al. [51] employed CCC technology based on the differential GPS system to furnish the roller operator with a straightforward visual aid, indicating the achieved compaction, the number of compaction passes, and other pertinent construction site data.

Moreover, GPS-integrated interpolation and grid partitioning have been employed in several research studies to ascertain lift thickness. A relatively straightforward method involved acquiring elevation (Z-value) measurements from the RTK-GPS on a CCC compactor for one lift and comparing these values to the elevation measurements from the preceding lift, thereby determining the lift thickness by calculating the difference between the two elevation measurements at the same location [52].

The CCC equipment does not record data at the same spatial locations (i.e., identical X and Y coordinates) from lift to lift. To circumvent this challenge, a fixed-position (X, Y) coordinate grid was established alongside the RTK GPS-measured elevation values for a given lift, supplemented by geospatial interpolation methods to predict the corresponding elevations at each grid point. Once the elevation at each grid point has been established for two adjacent lifts, the lift thickness at each grid point can be calculated by taking the elevation difference between the successive layers [13].

Additionally, the ordinary Kriging method is utilized for the spatial interpolation of roller-measured values (MV) of elevation to predict unknown values on a uniform grid for each lift of the constructed embankment, facilitating the determination of lift thickness [12]. An alternative interpolation technique for estimating elevation values at each grid point location is the inverse distance weighting (IDW) method. The final compacted thickness of each soil lift can then be determined by comparing the interpolated elevation value at each grid point with the interpolated value at the same grid point for the underlying lift [12, 53].

Several roller manufacturers and third-party companies, including Ammann [51], Geodynamik [33], Caterpillar [54], Bomag [20], and Trimble [19], have developed corresponding CCC/RICM/IC devices integrated with GPS technology to monitor compaction parameters throughout the construction process. Ammann's GPS-based compaction equipment was employed to track the number of compaction passes, while a combined DGPS/laser system achieved height precision within millimeters, facilitating the determination of lift thickness through the application of three-dimensional digital coordinate grid technology [55].

Caterpillar provided IC equipment equipped with an RTK-GPS system, which was utilized to monitor the thickness of compacted soil lifts during the construction of a roadway embankment. Both simple and advanced spatial analysis techniques were employed to interpolate the measured field elevation data onto a uniform grid for assessing lift thickness [12, 56]. Additionally, the Trimble intelligent compaction system, featuring on-board high-precision GPS, was utilized to continuously monitor the number of compaction passes, compaction trajectory, and lift thickness of the railway subgrade [57].

Building upon previous studies, we have further advanced the unmanned rolling compaction system (URC) to enhance the compaction quality of rockfill materials in earth-rock dams. This system comprises an unmanned roller, an RTK-GPS system, a wireless communication network, and a remote monitoring center. The unmanned roller autonomously collects real-time status information, relays it to the remote monitoring center, receives navigation instructions from the center for automatic navigation and driving, and executes control commands for remote operation.

The remote monitoring center establishes the navigation path, receives various status updates and emergency requests from the unmanned roller in real time, and dispatches immediate control instructions accordingly. Additionally, the center facilitates comprehensive monitoring of compaction parameters. The RTK-GPS system consists of both a base station and a moving station, which work together to ascertain the current positional coordinates of the vibratory roller. Specifically, the base station receives GPS satellite signals and determines the phase difference of the transmitted carrier in real time, while the moving station calculates the high-precision position of the vibratory roller based on the GPS signals it receives, along with phase difference signals transmitted by the base station.

The wireless communication system supports not only interactions between slow-moving construction equipment but also the transmission of remote control and on-site monitoring information. The framework of the URC system is depicted in Fig. 8.1.

The proposed URC system offers several significant advantages. Firstly, it facilitates unmanned operation of the rolling process based on Automatic Data Transmission (ADT), allowing for precise control of compaction parameters—such as the number of compaction passes, driving speed, vibration frequency, compaction trajectory, and lift thickness—thereby markedly enhancing the compaction quality of earth-rock dam materials during the rolling process.

Secondly, the system provides continuous monitoring throughout the entire rolling operation, enabling timely feedback for quality control. Thirdly, it alleviates the burden of complex operations from personnel, thereby mitigating the influence of human factors on compaction quality. Fourthly, it meets the demands of all-weather, high-strength filling construction on-site.

Fifthly, it effectively addresses issues of unrolled, cross-rolling, and repeated rolling that often occur between adjacent work faces. Ultimately, these advantages

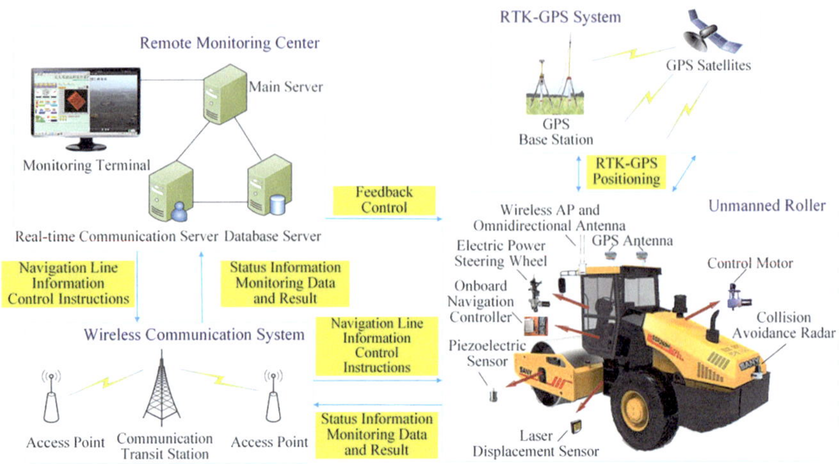

Fig. 8.1 Framework of the URC system

contribute to improved compaction quality and efficiency of rockfill materials in earth-rock dams, thereby enhancing overall project performance.

8.1.1 *Unmanned Roller*

The unmanned roller primarily comprises an on-board navigation controller equipped with a built-in navigation control program (NCP), a GPS receiver and antennas, a wireless access point (AP), omnidirectional wireless APs, an inter-vehicle communicator, ground speed radar, collision avoidance radar, a control motor, an electric power steering system, and a piezoelectric sensor (as illustrated in Fig. 8.1).

As a critical component of the embedded system, the NCP is responsible for both receiving and transmitting control information, as well as processing commands and inputting target metrics relayed by the remote monitoring center in real time. Furthermore, it concurrently receives RTK-GPS positional data, ground speed radar data, ranging radar data, and steering angle sensor data, ensuring comprehensive situational awareness.

The NCP initially executes remotely controlled instructions to power on, start, or shut down the vibratory roller. Subsequently, it invokes the steering control subroutine, wherein it assesses the deviation between the navigation target input and the positioning data to ascertain the desired steering angle. The NCP then compares the difference between this desired steering angle and the actual steering angle of the wheel, automatically adjusting the steering mechanism based on the output current of the PID algorithm. Consequently, the vibratory roller is directed in a predetermined orientation.

Additionally, the NCP engages the automatic acceleration subroutine, which compares the expected driving speed input values with the speed measured by the ground speed radar and evaluates the conditions for automatic braking. It regulates the automatic throttle actuator through the PID algorithm's output current, taking into account the status of surrounding vehicles, thus ensuring that the vibratory roller operates at the desired speed.

Furthermore, the NCP invokes an automatic brake subroutine to assess the degree of range variations detected by the ranging radar, thereby determining the movement of nearby vehicles through inter-vehicle communication. It then regulates the brake pedal based on the output current of the PID algorithm, effectively slowing down or halting the vibratory roller. Subsequently, the NCP transmits information—including vibration status, positional data, driving speed, automatic control status, and environmental video—to the remote monitoring center in real time.

Additionally, the on-board navigation controller continuously monitors the distance between adjacent vehicles and appropriately adjusts the throttle and brake in accordance with a predefined safe distance and speed. As a result, it successfully achieves an automatic driving capability that is responsive to the surrounding conditions.

(1) On-board automatic control system

The flowchart of the on-board automatic control system is illustrated in Fig. 8.2. Upon activation of the on-board navigation controller, the NCP commences an initialization process. Subsequently, it establishes a connection with the remote control device via wireless communication, connects to the RTK-GPS, and initializes the navigation control cycle, PID parameters, and other relevant state parameters.

Following this, the NCP initiates the monitoring of the operational status of each sensor and actuator within the entire automatic control system. Should any component fail to function properly, the NCP records its status parameters and transmits pertinent information to the remote monitoring center, where it will await resolution of the identified issues; otherwise, the NCP will refrain from executing automatic navigation operations.

If the NCP successfully passes the system self-test, it will proceed to enter automatic navigation mode, which is delineated into a series of automatic navigation processes in accordance with the navigation control cycle. Each process comprises three sub-processes: automatic steering control, automatic speed regulation, and automatic braking.

① The automatic steering control sub-process begins by reading the navigation target value and direction, followed by the acquisition of the latitude and longitude coordinates of its current position, which are then converted into plane coordinates. Next, the automatic steering control calculates the distance between the current position and the navigation direction, yielding the horizontal deviation value, and subsequently converts the lateral deviation into the desired automatic steering expectation angle for the current control cycle based on the current speed.

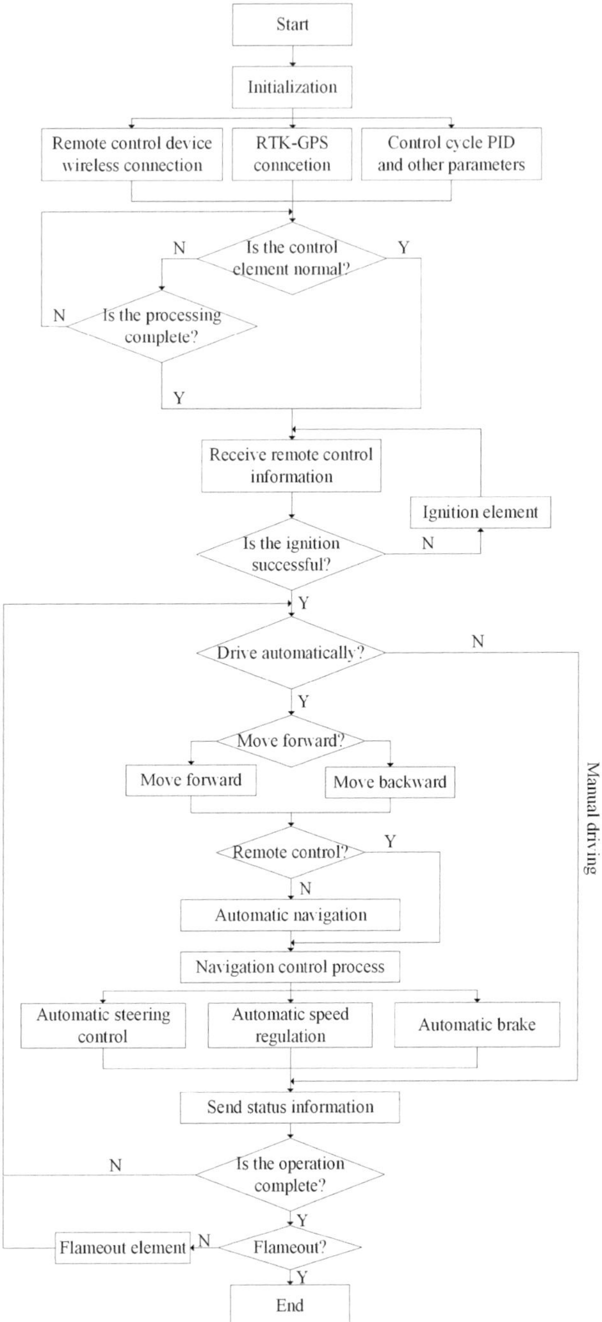

Fig. 8.2 Flowchart of on-board automatic control system

Following this, the automatic steering control retrieves the current steering angle from the steering wheel angle sensor to ascertain whether it has reached the limit established by the system or if it remains below the desired automatic steering angle. The steering PID control subroutine is invoked only when the current steering angle is less than the predetermined limit.

Following the comparison of the current steering angle with the expected automatic steering angle within the PID control subroutine, the PID adjustment is executed to address the rotation angle deviation. Subsequently, the control current signal is calculated and transmitted to the electric steering system, facilitating the automatic steering of the roller and thereby completing one cycle of the automatic steering control process.

② The automatic speed regulation sub-process begins by reading the expected input value for navigation speed along with the collision avoidance radar signal. It then retrieves the ground speed radar measurement of the roller to determine whether the driving speed meets the system-set speed limit. If the driving speed is below this limit, the PID control subroutine for speed adjustment is invoked.

After comparing the current speed value with the expected input speed value, the PID control subroutine modifies the speed deviation, calculates the control current signal to the electric throttle device, and subsequently adjusts the driving speed of the vibratory roller, thereby completing one cycle of the automatic speed regulation process.

③ The automatic brake sub-process is categorized into three distinct braking scenarios: parking, work, and emergency. In the parking scenario, which occurs at the conclusion of the navigation task, the sub-process first retrieves the coordinates of the navigation target position. It then reads the positional data acquired via the RTK-GPS. By comparing the navigation target position with the current position, the system determines whether the position deviation falls within an acceptable error range. If it does, the current brake signal is transmitted to the electric brake device to execute the braking and parking tasks.

For acceleration and deceleration work brake conditions, when the throttle device is in a closed-valve state, the PID brake subroutine is invoked based on the rate of distance reduction between the roller and any obstacles. The PID program then adjusts and regulates the brake current signal, which is transmitted to the electric brake device, thus facilitating the adjustment of driving speed. In emergency braking situations, upon receiving an unexpected stop signal, the maximum current signal is immediately output to the brake device, ensuring a rapid halt and completing a cycle of the automatic driving brake process.

(2) Three-dimensional navigation map

In path planning and navigation, it is essential to establish a rational data model. This model employs object-oriented concepts and methodologies to abstract spatial points and roads into node and unit classes, allowing for comprehensive management of the entire model while facilitating separate oversight of each class.

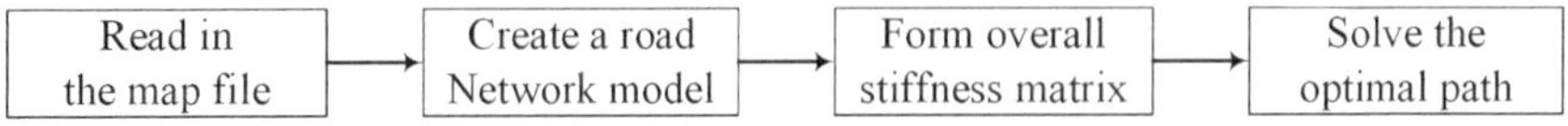

Fig. 8.3 Flowchart of the 3D navigation map data model

The design principles for the three-dimensional (3D) navigation map data model should adhere to a straightforward graphical structure, minimal redundancy, uncomplicated topological relationships, rapid spatial information queries, and an open data interface. The flowchart of the data model is depicted in Fig. 8.3.

① Import the map file, ensuring that it adheres to specific format requirements. These requirements include the output filename for the optimal path, the type of road network model, the domain type, the number of nodes, the number of units, the number of sections, the spatial coordinates of each node, the node number contained within each unit, and the relevant section information.
② Construct a road network model by saving the domain, node, unit, and section information from the read map file into their respective classes, including the Domain class, DofManager class, Element class, and CrossSection class.
③ Assemble a comprehensive stiffness matrix. Utilizing the node and element information from the road network model, construct a one-dimensional overall stiffness matrix with variable bandwidth storage, thereby establishing a foundational basis for path planning.
④ Determine the optimal path. Utilizing the node connection information contained within the one-dimensional global stiffness matrix with variable bandwidth storage, the Dijkstra algorithm is employed to address the shortest path problem in directed graphs. The A* algorithm is utilized for finding the shortest path in a static road network, while the D* algorithm is applied to resolve the shortest path in a dynamic environment. This approach effectively identifies the optimal path for both comprehensive and localized path planning, facilitating obstacle avoidance.

(3) Overall path planning

The Dijkstra, A*, and D* algorithms are all applicable for overall path planning. Among these, Dijkstra is a highly classical method; however, it lacks a heuristic component and necessitates substantial computational resources. In contrast, both A* and D* exhibit superior intelligence, enabling them to swiftly ascertain the optimal path. Furthermore, D* is a dynamic algorithm capable of replanning the optimal path in response to changes in road conditions.

To ascertain the most suitable algorithm for overall path planning, we utilize a complex road network model (path.in) to evaluate the accuracy and computational efficiency of the Dijkstra, A*, and D* algorithms. The road network model comprises a total of 2,500 nodes and 9,702 cells, with each internal node connected to eight surrounding nodes, maintaining an interval of 1 between horizontal and vertical nodes.

Table 8.1 Time utilised by the three algorithms for the path.in model and the optimal path calculation

Algorithm	Time (s)				Optimal path (m)			
	1–2500	2–2500	50–2451	100–2451	1–2500	2–2500	50–2451	100–2451
Dijkstra	0.26	0.25	0.26	0.25	69.3	68.9	69.3	68.9
A*	0.09	0.09	0.09	0.09	69.3	68.9	69.3	68.9
D*	0.09	0.09	0.09	0.09	69.3	68.9	69.3	68.9

Leveraging the road network model (path.in), we employ the three algorithms to compute the optimal distances for paths with various starting and ending points, specifically 1–2500, 2–2500, 50–2451, and 100–2451. The results of these optimal path calculations, along with the elapsed time for their resolution, are summarized in Table 8.1.

According to the data presented in Table 8.1, both A* and D* demonstrate a significant improvement in computational efficiency compared to Dijkstra. Moreover, due to D*'s capability to re-plan the path in response to changing road conditions, we ultimately select the D* algorithm to facilitate overall path planning for the URC system.

(4) Obstacle avoidance in local path planning

Building upon the preceding analysis, it is evident that the D* algorithm offers distinct advantages in overall path planning. When the roller encounters obstacles during its movement, the D* algorithm leverages previously planned information to re-evaluate the route, thereby avoiding redundant calculations of the same data and enhancing the efficiency of secondary planning.

Nevertheless, the D* algorithm is not without its shortcomings. As the roller inevitably navigates corners during its movement, D* searches in four or eight directions around the current position with each turn. Furthermore, being a length-first algorithm, the resultant path may include numerous extraneous turns. During the pathfinding process, the D* algorithm initially identifies obstacles before seeking alternative adjacent shortest feasible paths. As a result, the generated path may cling to obstacles, potentially leading to encounters with two adjacent barriers.

To address these issues and in light of previous research [58], we have opted for an improved D* algorithm based on the CA model as the local path planning algorithm for obstacle avoidance. This approach ensures that the roller navigates along an effective path, avoiding collisions while maintaining low operational costs.

8.1.2 RTK-GPS System

The RTK-GPS system is a positioning and navigation system founded on real-time kinematic technology, capable of achieving centimeter-level positioning accuracy.

This system has been employed in various studies for real-time data acquisition, precise positioning of construction machinery, and navigation at construction sites. Su et al. [59] proposed an enhanced boundary condition method to control the accuracy of the construction location system integrated with the RTK-GPS and RFID technologies. Robert et al. [60] utilized RTK-GPS technology to facilitate dynamic control and autonomous guidance for construction plants. Liu et al. [28] employed it to achieve real-time spatial visualization of compaction vehicles and quality assessments, as well as to visually indicate the locations of quality defects. Furthermore, our prior research [61] advocated for the use of roller-integrated acoustic wave detection techniques for rockfill materials based on RTK-GPS technology. In light of these studies, we have selected RTK-GPS technology to realize the high-precision positioning and navigation capabilities of the URC system.

The RTK principle entails that the GPS mobile station not only receives information from the base station but also from satellites. Subsequently, it performs differential processing on these two types of coordinate information to derive high-precision data. Once the optimal path between two points on the road network has been established, autonomous navigation can be executed using the ADT. The essence of automatic driving navigation lies in utilizing the coordinate information provided by the RTK-GPS to calculate the distance deviation, e_1, and the angular deviation, e_2, between the vehicle body and the optimal path in real-time. This information is then employed to compute the steering angle, facilitating automatic driving. The specific processes of navigation are outlined as follows:

① Transform the GPS latitude and longitude information into construction coordinates through coordinate conversion;
② Compute the rear body azimuth, α_h, using the left and right GPS coordinates;
③ Utilize the Controller Area Network (CAN) communication to obtain the actual steering angle, α_f, of the steering wheel, and subsequently calculate the azimuth of the front body, α_q (where $(\alpha_q = \alpha_h + \alpha_f)$);
④ Calculate the coordinates (x_q, y_q) of the front or rear wheel center by integrating the GPS data with the azimuth;
⑤ Calculate the distance deviation, e_1, and the angular deviation, e_2, between the vehicle body and the navigation route;
⑥ Calculate the steering angle, $t\alpha_f$, based on the deviations e_1 and e_2, thereby facilitating automatic driving navigation.

(1) Coordinate transformation

By configuring the GPS mobile station using the TestSerial software, data consisting solely of the GPGGA statement can be transmitted every 0.1 s. After extracting the longitude and latitude information from the GPGGA statement, geodetic coordinates can be derived through Gaussian projection. Subsequently, a four-parameter transformation (x_0, y_0, R_1, R_2) can convert the geodetic coordinates into local coordinates using the following conversion formula:

$$\begin{bmatrix} x \\ y \end{bmatrix} = \begin{bmatrix} x_0 \\ y_0 \end{bmatrix} + \begin{bmatrix} R_2 & R_1 \\ -R_1 & R_2 \end{bmatrix} \times \begin{bmatrix} x' \\ y' \end{bmatrix} \tag{8.1}$$

where x and y represent the coordinates of the target coordinate system, x' and y' denote the coordinates of the source coordinate system, while x_0 and y_0 are the translation parameters, and R_1 and R_2 are the rotation parameters.

(2) Vehicle body azimuth

The rear body azimuth, α_h, can be determined using the coordinates from the left and right sides of the GPS rover; the calculation formula is presented in Eq. (8.2). The front body azimuth can then be calculated from the rear body azimuth in conjunction with the actual steering angle of the steering wheel, as demonstrated in Eq. (8.3).

$$\alpha_h = \arcsin\left(\frac{|x_1 - x_r|}{\sqrt{(x_1 - x_r)^2 + (y_1 - y_r)^2}}\right) \tag{8.2}$$

$$\alpha_q = \alpha_h + \alpha_f \tag{8.3}$$

where (x_1, y_1, z_1) and (x_r, y_r, z_r) are the coordinates of the left and right sides of the GPS rover, respectively.

(3) Elevation correction formula

Given that the actual road possesses a certain slope, the left and right sides of the GPS rover exhibit specific tolerances, which will significantly impact the coordinates of the midpoint between the left and right sides of the GPS rover projected onto the horizontal plane. Therefore, the elevation must be corrected using the following formulas:

$$EC = \sqrt{(x_1 - x_r)^2 + (y_1 + y_r)^2} \tag{8.4}$$

$$x = \frac{x_1 + x_r}{2} - h\frac{z_1 - z_r}{EC}\sin\alpha_h \tag{8.5}$$

$$y = \frac{y_1 + y_r}{2} + h\frac{z_1 - z_r}{EC}\cos\alpha_h \tag{8.6}$$

where EC represents the elevation correction, (x_1, y_1, z_1) and (x_r, y_r, z_r) are the coordinates of the left and right sides of the GPS rover, respectively; h denotes the height of the vehicle body; and (x, y) are the coordinates of the midpoint between the left and right sides of the GPS rover.

(4) Front and rear wheel coordinate formulae

The body composition of the roller is illustrated in Fig. 8.4. By determining a (the horizontal distance from the center of the front wheel to the intermediate hinge), b

(the horizontal distance from the center of the rear wheel to the intermediate hinge), and c (the horizontal distance from the center of the rear wheel to the coordinates of the midpoint between the left and right sides of the GPS rover), the coordinates (x_{fq}, y_{fq}) of either the front wheel center or the rear wheel center can be derived. The front wheel coordinates can be calculated using Eqs. (8.7) and (8.8), while the rear wheel coordinates (x_{fq}, y_{fq}) can be obtained through Eqs. (8.9) and (8.10).

$$x_{fq} = x + (b + c)\cos\alpha_h + a\cos\alpha_h \tag{8.7}$$

$$y_{fq} = y + (b + c)\sin\alpha_h + a\sin\alpha_h \tag{8.8}$$

$$x_{rq} = x + c\cdot\cos\alpha_h \tag{8.9}$$

$$y_{rq} = y + c\cdot\sin\alpha_h \tag{8.10}$$

(5) Deviation and steering angle

Utilizing the coordinates of either the front wheel center or the rear wheel center, the distance deviation, e_1, and the angular deviation, e_2, of the vehicle body from the navigation route can be computed. Furthermore, the steering angle, t_{af}, can be determined in accordance with the actual steering angle, α_f, as well as the deviations e_1 and e_2 of the steering wheel. Additionally, the steering wheel is controlled in real-time based on t_{af}; thus, the roller navigates along a predetermined trajectory. Consequently, automatic driving is fully realized.

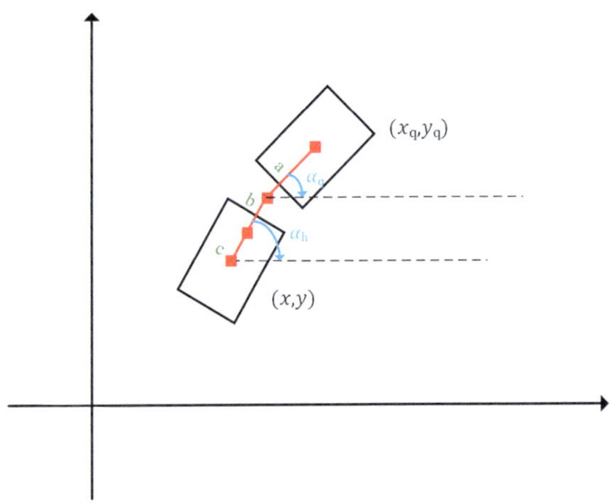

Fig. 8.4 Schematic of the vehicle body composition

8.1.3 Wireless Communication System

Communication between the remote monitoring center and the unmanned roller is primarily facilitated by a wireless communication system, employing a full-duplex serial mode. The wireless communication system (refer to Fig. 8.1) predominantly comprises the AP situated on the roller, the communication transit station, and the AP located at the remote monitoring center. The AP on the roller is tasked with transmitting information regarding compaction parameters (including the number of compaction passes, compaction trajectory, vibration frequency, elevation data, and driving speed), as well as video images of the working environment and status information (such as the distance between the roller and obstacles, along with the steering angle of the steering wheel) to the remote monitoring center. The communication transit station plays a crucial role in relaying this information to enable long-distance communication. Conversely, the AP at the remote monitoring center dispatches control commands (including ignition, power activation, throttle adjustment, braking, and shutdown), navigation line data, and remote-control commands in emergencies or special situations to the unmanned roller.

8.1.4 Remote Monitoring Centre

The remote monitoring center encompasses a real-time video surveillance (VS) system for the construction site, a remote monitoring system for the roller, and a visualized management system for compaction quality information. Adopting a client/server model, the remote monitoring center employs a four-layer architecture for data communication with controlled objects. This structure comprises the display layer, business logic layer, data layer, and remote-control layer, as illustrated in Fig. 8.5. Users access the application server via a self-developed monitoring platform client. As the business logic processing occurs on the server side and data processing is managed by the database server, both long-distance and local data access can be facilitated through the wireless network.

(1) Real-time video surveillance system

The real-time video surveillance system for the construction site primarily facilitates the monitoring of working conditions, including the material truck, roller, and bulldozer throughout the entire filling and rolling construction area, as well as the overall state of the construction site. This system (refer to Fig. 8.6) consists mainly of the server, industrial computer, camera, and wireless transmission station. The server, in conjunction with the industrial computer and camera, employs wireless communication to transmit data. The camera relays live, real-time images to the server, while the industrial computer provides real-time information regarding the working conditions of the material truck, roller, and bulldozer to the server.

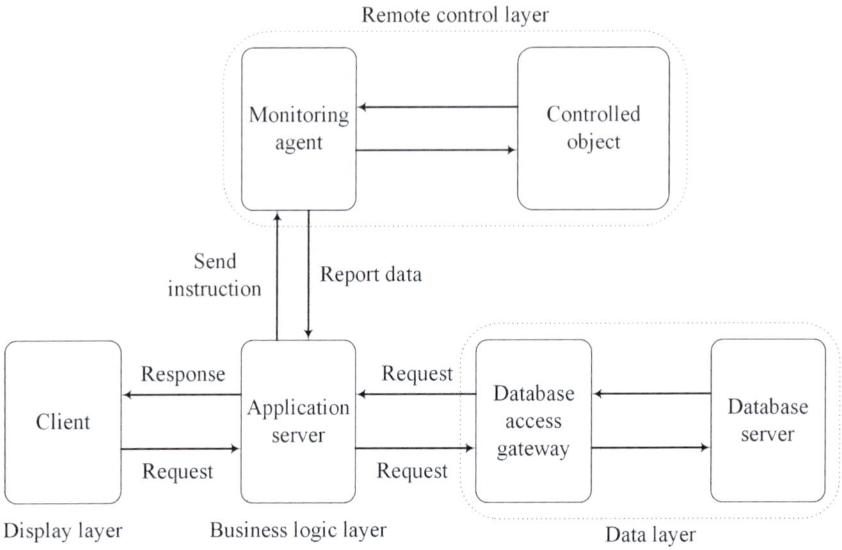

Fig. 8.5 Schematic diagram of remote monitoring centre

Fig. 8.6 Schematic of the real-time video surveillance system

(2) Remote monitoring system for the roller

The remote monitoring system for the roller primarily consists of the control computer and a wireless communication antenna. The control computer houses a monitoring program, a 3D digital model of the construction project, a construction schedule program, and a navigation planning program. The monitoring program is capable of simultaneously commanding and controlling multiple rollers during rolling operations. The flowchart illustrating the remote monitoring system for the roller is depicted in Fig. 8.7.

As illustrated in the figure, upon activation of the remote monitoring system for the roller, the monitoring program is first initialized to display the status of the monitoring equipment system and the monitored information of the rollers. Next, the monitoring program successively issues a power-on command remotely, following the order number of the roller. Subsequently, the monitoring program establishes a wireless communication connection with the on-board automatic control system of the rollers. The remote monitoring system then receives and displays video transmitted by the on-board automatic control system, along with spatial position information obtained through RTK-GPS technology, data from various sensors, and any emergency processing requests within a designated display frame. Following this, based on the real-time conditions of the rollers, the monitoring program remotely sends ignition and start commands to the rollers. Finally, the monitoring program transitions into the remote 3D digital model navigation control or human–computer interaction remote subroutine.

(3) Visualised management system for compaction quality information

We utilized AutoCAD, LabVIEW, and other software to establish a 3D digital model of the earth-rock dam, subsequently developing a compaction quality information management system based on a client/server architecture. Additionally, we integrated the 3D model into the server database, creating data table information based on rolling time and compaction quality metrics. This enabled the visualization of compaction quality for the rockfill materials along with a subsequent evaluation function, and facilitated the dynamic updating of compaction quality information based on the database file.

8.1.5 Case Study

(1) Testing site and materials

The earth-rock dam of the Qianping Reservoir, featuring a clay core, is situated in Henan Province, China. This dam, with a height of 90.3 m, encompasses six construction zones and utilizes four distinct types of dam filling materials, boasting a total earthwork filling capacity of 1.3×10^7 m^3. Our focus was on the rockfill materials

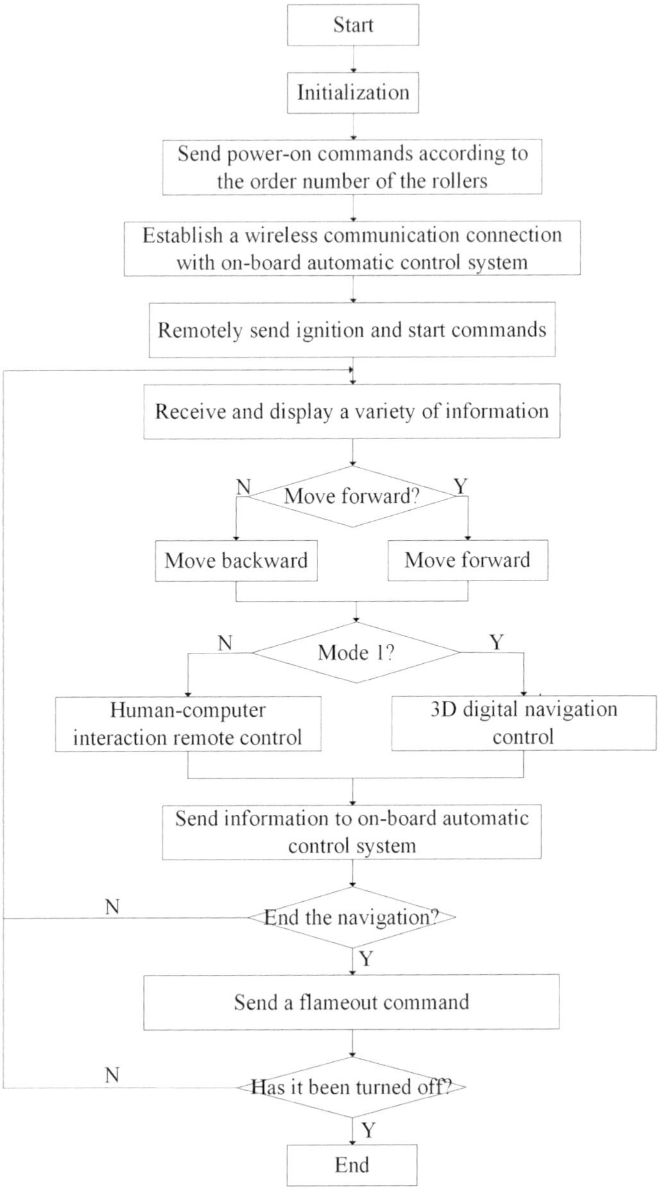

Fig. 8.7 Flowchart of the remote monitoring system for the roller

found in the major rockfill areas located at the downstream section of the dam. These rockfill materials, based on conservative estimates, have a maximum particle size of 1000 mm, with the proportion of particles smaller than 5 mm controlled to within 15%. The average moisture content is recorded at 4.8%, derived from the average results of 114 test groups. For the rockfill materials, the designed standard for the number of compaction passes is set at eight.

(2) Testing program

To assess the actual effectiveness of the URC system proposed in this section, several sets of on-site experiments were conducted. In this analysis, areas A and B were selected as focal points for evaluating the system's operational results. Figure 8.8 presents a photograph of the URC system test conducted in the downstream rockfill zone, while the experimental scheme is summarized in Table 8.2. As illustrated in Fig. 8.8, areas A and B are situated in the rockfill zone at a measured distance from the right dam abutment, with several unmanned rollers, including unmanned rollers 1, 2, and 3, participating throughout the entire test plan.

The vibratory rollers utilized in this field test were equipped with smooth drums specifically designed for the compaction of rockfill materials. In accordance with the design requirements, we conducted a total of 10 compaction operations, comprising two static and eight vibratory compaction processes (refer to Table 8.2) on the rockfill materials. Throughout the entirety of the test period, area A was filled with 9 layers, while area B was filled with 14 layers. With the implementation of the URC system and its associated devices at the Qianping project, we were able to gather data on compaction parameters, usage frequency, and the fault rate of the unmanned roller during compaction. Figure 8.9 illustrates the user interface of the URC system for this project. We obtained a comprehensive set of field-sampling

Fig. 8.8 Photograph of the URC system testing at the rockfill zone

Table 8.2 Test scheme for the URC system at the rockfill zone located downstream

Area	Date	Range (m)	Number of layers	Pile number	Number of compaction passes
A	7/11/2017–25/11/2017	0 + 200–0 + 400	9	A/B/C/D/E/F/G/H/I	2/8
B	23/9/2017–24/11/2017	0 + 400–0 + 680	14	a/b/c/d/e/f/g/h/i/j1/j2/k/l/m/n/o/p/q/r/s/t/u/v/w/x/y/z/aa/bb/cc/dd	2/8

data, including compaction parameters (such as the number of compaction passes, compaction trajectory, vibration frequency, lift thickness, and driving speed), the usage frequency of the unmanned roller, the fault rate of the unmanned roller, and the elevation of the compacted layers. This data was utilized for the actual effect analysis of the unmanned rolling compaction system proposed in this section.

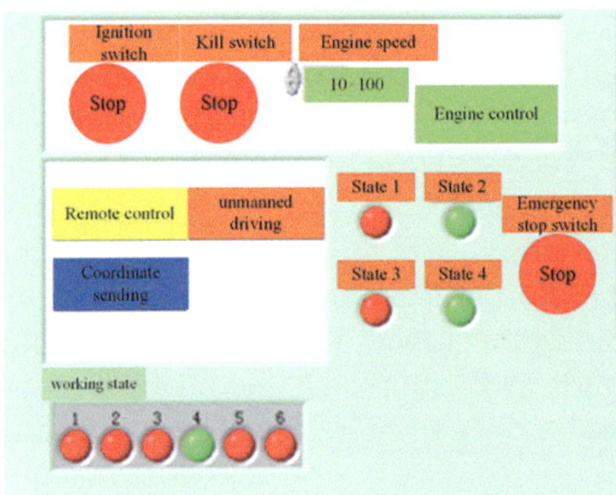

Remote monitoring system for the roller

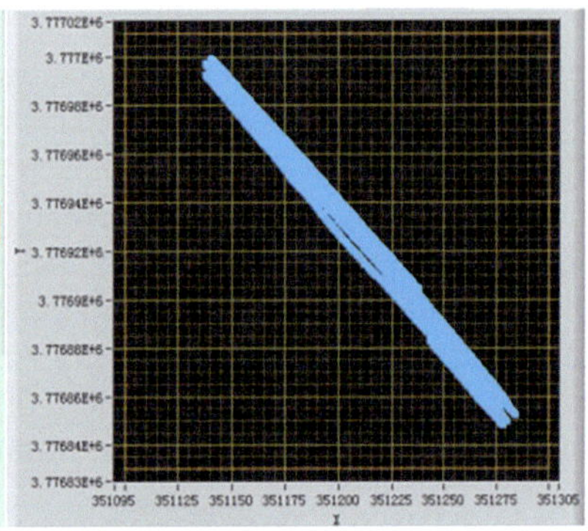

Real-time rolling trajectory

Real-time video surveillance system

Fig. 8.9 User interface of the URC system consisting of remote monitoring system, real-time rolling trajectory interface which are part of on-board monitoring system, and real-time video surveillance system for Qianping project

8.1.6 Result Analysis

(1) Result analysis of the compaction parameters

In accordance with the test plan outlined in Table 8.2, this section presents the statistically analyzed results concerning the compaction trajectory and the number of compaction passes for the filling construction areas A and B. Figure 8.10 depicts the qualified rate of the number of compaction passes for area A from November 7, 2017, to November 25, 2017, while Fig. 8.11 illustrates the same for area B from September 23, 2017, to November 24, 2017. The statistical results indicate that when the URC system is employed to fill each layer in both areas, the qualified rates for the number of compaction passes in filling areas with varying pile numbers consistently exceed 95%, with many surpassing 98%. The control accuracy of the compaction trajectory is maintained within 5 cm, while the control accuracy for vibration frequency and driving speed is 0.5 Hz and 0.05 m/s, respectively. Furthermore, the amplitude of the modified roller aligns with the parameters of the original roller and can be regarded as a constant. In certain areas, the number of compaction passes may not have met the required standards due to instances of several unmanned rollers being overloaded during the initial operation of the URC system. Consequently, this may result from various factors, including failure to adhere to work schedules, inadequate on-site management of the unmanned rolling area, a lack of coordination between the management personnel of the remote monitoring center and those present on-site, disruptions in wireless communication caused by inclement weather, or other uncontrollable factors occurring on-site.

According to the specifications for existing earth-rock dam construction [62], compaction parameters significantly influence the compaction quality of earth-rock dams. To visually demonstrate the control effectiveness of the URC system on compaction quality, this section presents the compaction parameter monitoring diagram for area A from November 15, 2017, as depicted in Fig. 8.12. As observed from Figs. 8.10, 8.11, and 8.12, the URC system exhibits clear advantages over manual operation. Specifically, the URC system can accurately regulate the compaction trajectory, effectively addressing issues such as unrolled, cross-rolling, and repeated-rolling, thereby ensuring a high qualified rate of compaction passes. Additionally, the URC system can precisely control the compaction trajectory, vibration frequency, and driving speed, significantly minimizing the influence of human operators on the compaction quality of the rockfill materials. The URC system also facilitates real-time visualization of compaction parameters, allowing for comprehensive monitoring of rolling compaction quality under all weather conditions. Furthermore, the unmanned characteristics of the URC system render it exceptionally suitable for rolling construction in hazardous environments, nighttime operations, or extreme conditions, such as those experienced in plateau areas. Based on this analysis, the URC system effectively enhances the control of compaction quality for rockfill materials.

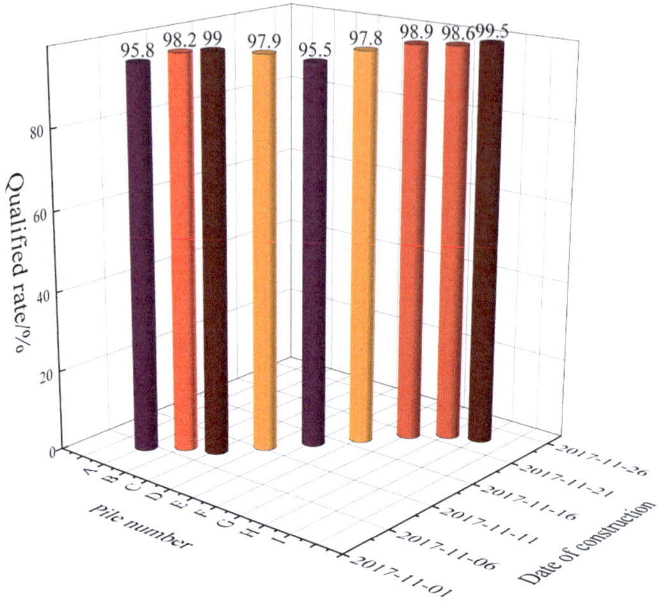

Fig. 8.10 Qualified rate of the number of compaction passes for area A

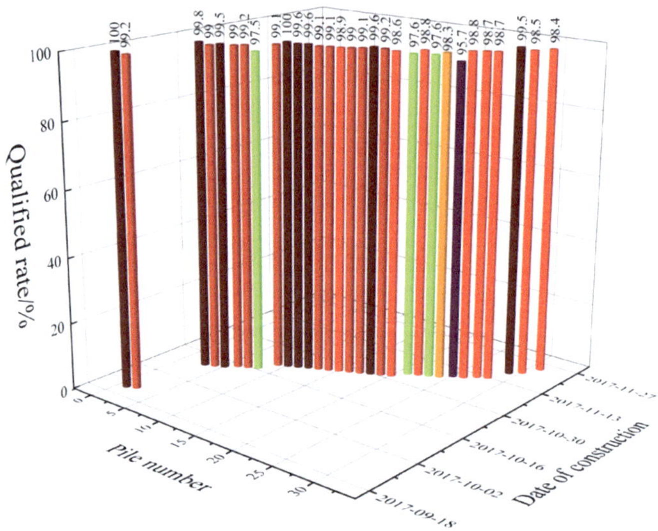

Fig. 8.11 Qualified rate of the number of compaction passes for area B

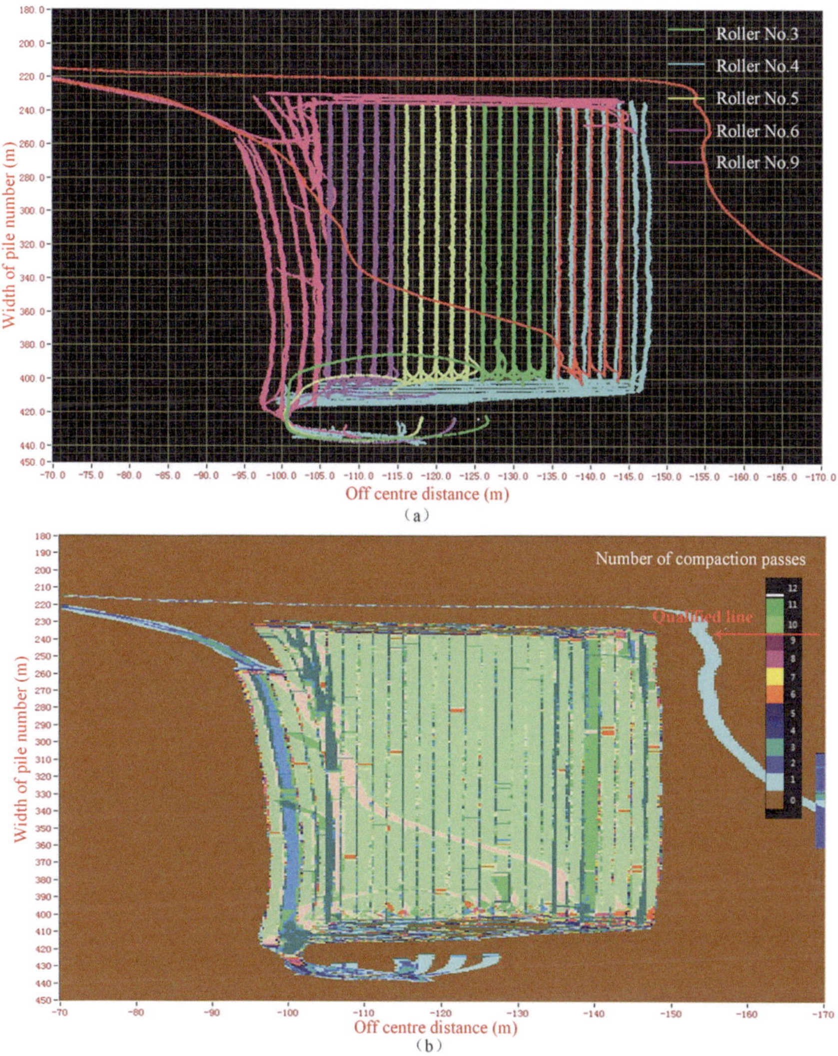

Fig. 8.12 Compaction parameter monitoring diagram of area A on 15 November 2017: **a** Compaction trajectory; **b** the number of compaction passes

(2) Statistics and analysis of the use frequency of unmanned rollers

This section outlines the usage frequency of unmanned and manned rollers during the filling construction of areas A and B, respectively. Statistical results reveal that unmanned rollers are utilized at a significantly higher frequency than their manned counterparts. Throughout the entirety of the testing process, unmanned rollers primarily undertook the rolling tasks in the main construction area (refer to Fig. 8.12), while manned rollers were tasked with rolling at corners and seams (also shown in Fig. 8.12). In comparison to manned rollers, the URC system, which utilizes unmanned rollers, plays a vital role in the filling process of rockfill materials situated downstream of the earth-rock dam. This analysis indicates that the URC system effectively liberates roller operators from complex tasks, thereby greatly diminishing the impact of human factors on compaction quality and significantly enhancing the efficiency of filling construction.

(3) Statistics and analysis of the fault condition of unmanned rollers

This section presents a comprehensive analysis of the fault conditions encountered by unmanned rollers during the rolling construction in areas A and B. Statistical results indicate that the number of faults associated with unmanned rollers is significantly lower than the overall fault count, reflecting the high reliability of these unmanned systems. Based on the statistical analysis, the fault rates of unmanned rollers in areas A and B during construction are approximately 8.8% and 7.9%, respectively. These findings suggest that the URC system, built upon unmanned rollers, exhibits characteristics of stable operation, strong adaptability, convenient maintenance, and exceptional safety.

(4) Statistics and analysis of lift thickness of compacted layer

① Discussion regarding the accuracy of RTK-GPS equipment

Given that the construction site is relatively expansive and the GPS base station is situated on an open platform atop the left bank abutment, not far from the construction area (refer to Fig. 8.8), communication between the base station and the construction site is unobstructed by any obstructions or trees. We selected a set of location points within the test site and verified the accuracy of the RTK-GPS equipment by comparing its measurements with those obtained from static GPS and leveling instruments. A total of 32 test points were utilized for the accuracy analysis of the RTK-GPS equipment. Table 8.3 presents the test results concerning the accuracy of the RTK-GPS equipment, while the relative error of the test results obtained from both RTK-GPS and conventional methods is illustrated in Fig. 8.13.

The data presented in the table indicates that the RTK-GPS equipment exhibits greater accuracy in the horizontal plane (x, y) compared to its elevation measurements z. The average absolute error for the plane is less than 1 cm, while the average absolute

Table 8.3 Test results of RTK-GPS equipment accuracy

Coordinate axis	Minimum absolute error/cm	Maximum absolute error/cm	Average absolute error/cm	Minimum relative error/%	Maximum relative error/%	Average relative error/%
x	5.3E-08	0.4	0.2	6.7E-8	0.5	0.2
y	7.6E-09	2.0	0.7	2E-8	3.1	1.3
z	1.3	3.0	2.3	3.8	8.9	6.9

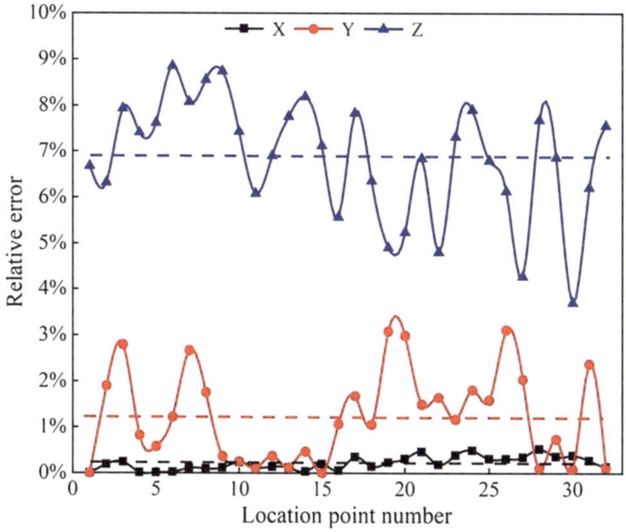

Fig. 8.13 Relative error of the test results obtained from RTK-GPS and conventional methods

error for elevation is under 2.5 cm. As illustrated in Fig. 8.13, the average relative error increases progressively from the x direction to the y direction and subsequently to the z direction. The average relative error for the plane is less than 1.5%, whereas the average relative error for elevation is less than 7%. Based on this analysis, the results from multiple comparison experiments confirm that the measurement accuracy of the RTK-GPS equipment sufficiently meets the operational requirements of the URC system and on-site construction.

② Statistics and analysis of the elevation of the compacted layer

Area A comprises 9 layers of rockfill materials that were filled between November 7, 2017, and November 25, 2017, while area B consists of 14 layers filled from September 23, 2017, to November 24, 2017. Utilizing the monitoring data obtained from the URC system in areas A and B, this section presents an evaluation of the elevation data for the compacted layers. The results demonstrate that the URC system can maintain the elevation differences at each position of the compacted layer within 15 cm, given that the design standard for filling layer thickness is 80 cm. The monitoring data from areas A and B on November 12, 2017, are selected for illustration purposes to more effectively convey the control impact on elevation, as depicted in Fig. 8.14. The statistical results, along with Fig. 8.14, indicate that the URC system successfully ensures uniform settlement of the compacted layers, thereby regulating the compaction quality of earth-rock dams through precise control of lift thickness.

③ Uncertainty in the computed average lift thickness

To illustrate the uncertainty in the computed average lift thickness, area E in Fig. 8.14 is designated as the calculated test region. Area E is subdivided into 3360 uniform

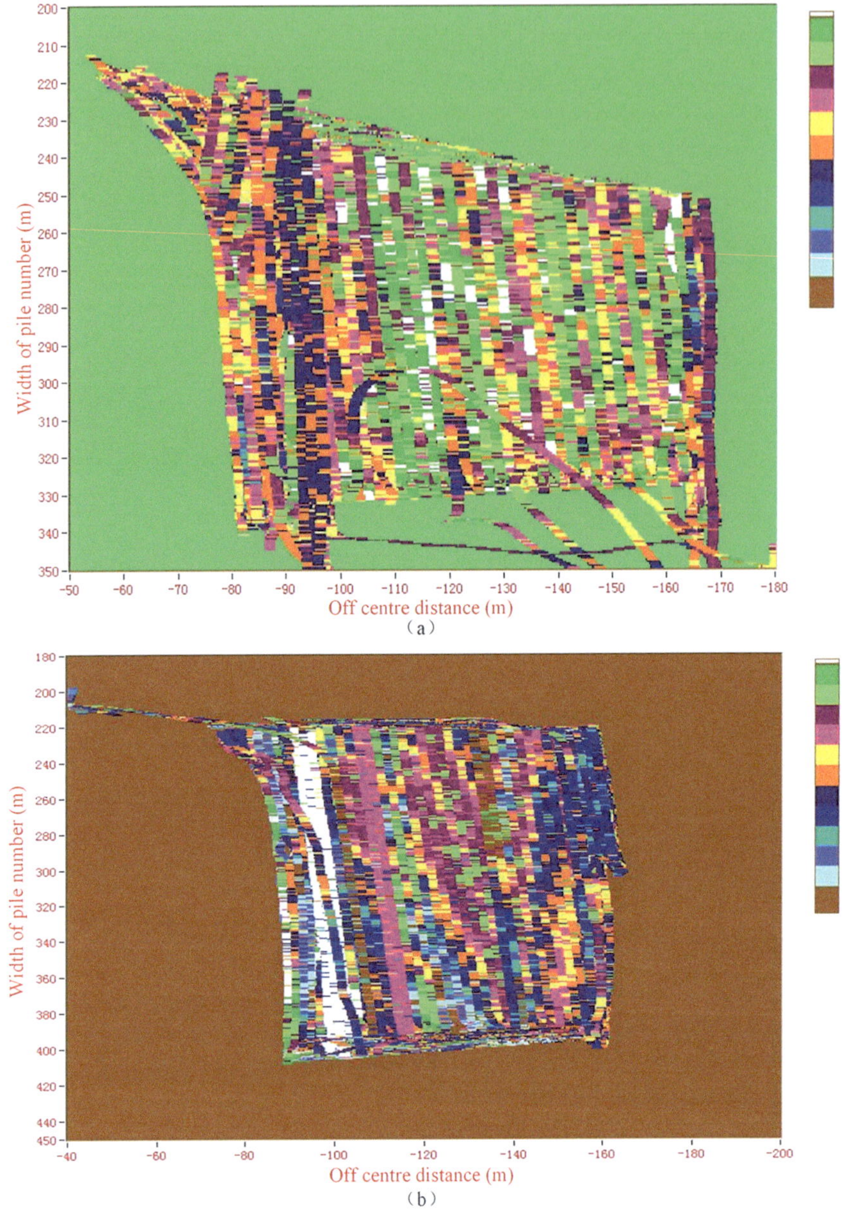

Fig. 8.14 Elevation information of the work surface after completion of the rolling operations: **a** area A and layer 4; **b** area B and layer 12

Table 8.4 Lift thickness and the uncertainty of layer 12 in area E

Number of grids	Designed thickness/cm	Maximum lift thickness/cm	Minimum lift thickness/cm	Average lift thickness/cm	Uncertainty/cm
3360	80	81	66	76	0.04

grids, each measuring 2 m by 1 m. Following the rolling operations conducted with the URC system according to the specified number of compaction passes, this section undertakes a statistical analysis of the lift thickness data from these grid points. The calculation of the lift thickness, denoted as H_k, in grid k is represented through Eqs. (8.11) and (8.12) [3]. The lift thickness h_k, t in grid k at time t is determined using Eq. (8.11).

$$h_{k,t} = z_t - Z_{k,0} \tag{8.11}$$

hence, the lift thickness in the k th gird can be calculated using Eq. (8.12).

$$H_k = \max_t \{h_{k,t}\} = Z_{k,1} - Z_{k,0} \tag{8.12}$$

z_t represents the elevation at time t; $z_{k,o}$ denotes the initial elevation of grid k; and $z_{k,o}$ aligns with the design thickness of the filling layer. The elevation of the grid following compaction is captured in Eq. (8.13).

$$Z_{k,1} = \max_t \{z_t | (x_t, y_t) \in G_k\} \tag{8.13}$$

where, $(x_t, y_t)\varepsilon G_k$means that in the t th time interval, the compaction trajectory point (x_t, y_t) is in the k th gird G_k.

Utilizing the aforementioned formula and the Type A evaluation method of standard uncertainty, the required lift thickness, maximum lift thickness, minimum lift thickness, average lift thickness, and the uncertainty in the computed average lift thickness for the filling layer of area E are summarized in Table 8.4. As evidenced by Table 8.4, the measurement results exhibit high quality, indicating that the lift thickness can be effectively monitored throughout the construction process.

8.2 Intelligent Rolling Compaction System

In recent years, the rapid advancement of China's hydropower industry has led to an increasing demand for the application of artificial intelligence technology in quality control (QC) for earth-rock dam construction. To enhance the compaction quality of dam materials during such construction, Liu et al. [63] developed an automatic control and real-time monitoring system specifically focused on the QC of rockfill materials in earth-rock dams. Meanwhile, Ranasinghe et al. [64, 65] introduced a novel predictive tool utilizing artificial neural networks (ANNs) to forecast the effectiveness of rolling compaction and to enhance ground compaction capabilities. Imran et al. [66] created an ANN-based intelligent control system for estimating the stiffness of the subgrade during construction. Taghavifar et al. [67] employed a hybridized ANN and imperialist competitive algorithm to predict soil compaction outcomes. Isik and Ozden [68], along with Günaydin [69], developed ANN prediction models for estimating soil compaction parameters. Furthermore, Commuri et al. [44] designed a neural network-based intelligent control analyzer for assessing the compaction quality of hot asphalt mixes. Lastly, Wang et al. [84] sought to manage the assessment and control of compaction quality in earth-rock dams by developing a three-dimensional, real-time monitoring system that integrates a smart bacteria-foraging algorithm with customized kernel-support vector regression and enhanced probabilistic neural networks.

Despite the significant advancements and widespread application of AI technology, numerous pressing challenges remain to be addressed. Firstly, the current CMS-based framework of material-machine-information-human decisions can result in control errors, with decision-makers exerting a considerable influence on the management of compaction quality, thereby introducing uncertainties into the construction schedule. Secondly, the uniformity of compaction in filling materials is susceptible to human factors. Thirdly, in terms of QC, the non-circular closed-loop characteristics of detection, analysis, decision-making, and control contribute to a lack of intelligence, leading to low efficiency, accuracy, and performance. Lastly, the application of AI technology in the realm of construction QC for earth-rock dams faces numerous obstacles, including technical realization, engineering application, and adherence to construction standards.

In response to these challenges, this section introduces an IRC method and system designed for the QC of earth-rock dams during rolling. By employing a proposed system integration approach featuring a three-layer structure, the IRC system is realized through the integration of the CAWD system, URC system, IDM system, and a remote monitoring center. This innovative IRC method and system effectively mitigate the influence of human factors, resolve issues of under- or over-compaction, ensure the uniformity of compaction in filling materials, meet construction requirements under various weather conditions or extreme circumstances, enhance compaction quality and construction efficiency, and propel rolling construction technology towards greater intelligence.

Furthermore, the dynamic optimization of compaction parameters serves as an effective approach to enhance both the quality and efficiency of earthwork compaction [71–73]. Recently, research has predominantly concentrated on two models. The first type employs a compaction dynamics model that simulates the vibration compaction process to optimize compaction parameters. The second type utilizes a prediction model for compaction parameters, grounded in AI algorithms, to adaptively optimize and adjust these parameters throughout the rolling process [73, 74].

The fundamental concept of a compaction dynamics model is to accurately simulate the vibration compaction process, calculating the dynamic response of soil-roller interactions during this process, and determining the elastic–plastic deformation of the filling materials, along with the influence of various parameters on the compaction outcome and other relevant factors. Based on these simulations, the optimization of compaction parameters can be pursued [75–77]. The origins of research related to the compaction dynamics model can be traced back to the 1950s [78]. Drawing upon the theory of complete elastic vibration, Yoo and Selig [74, 79] introduced a two-degrees-of-freedom (2DOF), linear, lumped-parameter, spring-damping system model to represent the vibration system of vibratory rollers. Following a dynamic theoretical analysis of the vibratory compaction system, Machet and Morel [80] proposed their own 2DOF dynamics model. Taking into account the viscoplastic and elastoplastic deformation characteristics of soil, Pietzsch and Poppy [81] established a four-degrees-of-freedom dynamics model for the frame-vibration wheel-soil system. Additionally, Beainy et al. [72] developed a viscoelastic-plastic model for asphalt pavement capable of accurately reflecting the dynamic properties of the asphalt mat during compaction. Acknowledging that the equivalent dynamic shear modulus diminishes and the equivalent damping ratio increases under conditions of high shear strain, Wersäll et al. [82] proposed a nonlinear 2DOF dynamics model that incorporates strain softening.

Due to the dynamic interactions between the roller and the soil during vibration compaction, the working parameters of the roller in the model are selected and optimized immediately following the establishment of an appropriate compaction dynamics model for the filling materials. To this end, numerous scholars have conducted extensive research on the selection and optimization of these working parameters. In the initial stages, investigations focused on the relationship between the vibration frequency of the roller and the compactness of the soil, emphasizing the selection of an optimal vibration frequency for the filling materials [83].

Pietzsch and Poppy [81] utilized the established compaction dynamics model to conduct a theoretical analysis of the compaction effect and vibration characteristics of the vibratory roller, determining the optimal working parameters based on field test results. Employing a method to ascertain vibration parameters that can be automatically adjusted alongside simulation test results, Ding and Zhang [84] posited that the optimal vibration parameters should be contingent upon the characteristics of the filling materials. Zhou [85] explored the dynamics model for a vibrating roller, identifying ideal operating parameter controls and optimizing these parameters to achieve the best compaction results. Utilizing the 2DOF dynamics model, Zhang [86] executed a dynamic analysis and calculation of the YZ30 full hydraulic, double-drive, self-propelled vibratory roller to determine the vibration parameters for the model. These parameters were subsequently simulated and optimized using Matlab, resulting in the identification of the optimal excitation frequency and amplitude values of the roller under varying working conditions.

Furthermore, several researchers have employed finite element and discrete element methods to simulate the compaction process, aiming to determine and optimize the working parameters of the roller [87, 88]. However, this approach utilizing the compaction dynamics model is fraught with several limitations. For instance, the mechanisms of vibration compaction are intricate and challenging to accurately simulate within a model. Many existing models inadequately consider the plastic deformation that occurs during the compaction process and fail to account for variations in model parameters, leading to an inability to accurately depict the vibration compaction process. Additionally, the compaction dynamics modeling process is complex and protracted, exhibiting poor timeliness that does not meet the demands of large-scale rapid earthwork construction, particularly in the context of earth-rock dams. Moreover, current models primarily operate at a qualitative analysis level, with limited theoretical and applied research on intelligent decision-making concerning compaction parameters, highlighting the necessity for further investigation.

Other models have sought to adaptively adjust compaction parameters through a predictive model based on AI algorithms, thereby optimizing the compaction process in a different manner. Beainy et al. [66, 89] and Commuri et al. [44, 90] developed a neural network-based intelligent control analyzer for estimating the compaction quality of hot asphalt mixes, enabling operators to modify the rolling strategy according to these estimations to avert both over-compaction and under-compaction. After identifying the primary influencing factors on compaction parameters, Li [91] employed a genetic algorithm (GA) to predict the working parameters of the roller, enhancing the backpropagation GA to automatically adjust the proportional-integral-derivative (PID) control parameters in real time. Zhang et al. [92] advanced the compaction process and the intelligence of the roller's automatic control by constructing an intelligent control system based on the CAN bus and utilizing a GA method to optimize the fuzzy control strategy within the controller. Wang et al. [70] introduced a smart bacteria-foraging algorithm combined with customized kernel-support vector regression and enhanced probabilistic neural networks for compaction quality assessment and feedback control in earth-rock dams. Isik and Ozden [68] presented ANN prediction models for estimating compaction parameters applicable to both coarse- and fine-grained soils. Additionally, Xu and Chang [93] developed an innovative integrated material-machine-information and human-decision system that provides feedback to construction personnel, facilitating the optimization of compaction efforts and enhancing the adaptive control of pavement material density.

Despite these advancements in optimizing compaction parameters and QC of earthwork, several challenges remain unresolved. Presently, relevant research predominantly focuses on road construction, with a notable deficiency in the application of AI algorithms for optimizing the compaction process in earth-rock dams. The system design continues to operate within the material-machine-information and human-decision framework, thereby failing to mitigate the impacts of human factors on the optimization of the compaction process. Moreover, there has been a lack of effective integration and fusion studies aimed at the continuous detection of compaction quality and the precise control of compaction parameters. The exploration of AI technology for machine learning decision-making regarding compaction parameters, as well as the optimization of the compaction process, remains essential for enhancing the intelligent construction of earth-rock dams. This paper proposes an intelligent decision-making (IDM) system founded on a global compaction process-dynamic optimization method (GCPDOM) tailored for earth-rock dams, addressing these challenges and the shortcomings of existing QC methodologies. This system is designed to effectively control the compaction quality of earth-rock dams and to further the exploration of intelligent construction practices in this domain.

The framework and architecture of the IRC system are illustrated in Fig. 8.15. Within the framework of intelligent material-machine-information-machine decision-making, the roller effectively compacts the fill materials, which respond to compaction through the evolution of material properties, including stiffness, deformation, relative density, and compactness. The information management module receives and statistically displays data concerning the CCI, specifically the *SCV* measurements, as well as the machine's operational state parameters, such as vibration frequency and driving speed, alongside other compaction metrics, including the number of compaction cycles.

This pertinent information is then input into the predictive model within the intelligent decision module, facilitating the determination of the roller's working parameters for the subsequent rolling cycle. Thereafter, the target values are relayed back to the roller to adaptively adjust the compaction efforts and control material properties. Upon the completion of construction, the material properties—such as relative density or compactness—covering the entirety of the construction area can be predicted using regression models and geostatistical methods for the benefit of quality inspectors and contractors. Moreover, an in-depth analysis of the compaction uniformity of the filling materials and the reliability of the IRC system can be conducted, providing valuable insights to guide the enhancement of subsequent compaction operations.

As depicted in Fig. 8.15, the proposed IRC system is fundamentally comprised of the CAWD, IDM, and URC systems, along with a remote monitoring center. The

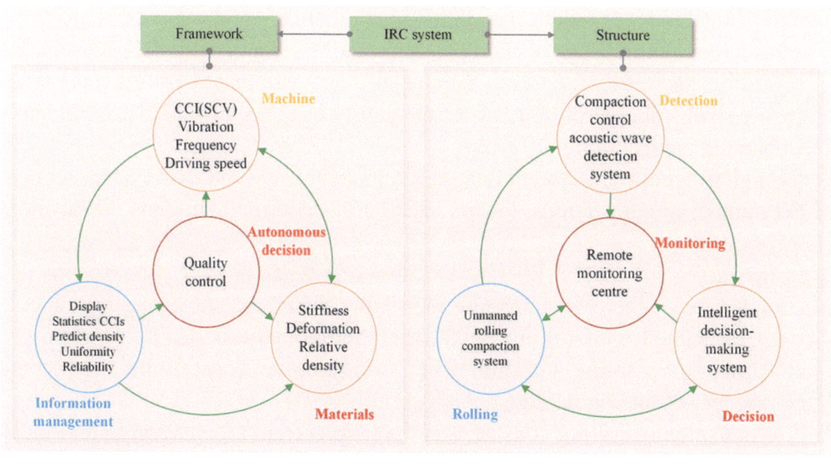

Fig. 8.15 Framework and structure of the IRC system

wireless communication system facilitates interactions between the subsystems and between these subsystems and the central monitoring facility.

The CAWD system is responsible for the continuous, real-time collection of compaction quality information characterized by the CCI for the fill material. The IDM system receives data from the CAWD, as well as information regarding compaction parameters obtained through monitoring the rolling operations. This information is then input into the GCPDOM to ascertain the roller's working parameters for the forthcoming rolling cycle.

Subsequently, the URC system acquires these working parameters from the IDM system, along with operational planning information, the navigation path, and scheduling details provided by the remote monitoring center, thereby executing the rolling construction task utilizing the modified roller. Upon the conclusion of each rolling cycle, the IRC system reiterates the information-gathering process and initiates a new detection cycle, thereby establishing a closed-loop system.

The remote monitoring center receives information regarding compaction quality, the roller's working parameters, real-time compaction data, and live monitoring information pertaining to the in situ construction. Moreover, the remote monitoring center, structured upon a triangular architecture, stores rolling construction data in real time and provides instantaneous communication services for the construction site. This paper proposes a system integration methodology featuring a three-layer structure based on fieldbus, Ethernet, and Internet protocols. Through this integration approach, the IRC system is effectively realized by amalgamating the CAWD, URC, IDM systems, and the remote monitoring center.

The IRC system delineated in this section offers numerous advantages:

① In comparison to existing technologies and the traditional framework of human decision-making, it embodies characteristics of intelligence, high efficiency, real-time information processing, and superior control accuracy, effectively mitigating the influence of human factors during rolling construction.
② the system boasts excellent scalability, high reliability, and ease of maintenance.
③ continuous monitoring of compaction quality, intelligent decision-making regarding the roller's working parameters, and automated rolling operations ensure stringent control of compaction quality in earth-rock dams, significantly enhancing compaction uniformity.
④ through the proposed system integration methodology, the IRC system reduces the number of connections, lowers installation costs, and shortens the development cycle.
⑤ furthermore, this system liberates construction workers from monotonous and complex tasks, thereby enhancing their labor value and substantially reducing risks associated with rolling operations. Additionally, it facilitates improved construction flexibility and diminishes working time, constituting an essential component of intelligent dam construction [94, 95].
⑥ The transition of rolling construction operations from automation to an intelligent level presents substantial benefits for earth-rock dams and can serve as a

reference for establishing construction standards within the realm of intelligent dam construction.

⑦ Collectively, these advantages enhance both the compaction quality and overall construction efficiency of the earth-rock dam, thereby bolstering the strength and stability of the entire project.

8.2.1 Basic Components of the System

(1) CAWD system

The CAWD system mainly includes an on-board controller with a built-in data acquisition and processing program, GPS receiver and antennas, GPS satellites, GPS base station, AP and antenna, power unit, acoustic sensor, remote wireless device, and a remote monitoring centre, as shown in Fig. 8.16. As an essential part of the software module in the embedded system, the built-in program is responsible for synchronously collecting and processing the acoustic and GPS signals and transmitting the compaction quality information to the remote monitoring centre using the wireless device. Moreover, it also receives commands sent by the remote monitoring centre in real-time. The remote monitoring centre, based on triangle architecture, is in charge of displaying the real-time compaction quality information of the construction area and comprehensively storing the in situ information and the spatial position of the roller.

The signal conditioning module of the onboard controller is utilized to process the time-domain signals from the acoustic field, as provided by the acoustic sensor,

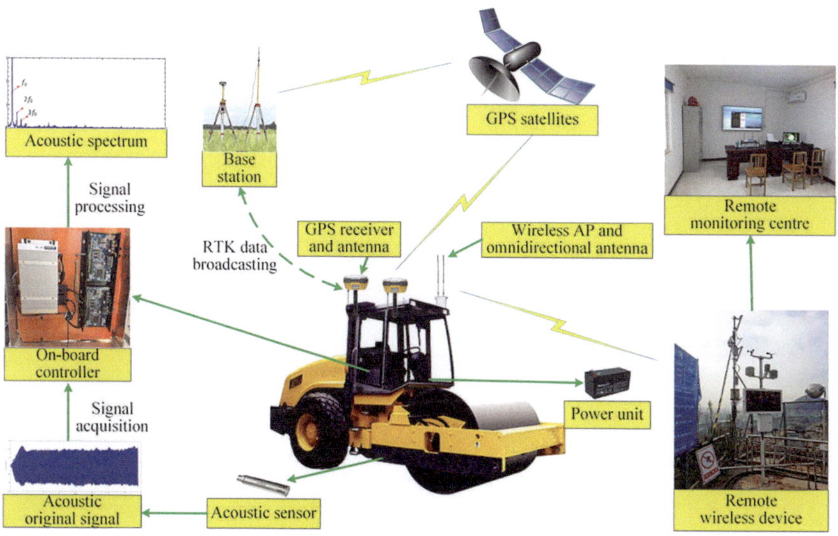

Fig. 8.16 Diagram of the CAWD system

alongside the spatial signals indicating the roller's position, as supplied by the GPS receiver, thereby facilitating effective signal transmission. The data acquisition module is designed to synchronously collect the signals relayed by the signal conditioning module and to relay these collected signals to the analysis processing module without delay.

For the roller's spatial signals, the analysis processing module employs a difference algorithm to compute the current position coordinates of the vibratory roller with centimeter-level accuracy. Leveraging this information, the compaction parameters—including lift thickness, the number of compaction cycles, and driving speed—can be calculated in real time, while the compaction trajectory is recorded instantaneously.

For the time-domain signal from the acoustic field, the analysis processing module sequentially filters, analyzes the spectrum, and conducts logarithmic processing on the signal in order to derive the second harmonic amplitude of the effective acoustic signal. This is subsequently calibrated according to the type of filling material to obtain the CCI, specifically the SCV. Following this, a CCI data analysis methodology is proposed, as illustrated in Fig. 8.17. Prior to further processing, the CCI measurements are transmitted in real time to the IDM system for utilization.

A correlation analysis of the SCV, gradation, moisture content, and relative density of the filling materials is conducted. Should a correlation be identified, the relevant variables may be incorporated into a statistical regression model. In addition to univariate linear regression, multivariate linear or nonlinear regressions between the CCI measurements (SCV) and in situ measurements have also been explored. Subsequently, the compaction quality encompassing 100% of the compaction area can be estimated using a geostatistical analysis method, such as Kriging interpolation.

Based on this analysis, a color-coded SCV map can be generated for a unit within the construction area, visually representing the compaction quality across the site.

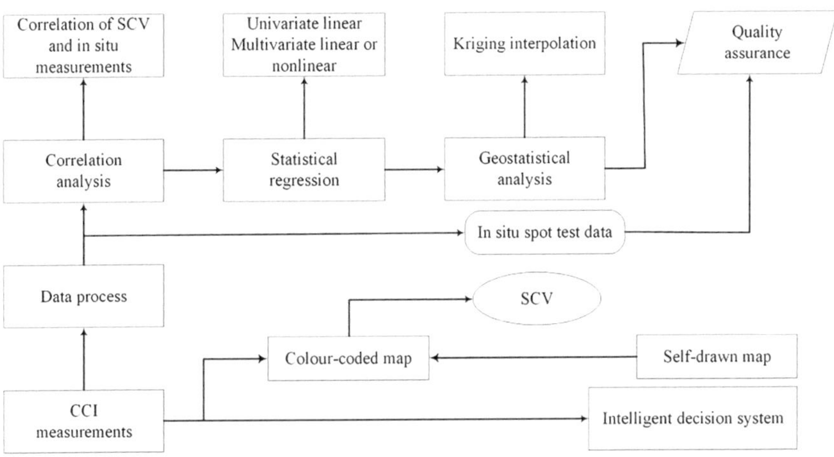

Fig. 8.17 Flow chart of CCI data process and analysis

This map enables the identification of areas exhibiting poor quality based on the specified number of compaction cycles, which can serve as a reference to prevent both over-compaction and under-compaction, and can subsequently be utilized for quality assessment.

(2) URC system

The URC system for earth-rock dams comprises thirteen components: an onboard navigation controller, an electric power steering wheel, a laser displacement sensor, a piezoelectric sensor, collision avoidance radar, a control motor, a power unit, a wireless access point with an omnidirectional antenna, a base station, a GPS receiver and antenna, GPS satellites, a remote wireless device, and a remote monitoring center, as depicted in Fig. 8.18.

The onboard navigation controller, equipped with a built-in NCP, is designed to receive control commands and navigation planning information dispatched by the remote monitoring center to facilitate automatic navigation and URC operations. Concurrently, the controller gathers real-time status information of the roller and transmits this data to the remote monitoring center for analysis and storage via the wireless AP and omnidirectional antenna.

The status information encompasses the current position of the roller, vibration frequency, driving speed, distance from obstacles, the number of steering wheel

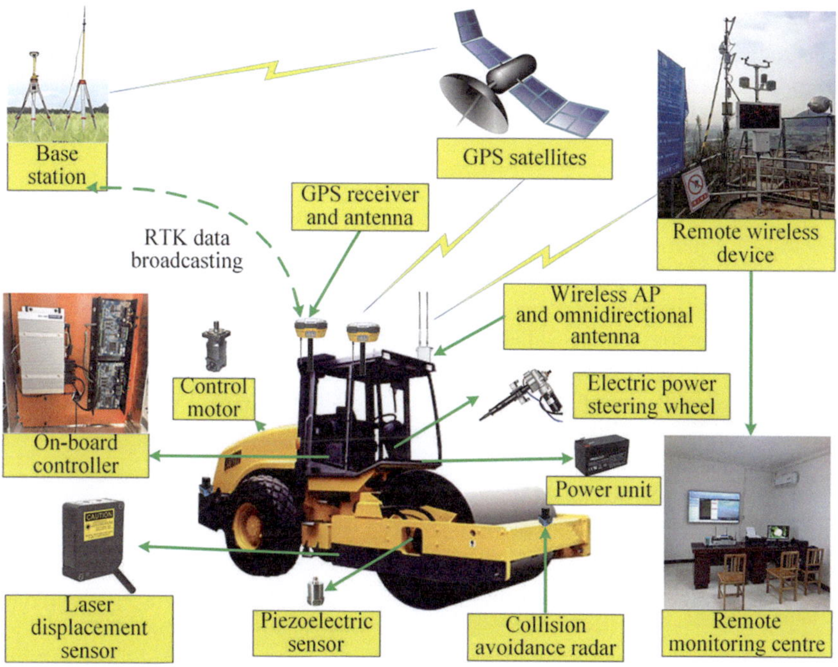

Fig. 8.18 Diagram of the URC system

angles, and real-time obstacle avoidance data pertinent to the construction site. The control commands encompass not only the fundamental operational directives for the roller-such as power-up, ignition, start, steering, throttle, brake, and shutdown commands-but also the working parameters of the roller, including driving speed and vibration frequency, for the forthcoming rolling cycle.

During the rolling construction operation, the remote monitoring center establishes a navigation line, receives various status updates and emergency request responses from the onboard navigation controller in real time, and dispatches control commands to the onboard navigation controller. The collision avoidance radar detects obstacles within the surrounding environment to ensure the safety of the URC system, and this data is transmitted to the remote monitoring center by the onboard navigation controller. In response, the remote monitoring center issues the appropriate control commands to facilitate timely deceleration or braking maneuvers, thereby avoiding potential obstacles.

In cases of emergency or unusual circumstances, construction management personnel situated at the remote monitoring center have the capability to switch to remote control mode, allowing for manual operation of the actuator of the URC system and the modified roller, thereby enhancing the safety of the URC system. Based on the specific construction requirements of the site and the standard number of modified unmanned rollers available, personnel at the remote monitoring center can opt for a single roller to independently complete the rolling task or deploy multiple rollers to work in coordination for task completion. Furthermore, personnel at the remote monitoring center can also manage the rolling tasks in parts of the construction area that are unsuitable for the URC system's operation, such as edges or corners.

(3) IDM system

Building upon previous studies, we have further advanced the IDM system to dynamically regulate the compaction process of rockfill materials during the construction of earth-rock dams. The framework of the IDM system is depicted in Fig. 8.19. The IDM system acquires real-time compaction quality data (i.e., static compaction value or SCV) and parameter information (i.e., current number of compaction cycles) for the rockfill materials, as provided by the CAWD system.

Utilizing the GCPDOM, the working parameters for the roller's subsequent rolling cycle are determined and communicated to the URC system. Upon receiving the working parameter information, the URC system executes the rolling operation within the designated construction area.

As demonstrated in our prior research [96], GCPDOM offers global optimization for the compaction process through a GA, aiming to minimize the remaining compaction time in contrast to the current local optimization method, which solely seeks to maximize compactness increments for the upcoming rolling cycle. Specifically, a nonlinear multiple regression model was established using SCV, relative density, and P5 content as variables, along with a relative density incremental function (RDIF), both derived from field compaction tests and subsequent regression analysis [96].

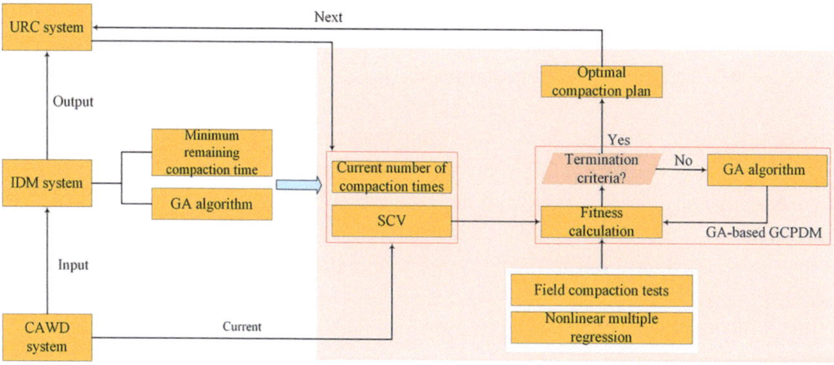

Fig. 8.19 Framework of the IDM system

With the nonlinear multiple regression model and *SCV* measurements, it becomes possible to calculate the current relative density. Once the current relative density and number of compaction cycles are input into the GCPDOM, global optimization of the compaction process is performed via the GA. An objective function serves to evaluate the advantages and disadvantages of individual factors within the GA. This objective function demands minimal computational load and robust function generality, forming the foundation for calculating the fitness function. The selected objective function, representing the remaining compaction time (RCT), is articulated in Eq. (8.14) [97].

$$\min RCT = \sum_{i=m+1}^{n} T_i = \begin{cases} \sum\limits_{i=m+1}^{n} \frac{1}{v_i} & \text{if } n \text{ is even} \\ \sum\limits_{i=m+1}^{n} \frac{1}{v_i} + \frac{1}{0.83} & \text{if } n \text{ is odd} \end{cases} \tag{8.14}$$

In this context, m represents the current relative density; D_n denotes the relative density at the n-th compaction cycle; $\Delta D_i = f(D_{i-1}, f_i, v_i)$ signifies the relative density increment for the i-th compaction cycle; D_{set} is the target relative density; D_{i-1} corresponds to the relative density at the $(i-1)$th compaction cycle; f_i indicates the vibration frequency of the i-th compaction cycle; and v_i represents the driving speed during the i-th compaction cycle. The iterative process concludes when the relative density reaches the target relative density after the n-th compaction cycle. The sequences comprising the working parameters (f_i, v_i) of the roller, where $i\varepsilon[m+1, n]$, constitute the solution to the optimal compaction plan, enabling the target construction area to be rolled in a manner that minimizes the RCT. Consequently, the working parameters of the rollers can be promptly adjusted to enhance the efficiency of the rolling operation in the subsequent rolling cycle, in accordance with the optimal compaction plan. Building on previous research, we have reaffirmed the efficacy of the GCPDOM prior to the implementation of the proposed IRC method and system (refer to Sect. 8.2.4).

(4) Remote monitoring center

The remote monitoring center comprises both software and hardware components. The software aspect primarily includes a remote monitoring and construction scheduling system tailored for in situ rolling construction, a three-dimensional visualization system, a real-time video surveillance system to track construction progress, and a construction quality information management system dedicated to the rockfill materials.

The core hardware components consist of a display, server, and remote wireless communication devices, all of which provide essential support for in situ rolling construction. The remote monitoring and construction scheduling system is employed to oversee the status information of the IRC system, along with the compaction quality data of the rockfill materials throughout the rolling construction process.

Utilizing a three-dimensional dynamic information model that reflects the actual construction progress, the system primarily simulates the three-dimensional visualization of the dam structure during construction, delivering valuable insights to the construction management personnel in the remote monitoring center. The real-time video surveillance system is utilized to receive images and alarm signals transmitted from the construction site, prompting personnel in the remote monitoring center to undertake various actions, such as manual operation of the remote control, camera functionality, and remote linkage, as well as facilitating remote video file downloads, image retrieval, digital video playback, and other related operations.

The construction quality information management system serves to store the construction data generated during the in situ rolling process, enabling effective tracking of this data and providing it to the relevant construction personnel for analysis and inquiry.

8.2.2 System Integration

As elaborated in Sect. 8.2.1, the IRC system for earth-rock dams comprises four relatively independent components: the CAWD system, the URC system, the IDM system, and the remote monitoring center. The primary challenge addressed in this section is how to facilitate organic and coordinated operation among these components through system integration, thereby maximizing overall benefits and achieving comprehensive optimization.

The proposed system integration methodology and structure are founded on fieldbus, LAN, and Internet technologies, aiming to harmonize detection, collection, control, execution, as well as information transmission, exchange, storage, and utilization. This approach enables equipment integration, information integration, and data integration [98, 99], thereby providing a solution to the aforementioned challenge.

The late twentieth century witnessed the advent of computer monitoring and control systems based on fieldbus technology, signifying the transition of dam construction into the era of automation [94]. In this digital age, fieldbus, LAN, and Internet technologies have primarily been employed to develop concrete temperature control systems [95, 100, 101]. Presently, dam construction is evolving towards an era characterized by intelligence [94, 95], during which innovative non-destructive testing methods [102], unmanned rolling compaction technologies [103], and intelligent decision-making techniques have been progressively implemented in the construction of earth-rock dams.

The realization of the IRC system necessitates effective integration among various subsystems and the remote monitoring center. To this end, we propose an integrated framework featuring a three-layer structure based on fieldbus, Ethernet, and Internet technologies, and apply this integrated approach to the compaction quality control of earth-rock dams.

The system integration structure diagram, based on fieldbus, Ethernet, and Internet technologies, is illustrated in Fig. 8.20. The framework is organized into a three-layer structure comprising the field device layer, the field monitoring layer, and the remote management layer.

The field device layer primarily consists of two CAN intelligent nodes. One node houses Controller 1# (slave), a series of sensors (1 to N), and GPS devices, while the other node contains Controller 2# (slave), actuators, and executing devices. In terms

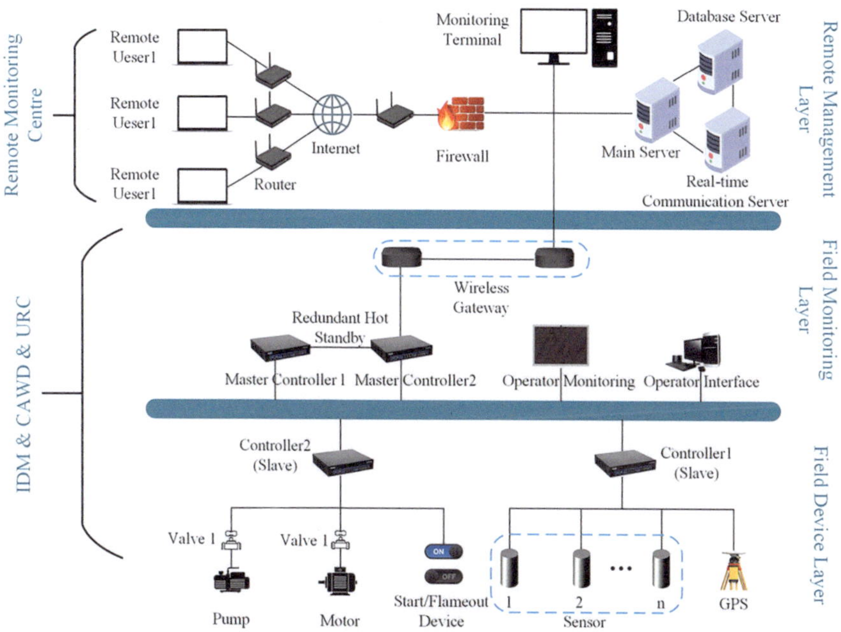

Fig. 8.20 System integration structure diagram based on fieldbus, Ethernet and Internet

of functionality, the field device layer interconnects various underlying devices on the roller through intelligent nodes, facilitating functions such as data acquisition, analog-to-digital conversion, digital filtering, and Proportional-Integral-Derivative (PID) control. The implementation of the CAN bus establishes the connection between the field device layer and the field monitoring layer.

The field monitoring layer encompasses the main controller, operator interface, operator monitoring, and a wireless gateway, serving several functions. It enables the seamless interaction of data, information, and control instructions between the field device layer and the remote management layer. In contrast to the basic control functions of the field device layer, it performs advanced control and optimized calculations, conducts status diagnosis and alarm analysis of devices, and provides operators with debugging interfaces for both devices and operating systems.

Moreover, the main controller, which represents the core component of the field monitoring layer, employs a dual-machine hot standby redundancy design to enhance reliability. The wireless gateway facilitates the connection between the lower-layer field control network (comprising the field device layer and field monitoring layer) and the upper-layer management network (the remote management layer). Utilizing Ethernet and Internet technologies, the upper-layer management network is tasked with real-time data storage and information management from the lower-layer field control network, enabling remote monitoring and control of in situ rolling construction, as well as allowing remote users to query the operational status of both the systems and the devices.

Utilizing the proposed integration framework and methodology, the CAWD, URC, and IDM systems are situated within the lower-layer field control network, while the remote monitoring center resides in the upper-layer management network. Specifically, various sensors and GPS devices associated with the CAWD and URC systems execute signal acquisition through Controller 1# (slave) at the field device layer. Meanwhile, a range of valves, pumps, motors, and start/flameout devices within the URC system are managed by Controller 2# (slave) for operational execution. The IDM system, integrated into the main controller of the field monitoring layer, facilitates optimized calculations and advanced logic control, whereas the remote monitoring center, structured in a triangular architecture, enables real-time data storage, analysis, and management.

Within the remote monitoring center, the triangular configuration consists of a main server, a database server, and a real-time communication server, with the monitoring terminal client overseeing in situ construction and data analysis. Remote users can query and browse construction data via the Internet. Through the application of the proposed system integration framework and methodology, the four relatively independent components—the CAWD system, URC system, IDM system, and remote monitoring center—can be seamlessly integrated into the IRC system.

The proposed integration method boasts several advantages. Firstly, it significantly reduces the number of connections, thereby lowering installation costs and shortening the development cycle, all while meeting diverse functional requirements. Secondly, it operates within an open network, endowing the IRC system with excellent scalability. Thirdly, this method not only enhances the reliability of the IRC system but also its adaptability to complex field environments. Fourthly, it allows for the interchangeability of devices with similar performance metrics from different manufacturers, simplifying maintenance of the IRC system. Ultimately, these advantages culminate in the optimization and enhancement of the overall performance of the IRC system.

8.2.3 Case Study

(1) Testing site and materials

The Qianping Reservoir, boasting a total storage capacity of 584 million m^3, is situated in the upper reaches of the Beiru River, a tributary of the Shaying River within the Huaihe River Basin. This significant project is recognized as one of the 172 major water conservancy initiatives identified by the State Council [104]. Upon the reservoir's completion, it is projected that the irrigated area will encompass approximately 339 km^2, with an annual water supply reaching 63 million m^3 and a power generation capacity of 18.81 million KWh.

As the pivotal structure of the reservoir, the primary dam is designed as a core wall earth-rock dam, featuring a maximum height of 90.3 m and a maximum length of 818 m. According to the design specifications, the primary dam, which consists of six construction areas, will be filled with four distinct materials, yielding a total earthwork filling capacity of up to 13 million m^3. The research presented in this section is concentrated on the rockfill materials employed for the construction of the rockfill body of the earth-rock dam.

Previous studies [105] have indicated that natural-graded rockfill materials sourced from the material yards of the upstream and downstream rivers possess a particle size of up to 1000 mm. Given that the rockfill materials utilized are derived from one of several material yards, it is essential to sample and analyze the particle size grading directly at the construction site. The results of the grain size analysis for field samples collected at the in situ test locations are presented in Fig. 8.21.

Point sampling tests, employing the water replacement method [106], conducted on the rockfill materials revealed that the maximum particle size was 400 mm. Furthermore, the proportion of particle sizes below 5 mm ranged from 11 to 18%, with the number of particles under 5 mm being controlled within a limit of 20%. In accordance with the construction quality requirements for the earth-rock dam of the Qianping Reservoir, it is imperative that the compactness of the core wall reaches 100%, and the relative density of the rockfill area exceeds 80%.

(2) Testing program

Multiple sets of field rolling experiments were conducted across various filling areas (A, B, and C) to evaluate the actual effects of the IRC system based on the optimization method. As depicted in Fig. 8.22, the material yard is situated in the upstream channel of the river. The test scheme for the IRC system at the rockfill zone located

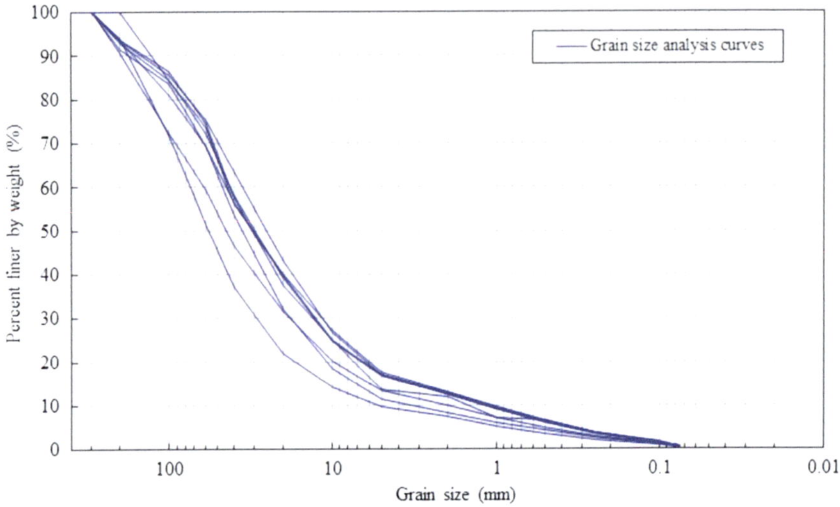

Fig. 8.21 Grain size analysis results for field samples taken from in situ test locations [107]

downstream is summarized in Table 8.5. Area C is dedicated to verifying the effectiveness of the proposed IDM system based on GCPDOM propagation, while areas A and B focus on assessing the performance and application of the IRC system.

Specifically, Area C comprises four piles—ee, ff, gg, and hh—designated for use with both conventional and modified rollers. The filling layer materials employed for testing in Area C were all sourced from the same material yard. According to the in situ construction plan during the test, nine filling layers were selected as test

Fig. 8.22 Photograph of the IRC system testing at the rockfill zone

Table 8.5 Test scheme for the IRC system at the rockfill zone located downstream

Area	Range (m)	Number of layers	Pile number	Object	Purpose
A	0 + 200–0 + 400	9	A/B/C/D/E/F/G/H/I	IRC system	Testing
B	0 + 400–0 + 600	14	a/b/c/d/e/f/g/h/i/j/k/l/m/n	IRC system	Testing
C	0 + 000–0 + 100	2	ee/ff/gg/hh	Optimization method	Verification

layers for Area A, while fourteen filling layers were designated for Area B. Given that the IRC method, along with its corresponding system, has been previously tested and implemented on the primary structure of the Qianping Reservoir, we were able to acquire compaction measurements, compaction parameters, the fault rate of the modified roller, and the actual rolling time required for each filling layer throughout the rolling process for this project.

We collected multiple sets of in situ rolling test data and sampling point test data, which included compaction measurements (such as *SCV* and relative density), compaction parameters (encompassing the number of compaction passes, compaction trajectory, vibration frequency, lift thickness, and driving speed), usage frequency of the modified roller, fault rate of the modified roller, and the elevation of the compacted layers. This comprehensive data set was utilized for the actual analysis of the IRC system.

8.2.4 Result Analysis

(1) Verification of effectiveness of GCPDOM

IRC system was designed and developed for the earth-rock dam within the Qianping Reservoir project. At the heart of the IDM system lies the GCPDOM, which aims to achieve dynamic optimization of compaction parameters—specifically, vibration frequency and driving speed—based on the current compaction state. This approach facilitates machine learning decision-making during compaction, mitigating the impact of human factors on compaction quality and construction efficiency.

This section aims to validate the effectiveness of GCPDOM through in situ comparative experiments. In alignment with the testing scheme presented in Table 8.5, verification was conducted on two filling layers in Area C, with each layer subdivided into two equally sized piles (each measuring 27 m × 24 m). Among these piles, ee and gg served as the comparative experimental groups, while ff and hh functioned as control groups. As illustrated in Table 8.6, the designated numbers of compaction cycles for static and vibratory rolling in the ee/ff piles were 2 and 8, respectively. For the gg/hh piles, the designated compaction cycles were 2 and x, where x represents the actual number of compaction cycles executed following the optimization of the compaction process for the gg/hh piles.

The comparative analysis focuses on the differences between the conventional compaction results and the optimized compaction parameters from the gg pile, alongside the compaction measurements of all four piles, to substantiate the efficacy of GCPDOM.

Table 8.7 presents the optimization results for the compaction parameters—namely, the number of compaction cycles and driving speed—for the gg pile, which involved a total of six compaction cycles. Upon reaching the sixth compaction cycle, the estimated average relative density was recorded at 86.5%. Furthermore, during the dynamic optimization process, the driving speed decreased from 0.83 to 0.57 m/s, while the vibration frequency increased from 28 to 30.1 Hz between the first and sixth passes.

Figure 8.23 illustrates the results of comparative experiments conducted between the conventional compaction process and the dynamically optimized compaction process. In the conventional method, test results from the sampling points indicate that the relative density surpasses 80% only after the seventh or eighth rolling pass, based on the average relative density derived from three sampling points for each test strip. In contrast, the results from the dynamic optimization demonstrate that the relative density can exceed 80% after just the fifth or sixth rolling pass.

According to the design requirements of eight rolling cycles, both the conventional and dynamically optimized compaction methods satisfy the construction quality standards; however, the dynamic optimization process is significantly more time-efficient. The verification results clearly demonstrate that GCPDOM offers substantial advantages over conventional compaction methods. Specifically, GCPDOM facilitates the dynamic optimization of the compaction process and enhances

Table 8.6 Verification test plan for the proposed GCPDOM

Pile number	Size (m)	Passes (static)	Passes (dynamic)	Driving speed (km/h)	Vibration frequency (Hz)
ee/ff	27m × 24	2	8	3	28
gg/hh	27m × 24	2	x	Optimisation	Optimisation

Table 8.7 Optimisation results of compaction process for the gg pile

Number of compaction passes	Driving speed (m/s)	Vibration frequency (Hz)	Estimated average relative density (%)
1	0.83	28	–
2	0.81	28.2	64.6
3	0.73	28.7	70.8
4	0.68	29.4	76.5
5	0.63	29.8	80.5
6	0.57	30.1	86.5

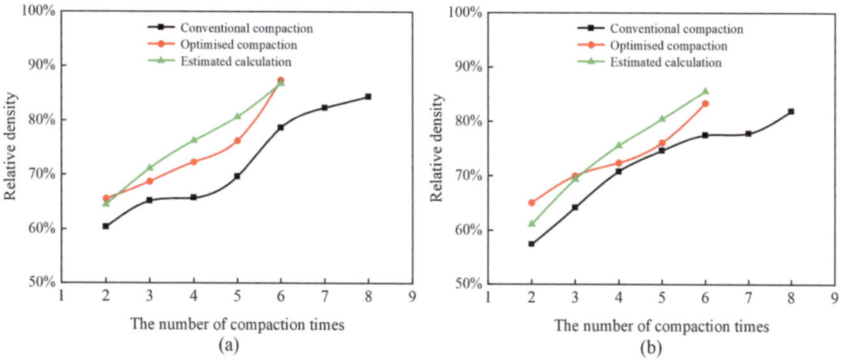

Fig. 8.23 Results of comparative experiments: **a** ee and gg; **b** ff and hh

machine learning decision-making for compaction strategies, effectively mitigating the influence of human factors on the quality of the compaction process.

The IDM system, underpinned by GCPDOM, establishes a closed-loop control system in conjunction with the CAWD and URC systems, thereby enabling the seamless integration of functional units such as continuous detection, intelligent decision-making, and URC throughout the compaction process. GCPDOM presents an intelligent solution for rolling compaction construction in dam projects, contributing significantly to the advancement of intelligent dam construction. In Area C, the test results substantiate the method's effectiveness for quality control of the two-layer materials and highlight its potential to enhance construction efficiency.

(2) Analysis of the compaction parameters results

Drawing upon previous studies, it has been established that the control accuracy of the vibration frequency, compaction trajectory, and driving speed of the modified roller can be precisely maintained within a specific range, with the amplitude of the modified roller aligning with the parameters of the original roller. After assessing the accuracy of the positioning equipment and calculating the uncertainty in the computed average lift thickness, it has been demonstrated that the lift thickness of the filling layers can be effectively monitored.

Moreover, statistical analysis of the number of compaction passes across various filling construction areas reveals that the quality ratings associated with the number of compaction passes in these areas are significantly superior to those achieved with conventional rollers. Consequently, this section presents only a selection of compaction parameters—namely, compaction trajectory and lift thickness—and provides an analysis of the results accordingly.

① Analysis of compaction trajectory results

As a vital component of the IRC system, the remote monitoring center is tasked with real-time oversight, data storage and acquisition, as well as the command and dispatch of in situ rolling operations. Within the remote monitoring center, the compaction

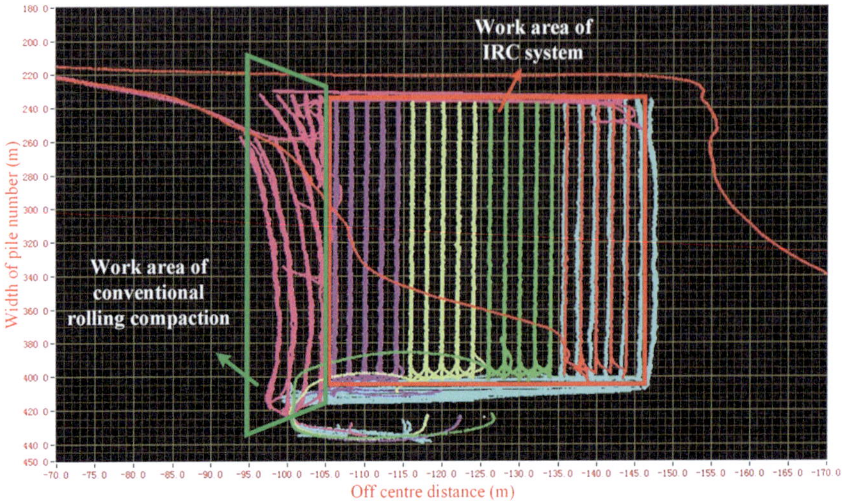

Fig. 8.24 Compaction parameter monitoring diagram of Layer 4 in area B: compaction trajectory

parameters during the rolling construction process are visualized in real-time. This section features the compaction parameter monitoring diagram for Layer 4 in Area B to illustrate the actual compaction trajectory, as depicted in Fig. 8.24.

These results reveal that the IRC system offers numerous significant advantages over conventional rolling compaction methods. Notably, the IRC system effectively mitigates the influence of human factors on compaction operations while maintaining precise control over the compaction trajectory of the roller. It enables real-time visual monitoring of compaction parameters and allows for instantaneous data storage in the remote monitoring center, facilitating subsequent data analysis and traceability.

With this system in place, five modified rollers can efficiently execute large-scale rolling construction tasks for the primary sections of Areas A and B, operating continuously to accommodate tight schedules. The benefits of the IRC system are especially pronounced under hazardous conditions, during nighttime, or in extreme environments. As demonstrated in the analysis results, the IRC system proficiently controls the compaction trajectory.

② Statistics and analysis of lift thickness of compacted layer

In accordance with the specifications for existing earth-rock dam construction [108] and the findings of previous research [3, 12, 28], regulating the lift thickness of each filling layer is crucial for maintaining the compaction quality of earth-rock dams. Controlling the lift thickness within a reasonable range enhances optimal compactor energy penetration and ensures a relatively uniform density of the final compacted materials. Consequently, this section presents the statistical analysis of the lift thickness for each compacted layer in Areas A and B following the completion of the rolling construction.

The monitoring data obtained from the IRC system for Layer 5 in Area A were utilized to visually represent the elevation of the compacted layer at various points post-rolling construction, as illustrated in Fig. 8.25. The color-coded map reveals that the elevation at each point within the rolling area of the IRC system falls within a specified range, with different colors indicating varying elevations, thereby facilitating analysis and assessment by management personnel in the remote monitoring center.

The resulting statistics for Areas A and B, derived from the calculation of lift thickness and the uniform grid method, are presented in Tables 8.8 and 8.9. These results demonstrate that lift thickness can be effectively monitored during construction, with the IRC system capable of controlling elevation discrepancies across the compacted layer within 15 cm (the designed lift thickness for the filling layer is 80 cm). Figure 8.25, along with the statistical findings from Tables 8.8 and 8.9, indicates that the IRC system promotes relatively uniform settlement across the various filling layers.

Based on the analysis presented, the IRC system can generate real-time color-coded maps of the lift thickness of the compacted layers. This capability is particularly advantageous, as it can be executed from the remote monitoring center without incurring the costs and delays typically associated with formal field elevation surveys. This system greatly benefits both the in situ construction personnel and the management staff in the remote monitoring center.

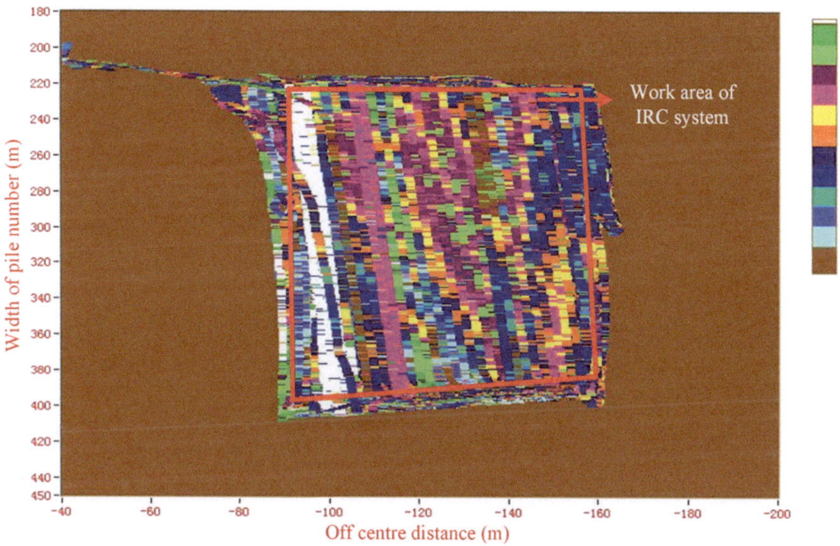

Fig. 8.25 Elevation information of the work area after completion of the rolling operations: area A and Layer 5

Table 8.8 Statistical results of lift thickness for area A

Layer number	Designed lift thickness (cm)	Maximum lift thickness (cm)	Minimum lift thickness (cm)	Average lift thickness (cm)	Varying range (cm)
1	80	83	67.9	76.5	14.8
2	80	82	66.9	75.2	14.7
3	80	79	64.3	74.8	15
4	80	79	64.6	74.6	14.4
5	80	80	64.7	75.0	15
6	80	80	66.1	74.8	14.2
7	80	82	67.0	76.1	14.8
8	80	78	63.2	73.9	14.5
9	80	80	64.8	74.7	14.6

Table 8.9 Statistical results of lift thickness for area B

Layer number	Designed lift thickness (cm)	Maximum lift thickness (cm)	Minimum lift thickness (cm)	Average lift thickness (cm)	Varying range (cm)
1	80	81	66	74	14.3
2	80	81	67	76	14.5
3	80	79	64	74	14.8
4	80	79	65	75	14.4
5	80	80	66	75	14.6
6	80	78	65	74	13.3
7	80	80	65	75	14.6
8	80	81	67	75	14.5
9	80	79	64	74	15
10	80	79	65	75	14.5
11	80	81	66	75	14.6
12	80	81	66	76	14.8
13	80	79	65	74	14.5
14	80	80	65	75	14.4

(5) Statistics and analysis of use frequency and the fault conditions

As evident from the preceding discussion, the IRC system consists of the CAWD, URC, and IDM systems, along with the remote monitoring center. Previous studies indicate that the triangular architecture of the remote monitoring center is less susceptible to failures, and the equipment and systems within the center exhibit commendable stability and reliability. The high-speed communication system linking the in situ construction machinery to the remote monitoring center is designed with redundancy to ensure the robustness and stability of its equipment and systems (refer to Figs. 8.16, 8.18, and 8.19).

In addition to the remote monitoring center, the primary functional modules of the remaining systems are integrated into the modified rollers within the IRC system. The frequency of usage for the modified rollers in the IRC system was documented separately from that of the conventional rollers in Areas A and B, as illustrated in Fig. 8.26. Based on the in situ mechanical conditions and the construction plan, a total of five modified rollers were allocated for dispatch within the IRC system, while three conventional rollers were designated for dispatch for traditional rolling compaction.

The statistical methodology dictates that if a roller is utilized throughout the entire construction process of an individual layer, it is classified as in an "intact state"; conversely, if it is not utilized entirely, it is deemed to be in a "fault state". The statistical results reveal that during the entire construction period for Areas A and B, the modified rollers in the IRC system were deployed one hundred times, whereas the conventional rollers were utilized forty-five times. Furthermore, the calculation of the utilization rate for the construction machinery is presented in Eq. (8.15) [109].

$$U_r = \frac{N_u}{N_t} \times 100\% \tag{8.15}$$

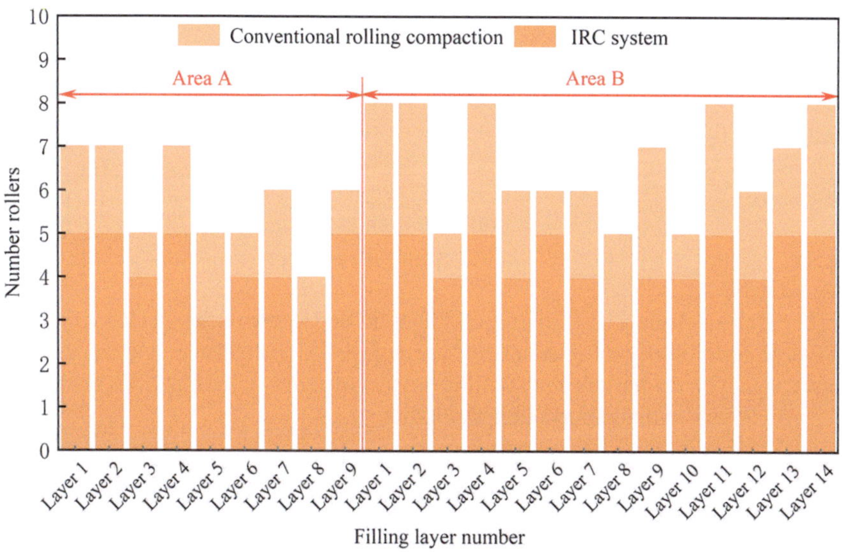

Fig. 8.26 Statistics of use frequency of the rollers in the IRC system and conventional rolling compaction: area A and area B

Table 8.10 Statistical results of the utilization rates and fault rates of the modified rollers in the IRC system for area A and area B

Area	Utilisation rate (%)	Fault rate (%)
A	87	14
B	89	10

where U_r denotes the utilization rate of the construction machinery; N_u represents the number of machines actually employed during the reporting period; and N_t indicates the total number of machines available during the reporting period. The frequency of equipment failure can be determined using the conventional definition presented in Eq. (8.16).

$$U_f = \frac{N_{fh}}{N_{fh} + N_{ah}} \times 100\% \tag{8.16}$$

where U_f signifies the fault status of the construction machinery; N_{fh} represents the machine-hours of failure downtime experienced by the machinery during the reporting period; and N_{ah} denotes the total expected machine-hours for the machinery throughout the reporting period.

Utilizing the aforementioned formulas and statistical criteria, the utilization rates for both the modified rollers in the IRC system and the conventional rollers, as well as the fault rates for each type, can be computed. The results of these calculations are presented in Table 8.10. The utilization rates of the modified rollers in Areas A and B are approximately 87.0% and 88.6%, respectively, while the fault rates for these rollers in Areas A and B are approximately 13.5% and 10.3%, respectively.

As illustrated in Fig. 8.26 and Table 8.10, the primary sections of Areas A and B have been effectively compacted using these modified rollers. The data reflects the outstanding working conditions and high utilization rates of the modified rollers, which correspond to a relative decrease in production costs. As a key performance index for machinery operations, the fault rate indicated in Table 8.10 demonstrates that the modified rollers are operating in excellent condition and are easily maintainable.

The statistical analyses further reveal that the IRC system exhibits characteristics of stable operations, strong adaptability, convenient maintenance, and remarkable safety. It plays a pivotal role in the filling process of earth-rock dams when compared to conventional rolling compaction methods.

(6) Statistics and analysis of construction progress of compacted layer

The subsequent step is to ascertain whether the IRC method and system can enhance the construction efficiency of earth-rock dams. This section presents statistical data and analyses regarding the scheduled versus actual construction times necessary for each of the filling layers in Areas A and B, thereby directly reflecting the effectiveness of the proposed method through the actual progress observed at the construction site.

Initially, the planned duration for each filling layer was established for manual rolling compaction. It is understood that the processes of dam material extraction, transportation, and paving can be adequately managed and controlled throughout the filling construction phase. Consequently, following the completion of preliminary preparations, the filling construction can proceed in a systematic manner according to the construction schedule. The IRC system is primarily employed to enhance the compaction quality of the filling materials and to improve the efficiency of rolling compaction.

The initially planned times and the actual durations required to complete the rolling operations for each layer in Areas A and B are depicted in Fig. 8.27. The results indicate that the actual time taken to finish the rolling operations for each layer in both areas is significantly shorter than the planned time. For Area A, the average duration for rolling construction was 4.87 h, while for Area B, it was 4.76 h. The average construction times for Areas A and B are relatively comparable, with only minor fluctuations in the actual times required to complete the rolling operations for each layer, underscoring the reliability of the IRC system.

The variance in the actual construction times for the completion of rolling in Areas A and B can be attributed to the complexities and diverse factors influencing the progress of the in situ rolling construction. These factors include the procurement of fuel for the rollers, shift changes among the management personnel in the remote monitoring center and the on-site construction teams, as well as the supplemental watering of the in situ filling materials.

In Area A, the results illustrated in Fig. 8.27 reveal that the construction efficiency for each filling layer has been enhanced in comparison to the initially planned

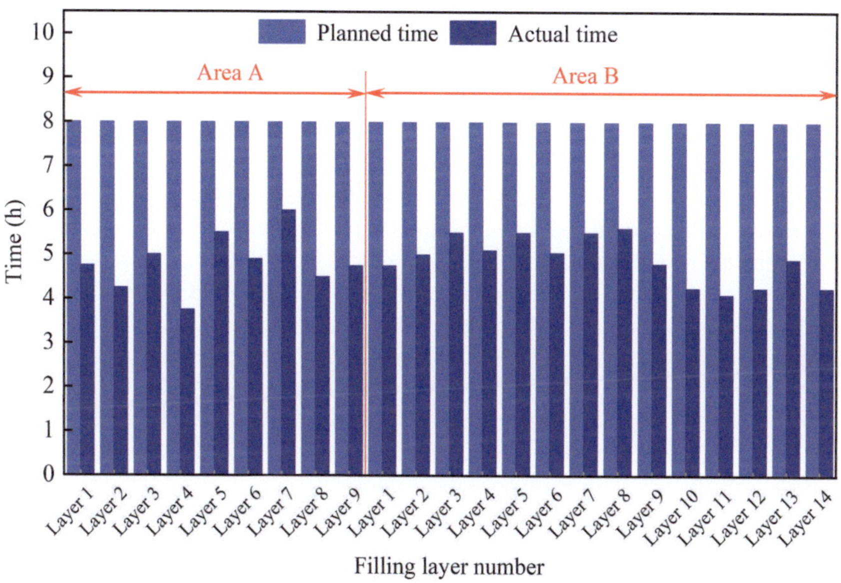

Fig. 8.27 Statistics of initial planned time and the actual time required for each layer of filling construction in areas A and B

rolling time, with improvements ranging from a minimum of 30.5% to a maximum of 103.6%, and an average increase of 67.2%. Similarly, in Area B, the construction efficiency for each filling layer has also shown corresponding improvements, with increases of at least 36.9%, a maximum of 102.5%, and an average enhancement of 70.7%.

When compared to the original plan for conventional rolling compaction, the implementation of the IRC system plays a crucial role in ensuring that rolling construction is completed punctually, with both quality and quantity assured according to the construction schedule, while simultaneously enhancing the rolling construction efficiency for each filling layer in Areas A and B.

The primary rationale for this significant difference lies in the multiple advantages of the proposed IRC system. Firstly, based on actual in situ conditions, the time required to complete the rolling operation for each layer exhibited considerable fluctuations during conventional rolling compaction, frequently exceeding the initial planned completion times, which undermined the assurance of construction progress. Thus, without a guarantee of progress, improvements in construction efficiency are unattainable.

Secondly, the system can operate continuously with adequate fuel supply, facilitating orderly rolling construction both day and night. Thirdly, the IRC can effectively reduce the number of compaction cycles while ensuring the quality of filling material compaction, thereby enhancing overall construction efficiency.

Fourthly, this method employs the URC system for rolling operations, which effectively mitigates issues arising from driver-related delays that can impact construction schedules, such as rest breaks due to fatigue, meal interruptions, or other physiological pauses.

Fifthly, the CAWD system conducts high-efficiency NDT and evaluates the compaction quality across the entire working surface of the filling materials. This capability provides timely feedback and controls the entire compaction area, thereby reducing the number of sampling points after rolling operations and minimizing delays associated with supplementary rolling compaction or other remedial actions due to delayed feedback in conventional rolling compaction, thus facilitating improvements in compaction efficiency.

Lastly, the closed-loop detection, analysis, decision-making, and control processes effectively integrate the relatively independent components of conventional rolling compaction, optimizing the QC process and accelerating rolling construction operations.

In conclusion, the IRC system, with its aforementioned advantages, positively influences both the assurance of construction progress and the enhancement of construction efficiency for earth-rock dams during the rolling construction phase.

(7) Compaction quality assurance for the entire filling construction area

① Compaction quality assessment using the IRC system

By conducting a compaction quality assessment based on measurements obtained from the CAWD system, as detailed in Chap. 4, it is evident that the overall compaction quality rate exceeds 95%. The proposed *SCV*-based compaction quality assessment method effectively evaluates the compaction quality across the entire work area. Consequently, this section integrates the *SCV*-based assessment method into the IRC system and employs it to evaluate the compaction quality of all filling layers in Areas A and B.

This section presents the statistics and analyses related to the compaction quality assessment results for all filling layers in both areas. The general process for the assessment is outlined as follows: (a) during the rolling construction of each filling layer in Areas A and B, compaction measurements were obtained in real-time using the IRC system; (b) a correlation analysis was subsequently performed to examine the relationships between the SCV, gradation, moisture content, and the relative density of the filling materials; (c) based on the results of the correlation analysis, key factors were selected to establish a regression model for compaction quality, which was then validated through a significance test; (d) from the established compaction quality regression models, the most suitable model was selected, and geostatistical analyses were conducted on the compaction quality of the filling layers throughout the work area; Utilizing steps (a) through (d), the overall compaction quality of the entire filling construction area could be assessed according to the predefined relative density control standard (where the relative density of the rockfill area must exceed 80%), allowing for the calculation of the overall passing rate of the entire work area, denoted as P_r, as follows [1]:

$$P_r = \frac{A_s}{A_f} \times 100\% \tag{8.17}$$

where P_r represents the overall passing rate for the entire work area, A_s denotes the area exhibiting a relative density greater than 80%, and A_f signifies the total area of the work unit.

As the previous study [110] has elaborated on the estimation process in detail, this section refrains from reiterating the specific steps involved in assessing the compaction quality of each filling layer in Areas A and B. The *SCV*-based assessment method, alongside the IRC system, has been employed for the compaction quality assessment of the entire filling construction area of each pile. Statistical results indicate that the overall passing rate of compaction quality for each pile exceeds 97%. These findings can assist the on-site construction personnel and the management teams at the remote monitoring center in providing timely feedback for quality control and improvement efforts.

In contrast to the IDM system, which primarily focuses on optimizing the compaction process to enhance rolling construction efficiency, the compaction quality assessment facilitated by the IRC system prioritizes the timely identification of poor-quality areas after rolling construction is completed. This approach allows for prompt feedback control to ensure the compaction quality of the filling construction area.

Based on the aforementioned statistics and analyses, several benefits arise from utilizing the IRC system to evaluate the compaction quality of the filling layers. The system ensures compaction quality while minimizing the need for costly rework, achieving a comprehensive compaction quality assessment that encompasses 100% of the filling construction area. This significantly reduces the uncertainty in quality assurance that is often associated with random point sampling. Furthermore, the assurance of compaction quality across the entire rolling construction area can effectively double the guarantees provided by traditional methods. Consequently, this system can deliver timely feedback and guidance to prevent quality defects, enhance the uniformity of the filling materials, and ultimately improve the overall performance of the earth-rock dam.

② Compaction quality assurance using the traditional method

The water replacement method yields traditional measurements of dry density, maximum dry density, minimum dry density, and moisture content for each test pit, with each measurement determined according to depth in accordance with the relevant standards. The relative density D_r of the filling materials for each test pit, which is primarily utilized in this section for the optimization of the compaction process and the assessment and assurance of compaction quality, can be calculated from the measured in situ dry density ρ_d and the maximum and minimum dry densities ($(\rho_d)_{\max}$ and $(\rho_d)_{\min}$, respectively), as follows[111]:

$$D_r = \frac{(\rho_d)_{\max}}{\rho_d}\left[\frac{\rho_d - (\rho_d)_{\min}}{(\rho_d)_{\max} - (\rho_d)_{\min}}\right] \times 100\% \tag{8.18}$$

Utilizing the water replacement method, several randomly selected test pits were designated for quality assurance during the compaction of rockfill materials, aimed at verifying that the *SCV*-based assessment method yields reasonable estimates of in situ compaction quality. In addition to the aforementioned test points, final quality assurance tests based on the water replacement method were conducted on the currently compacted layer prior to the commencement of further construction on the subsequent filling layer. A total of 132 sets of experimental samples were collected for in situ measurements, serving as the ultimate assurance for the compaction quality of the filling materials within each pile.

Figure 8.14 illustrates the relative density of test points across all piles in Areas A and B, alongside the absolute error between the relative density of each point and the control criteria. The results depicted in Fig. 8.28 reveal that the relative density of all test points exceeds the control threshold of 80%, indicating that all test piles in Areas A and B successfully meet the compaction assurance standards. Furthermore, the average absolute error between the relative density and the control criteria is slightly above 7%, reflecting a minor deviation of the overall measured values from the control standards and indirectly highlighting the effectiveness of the controls within the IRC system.

Consequently, these results not only demonstrate that the IRC system can effectively manage the compaction quality of the test piles throughout Areas A and B, but they also reaffirm that the traditional method can be integrated with the *SCV*-based compaction quality assessment approach. This integration reduces the likelihood

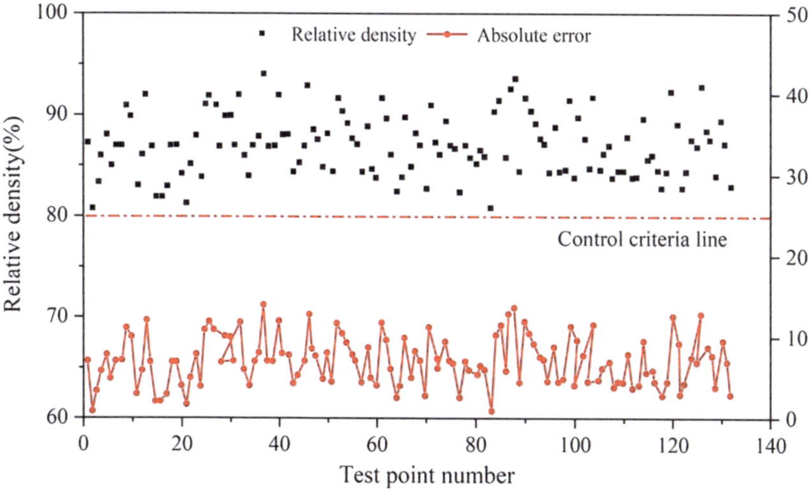

Fig. 8.28 Relative density and absolute error of test points for all piles in areas A and B

of poorly compacted areas and provides a dual assurance of quality control while enhancing the performance of the earth-rock dam.

According to the construction requirements, the acceptable compaction quality rate for the filling unit in the rockfill material area must not fall below 85%. When the IRC system is implemented in other areas of the Qianping project, each filling unit will also undergo compaction quality assurance via the traditional method. The findings indicate that each filling unit achieves an excellent rating that meets the specified requirements, with an impressive average compaction quality rate of 95.3%, which significantly surpasses the average compaction quality rate of 87.4% achieved through conventional rolling compaction methods.

Based on the preceding analysis, the IRC system proposed in this section can significantly diminish the extent of poorly compacted areas, adeptly manage the compaction quality of rockfill materials, and enhance the overall construction quality of earth-rock dams.

References

1. Liu DH, Sun J, Zhong DH et al (2012) Compaction quality control of earth-rock dam construction using real-time field operation data. J Constr Eng Manag 138(09):1085–1094
2. Hua T, Yang X, Yao Q, Li H (2018) Assessment of real-time compaction quality test indexes for rockfill material based on roller vibratory acceleration analysis. Adv Mater Sci Eng 2018:1–15
3. Liu DH, Lin M, Li S (2016) Real-time quality monitoring and control of highway compaction. Autom Constr 62:114–123
4. Peurifov RL, Schexnayder C, Shapira A (2010) Construction planning, equipment and methods, 6th edn. McGraw-Hill, N.Y., USA. ISBN: 9780073401126

5. China Electricity Council (2013) D/T5129-2013-specifications for rolled earth-rockfill dam construction. China Electric Power Press, Beijing, China
6. Tan YQ, Wang HP, Ma SJ, Xu HN (2014) Quality control of asphalt pavement compaction using fibre Bragg grating sensing technology. Constr Build Mater 54:53–59
7. Kassem E, Liu WT, Scullion T, Masad E, Chowdhury A (2015) Development of compaction monitoring system for asphalt pavements. Constr Build Mater 96:334–345
8. White DJ, Jaselskis EJ, Schaefer VR et al (2005) Real-time compaction monitoring in cohesive soils from machine response. Transp Res Rec 1936(1):172–180
9. Zhong DH, Li XC, Cui B, Wu BP, Liu YX (2018) Technology and application of real-time compaction quality monitoring for earth-rockfill dam construction in deep narrow valley. Autom Constr 90:23–38
10. Huang SX, Liu JN, Wu XM (2005) GPS real-time supervisory system and its preliminary application in the construction of face rockfill dam. Geomat Inf Sci Wuhan Univ 30(09):813–816
11. Zhong DH, Liu DH, Cui B (2011) Real-time compaction quality monitoring of high core rockfill dam. Sci China Technol Sci 54(07):1906–1913
12. Meehan CL, Khosravi M, Cacciola D (2013) Monitoring field lift thickness using compaction equipment instrumented with global positioning system (GPS) technology. Geotech Test J 36:755–767
13. Cacciola DV, Khosravi M, Meehan CL (2014) Using compaction equipment instrumented with global positioning system (GPS) technology to monitoring field lift thickness. In: Geo-congress 2014: geo-characterization and modelling for sustainability, Geotechnical Special Publication No. 234, ASCE, Reston, VA, 2014, pp 2630–2639
14. Huang SX, Zhang W, Wu G (2018) Research on real-time supervisory system for compaction quality in face rockfill dam engineering. J Sens 2018:6487405
15. Hu W, Huang BS, Shu X, Woods M (2017) Utilising intelligent compaction meter values to evaluate construction quality of asphalt pavement layers. Road Mater Pavement Des 18(4):980–991
16. Adam D (1997) Continuous compaction control (CCC) with vibratory rollers. Environmental Geotechnics. In: Proceedings of the 1st Australia-New Zealand conference on environmental geotechnics-geoenvironment 97, Melbourne, Victoria, Australia, 26–28 November 1997; A. A. Balkema, Rotterdam, The Netherlands, pp 245–250
17. Pistrol J, Villwock S, Völkel W, Kopf F, Adam D (2016) Continuous compaction control (CCC) with oscillating rollers. Procedia Eng 143:514–521
18. White DJ, Vennapusa PKR, Gieselman HH (2011) Field assessment and specification review for roller-integrated compaction monitoring technologies. Adv Civ Eng 2011(1):783836
19. White DJ, Vennapusa PK (2010) A review of roller-integrated compaction monitoring technologies for earthworks. Final Report ER10-04; Iowa State University, Ames, IA, USA, pp 1–36
20. Briaud J, Seo J (2003) Intelligent compaction: overview and research needs. Texas A&M University, College Station, TX, USA, pp 1–84
21. Anderegg R, Kaufmann K (2004) Intelligent compaction with vibratory rollers: feedback control systems in automatic compaction and compaction control. Transp Res Rec 1868:124–134
22. Thurner H, Sandström A (1980) A new device for instant compaction control. In: Proceedings of the international conference on compaction, Paris, France, vol 2, pp 611–614
23. Facas NMM (2010) Position reporting error of intelligent compaction and continuous compaction control roller-measured soil properties. J Test Eval 38(1):1
24. Meehan LC, Cacciola VD, Tehrani SF et al (2017) Assessing soil compaction using continuous compaction control and location-specific in situ tests. Autom Constr 73:31–44
25. Adam D, Brandl H (2003) Sophisticated roller integrated continuous compaction control. IN: Proceedings of 12th Asian regional conference on soil mechanics and geotechnical engineering-geotechnical infrastructure for the new millennium, Singapore, pp 427–430. 98-1238-559-2

26. Facas NW, Rinehart RV, Mooney MA (2011) Development and evaluation of relative compaction specifications using roller-based measurements. Geotech Test J 34(6):634–642
27. Vennapusa RKP, White JD, Morris DM (2010) Geostatistical analysis for spatially referenced roller-integrated compaction measurements. J Geotech Geoenviron Eng 136(6):813–822
28. Liu D, Li Z, Lian Z (2014) Compaction quality assessment of earth-rock dam materials using roller-integrated compaction monitoring technology. Autom Constr 44:234–246
29. Facas N, Mooney M (2010) Position reporting error of intelligent compaction and continuous compaction control roller-measured soil properties. J Test Eval 38(1):1
30. Vennapusa PKR, White DJ, Schram S (2013) Roller-integrated compaction monitoring for hot-mix asphalt overlay construction. J Transp Eng 139(12):1164–1173
31. Vennapusa PKR, White DJ, Siekmeier J, Embacher RA (2012) In situ mechanistic characterizations of granular pavement foundation layers. Int J Pavement Eng 13(1):52–67
32. White JD, Thompson JM (2008) Relationships between in situ and roller-integrated compaction measurements for granular soils. J Geotech Geoenviron Eng 134(12):1763–1770
33. Sandström Å, Pettersson CB (2004) Intelligent systems for QA/QC in soil compaction. In: Proceedings of the annual transportation research board meeting, Transportation Research Board, Washington, D.C., USA. 03-0909-461-5
34. White D, Vennapusa P, Thompson M (2007) Field validation of intelligent compaction monitoring technology for unbound materials
35. Thompson JM, White JD (2008) Estimating compaction of cohesive soils from machine drive power. J Geotech Geoenviron Eng 134(12):1771–1777
36. Komandi G (1999) An evaluation of the concept of rolling resistance. J Terrramech 36(3):159–166
37. Mooney M, Adam D (2017) Vibratory roller integrated measurement of earthwork compaction: an overview. In: Proceedings of the FMGM 2007: seventh international symposium on field measurement in geomechanics, Boston, MA, USA, 27 September 2017, pp 1–12
38. Scherocman J, Rakowski S, Uchiyama K (2007) Intelligent compaction, does it exist? 2007 Canadian Technical Asphalt Association (CTAA) Conference, Victoria, BC, 2007, pp 1–25
39. Mooney MA, Rinehart RV, Facas NE, Musimbi OM, White DJ, Vennapusa P (2010) Intelligent soil compaction systems. Final Report, NCHRP 676, Transportation Research Board, Washington, D.C. 978-0-309-15519-9
40. Dondi G, Sangiorgi C, Lantieri C (2014) Applying geostatistics to continuous compaction control of construction and demolition materials for road embankments. J Geotech Geoenviron 140(3):06013005
41. Kim SK, Seo J, Russell JS (2012) Intelligent navigation strategies for an automated earthwork system. Autom Constr 21:132–147
42. Kim SK, Russell JS (2003) Framework for an intelligent earthwork system: part I. Syst Arch Autom Constr 12:1–13
43. Rinehart RV, Mooney MA (2008) Instrumentation of a roller compactor to monitor vibration behavior during earthwork compaction. Autom Constr 17(2):144–150
44. Commuri S, Mai AT, Zaman M (2011) Neural network-based intelligent compaction analyzer for estimating compaction quality of hot asphalt mixes. J Constr Eng Manag 137(9):633–715
45. Ju Y, Lin G, Fan Y, Liu Z (2005) Intelligent compaction control based on fuzzy neural network. In: Sixth international conference on parallel and distributed computing applications and technologies (PDCAT'05). IEEE Computer Society, Dalian, China
46. Minchin RE, Swanson DC, Gruss AF, Thomas HR (2008) Computer applications in intelligent compaction. J Comput Civ Eng 22:243–251
47. White DJ, Vennapusa PKR (2009) Report of the workshop on intelligent construction for earthworks. EERC Publication ER09-02; Sheraton West Des Moines Hotel: West Des Moines, IA, USA, pp 1–156
48. Zhong DH, Cui B, Liu DH, Tong DW (2009) Theoretical research on construction quality real-time monitoring and system integration of core rock-fill dam. Sci China Ser E Technol Sci 52(11):3406–3412

49. Karray M, Lefebvre G, Ethier Y, Bigras A (2010) Assessment of deep compaction of the Peribonka dam foundation using "modal analysis of surface waves" (MASW). Can Geotech J 47:312–326
50. Xu Q, Chang GK (2013) Evaluation of intelligent compaction for asphalt materials. Autom Constr 30:104–112
51. Anderegg R, von Felten D A, Kaufmann K (2006) Compaction monitoring using intelligent soil compactors. In: Geocongress 2006: geotechnical engineering in the information technology age, pp 1–6
52. Cacciola DV (2013) Using continuous compaction control systems within an earthwork compaction specification framework. A thesis submitted to the Faculty of the University of Delaware in partial fulfillment of the requirements for the Degree of Master of Civil Engineering
53. Isaaks EH, Srivastava RM (1989) An introduction to applied geostatistics. Oxford University Press, New York. 01-9505-012-6
54. Thompson MJ, White DJ (2007) Field calibration and spatial analysis of compaction monitoring technology measurements. Transp Res Rec 2004(1):69–79
55. Kaufmann K, Anderegg R (2008) 3D-construction application III GPS-based compaction technology. In: 1st International conference on machine control & guidance, ETH Zurich, Switzerland, pp 1–10
56. Corcoran PT, Fernandez T (2001) Method and apparatus for determining the performance of a compaction machine based on energy transfer. United States Patent: US 6188942
57. Trimble Company Website (2018)
58. Shi J, Liu C, Xi H (2016) Improved D* path planning algorithm based on CA model. J Electron Meas Instrum 30(1):30–37
59. Su X, Li S, Yuan C et al (2014) Enhanced boundary condition-based approach for construction location sensing using RFID and RTK GPS. J Constr Eng Manag 140(10):04014048
60. Roberts GW, Dodson AH, Ashkenazi V (1999) Global positioning system aided autonomous construction plant control and guidance. Autom Constr 8(5):589–595
61. Zhang Q, Liu T, Li Q (2017) Roller-integrated acoustic wave detection technique for rockfill materials. Appl Sci 7(11):1118
62. ISO 17635: 2016 (2016) Non-destructive testing of welds-General rules for metallic materials, Switzerland
63. Liu DH, Cui B, Liu YG, Zhong DH (2013) Automatic control and real-time monitoring system for earth-rock dam material truck watering. Autom Constr 30:70–80
64. Ranasinghe RATM, Jaksa MB, Nejad FP, Kuo YL (2019) Prediction of the effectiveness of rolling dynamic compaction using artificial neural networks and cone penetration test data. Chin J Rock Mech Eng 38:153–170
65. Ranasinghe R, Jaksa M, Kuo Y, Nejad FP (2017) Application of artificial neural networks for predicting the impact of rolling dynamic compaction using dynamic cone penetrometer test results. J Rock Mech Geotech Eng 9:340–349
66. Asif Imran S, Barman M, Commuri S et al (2018) Artificial neural network–based intelligent compaction analyzer for real-time estimation of subgrade quality. Int J Geomech 18(6):04018048
67. Taghavifar H, Mardani A, Taghavifar L (2013) A hybridized artificial neural network and imperialist competitive algorithm optimization approach for prediction of soil compaction in soil bin facility. Measurement 46:2288–2299
68. Isik F, Ozden G (2013) Estimating compaction parameters of fine- and coarse-grained soils by means of artificial neural networks. Environ Earth Sci 69:2287–2297
69. Günaydin O (2009) Estimation of soil compaction parameters by using statistical analyses and artificial neural networks. Environ Geol 57:203–215
70. Wang J, Zhong D, Adeli H, Wang D, Liu M (2018) Smart bacteria-foraging algorithm-based customized kernel support vector regression and enhanced probabilistic neural network for compaction quality assessment and control of earth-rock dam. Expert Syst 35:e12357

71. Teng YN, Li XP, Ma H, Wen BC (2009) Study on vibration friction mechanism and vibration response analysis based on vibration compaction system. Appl Mech Mater 16–19:84–87
72. Beainy F, Commuri S, Zaman M, Syed I (2013) Viscoelastic-plastic model of asphalt-roller interaction. Int J Géoméch 13:581–594
73. Li QB, Zhang QL, An ZZ, Liu TY (2018) An intelligent rolling compaction system for earthwork. China Patent: CN201810720255.3
74. Yoo T-S, Selig ET (1979) Dynamics of vibratory-roller compaction. J Geotech Eng Div 105:1211–1231
75. Zhang QZ (2010) Study on the model and parameters of the vibration compaction system in the soil subgrade. PhD Thesis, Chang'an University, Xi'An, China
76. Wersäll C, Nordfelt I, Larsson S (2017) Soil compaction by vibratory roller with variable frequency. Géotechnique 67:272–278
77. Wersäll C, Nordfelt I, Larsson S (2018) Resonant roller compaction of gravel in full-scale tests. Transp Geotech 14:93–97
78. Хархута НЯ (1957) Theory and machinery for compacting soil. China Hydraulic press, Beijing (in Chinese)
79. Yoo TS (1975) A theory for vibratory compaction of soils. PhD thesis, State University of New York, Buffalo, New York
80. Machet JM, Morel G (1977) Vibratory compaction of bituminous mixes in France. In: Association of asphalt paving technologists proc, vol 46, pp 326–340
81. Pietzsch D, Poppy W (1992) Simulation of soil compaction with vibratory rollers. J Terramech 29:585–597
82. Wersäll C, Larsson S, Bodare A (2014) Dynamic response of vertically oscillating foundations at large strain. Computer Methods and Recent Advances in Geomechanics. In: Proceedings of the 14th international conference of international association for computer methods and recent advances in geomechanics, London, UK, 13 December 2014, pp 643–647
83. Fujii H (1992) Problems between soil and construction machinery with special reference to field compaction. J Terrramech 29(1):35–55
84. Ding SH, Zhang ZF (1992) Investigation into controlling scheme for new type oscillating roller & soil parameters on-line identifier. J Wuhan Univ Technol 14(3):8–16. Information & Management Engineering
85. Zhou CX, Zhou CL (2000) The dynamic model and operating parameter control of vibrating roller. J East China Shipbuild Inst 14(6):28–31
86. Zhang F (2003) Dynamic analysis, computer simulation and parameters optimization for the YZ30 type vibration roller. Master thesis, Xi'an University of Technology, Xi'an
87. Capraru C, Pistrol J, Villwock S, Völkel W, Kopf F, Adam D (2014) Numerical simulation of soil compaction with oscillatory rollers. In: XV Danube-European conference on geotechnical engineering (DECGE 2014), Vienna, Austria, pp 1–6
88. Roozbahany EG, Partl MN, Guarin A (2015) Particle flow during compaction of asphalt model materials. Constr Build Mater 100:273–284
89. Beainy F, Commuri S, Zaman M (2012) Quality assurance of hot mix asphalt pavements using the intelligent asphalt compaction analyser. J Constr Eng Manag 138(2):178–187
90. Commuri S, Mai A, Zaman M (2009) Calibration procedures for the intelligent asphalt compaction analyser. J Test Eval 37(5):454–462
91. Li J (2012) Study on control techniques of the vibratory roller based on soil compactness characteristic. PhD thesis, Xi'an University of Technology, Xi'an, China
92. Zhang Y, Zhang J, Shu XZ, Guo L, Shi Y, Liu XB (2009) Optimization of intelligent compactness control rule of vibratory roller based on genetic algorithm method. In: Proceedings of the 2009 fifth international joint conference on INC, IMS and IDC, Seoul, Korea, 25–27 August 2009, pp 1943–1947
93. Xu Q, Chang KG (2016) Adaptive quality control and acceptance of pavement material density for intelligent road construction. Autom Constr 6278–6288
94. Li QB, Lin P (2014) Demonstration on intelligent dam. J Hydroelectr Eng 33(1):139–146
95. Li QB, Shi J (2015) Dam construction 4.0. J Hydroelectr Eng 34(8):1–6

96. Röntgen WC (1896) On a new kind of rays. Science 3(59):227–231
97. An Z, Liu T, Zhang Z, Zhang Q, Huangfu Z, Li Q (2019) Dynamic optimization of compaction process for rockfill materials. Autom Constr 110:103038
98. Gan M, Li QF, Liu XJ, Wang ZA (2008) Fieldbus technology and its application. China Machine Press, Beijing, 2008 (ISBN: 9787111142690)
99. Ding EJ, Ma FQ (2003) Monitoring system and fieldbus. China University of Mining & Technology Press, Xuzhou (ISBN: 9787810707961)
100. Lin P, Li QB, Jia PY (2014) A real-time temperature data transmission approach for intelligent cooling control of mass concrete. Math Probl Eng 2014:1–11
101. Lin P, Li QB, Hu H (2012) A flexible network structure for temperature monitoring of a super high arch dam. Int J Distrib Sens Netw 2012:1–10
102. Hua T, Yang ZH, Yang X et al (2020) Assessment of geomaterial compaction using the pressure-wave fundamental frequency. Transp Geotech 22:100318
103. Yao Y-P, Ruan Y-Z, Chen J, Geng Y, Zhang X, Liu B-Y, Zong X-P, Yu G-Z (2018) Research on a real-time monitoring platform for compaction of high embankment in airport engineering. J Constr Eng Manag 144:04017096
104. Ministry of Transport, The State Council of the People's Republic of China (2018) (Accessed data: 10 October, 2019)
105. Huangfu ZH, Zhang ZX, Li CS, Yang YS (2017) Field compaction tests of dam gravel material for Qianping reservoir, China. Water Resour 12:25–26
106. ASTM (2013) D5030/D5030M-13a—standard test methods for density of soil and rock in place by the water replacement method in a test pit. Annual Book of ASTM Standards
107. Zhang QL, An ZZ, Liu TY et al (2020) Intelligent rolling compaction system for earth-rock dams. Autom Constr 116
108. I.O.W.R.A.H.R. (IWHR) (2001) D/T5129-2001-specifications for rolled earth-rockfill dam construction, China Electric Power Press, Beijing, China (Chinese)
109. Yin B (2010) Analysis on computational methods of availability and utilization ration of mechanical equipment. Yangtze River 41(5):83–87
110. Zhang QL, Liu TY, Zhang ZS, Huangfu ZH, Li QB, An ZZ (2019) Compaction quality assessment of rockfill materials using roller-integrated acoustic wave detection technique. Autom Constr 97:110–121
111. Chang MF, Yu G, Na YM, Choa V (2006) Evaluation of relative density profiles of sand fill at a reclaimed site. Can Geotech J 43:903–914

Uncited Reference

112. Lamb HI (1904) On the propagation of tremors over the surface of an elastic solid. Philos Trans R Soc Lond Ser A Containing papers of a mathematical or physical character 203(359–371):1–42
113. De Hoop AT (1960) A modification of Cagniard's method for solving seismic pulse problems. Appl Sci Res Sect B 8(1):349–356
114. Reissner E (1936) Stationäre, axialsymmetrische, durch eine schüttelnde Masse erregte Schwingungen eines homogenen elastischen Halbraumes. Ingenieur-Archiv 7(6):381–396
115. Quinlan P (1954) The elastic theory of soil dynamics. In: Symposium on dynamic testing of soils. ASTM International
116. Sung T (1954) Vibrations in semi-infinite solids due to periodic surface loading. In: Symposium on dynamic testing of soils. ASTM International
117. Biot MA (1962) Mechanics of deformation and acoustic propagation in porous media. J Appl Phys 33(4):1482–1498
118. Paul S (1976) On the displacements produced in a porous elastic half-space by an impulsive line load (Non-dissipative case). Pure Appl Geophys 11,4(4):605–614

119. Philippacopoulos AJ (1988) Lamb's problem for fluid-saturated, porous media. Bull Seismol Soc Am 78(2):908–923
120. Wang LZ, Chen YM, Wu SM, Ding HJ (1996) A solution to saturated elastic semi-space under harmonic force of low frequency. J Hydraul Eng 2:84–88
121. Cai Y, Meng K, Xu C (2004) Stable response of axisymmetric two-phase water-saturated soil. J Zhejiang Univ Sci A 5(9):1022–1027
122. Jin B, Liu H (2000) Horizontal vibrations of a disk on a poroelastic semi-space. Soil Dyn Earthq Eng 19:269–275
123. Kröber W, Floss EHR, Wallrath W (2001) Dynamic soil stiffness as quality criterion for soil compaction. In: Geotechnics for roads, rail tracks and earth structures. A. A. Balkema Publishers, Lisse, The Netherlands, pp 189–199. 90-2651-844-7
124. Mooney MA, Gorman PB, Gonzalez JN (2005) Vibration based health monitoring during earthwork construction. Struct Health Monit 2(4):137–152
125. Cacciola DV, Meehan CL, Khosravi M (2013) An evaluation of specification methodologies for use with continuous compaction control equipment. In: Proceedings of geo-congress 2013: stability and performance of slopes and embankments III, San Diego, CA, March 3–7, ASCE, Reston, VA, 2013, pp 413–416
126. Vennapusa PKR, White DJ (2016) Machine drive power based roller-integrated compaction measurements for cohesive embankment construction. Geo-Chicago 2016:571–580
127. Rinehart R, Mooney M, Facas N, Musimbi O (2012) Examination of roller-integrated continuous compaction control on Colorado test site. Transp Res Rec 2310:3–9

MIX
Papier aus verantwortungsvollen Quellen
Paper from responsible sources
FSC® C105338

If you have any concerns about our products,
you can contact us on
ProductSafety@springernature.com

In case Publisher is established outside the EU,
the EU authorized representative is:
Springer Nature Customer Service Center GmbH
Europaplatz 3, 69115 Heidelberg, Germany

Printed by Libri Plureos GmbH
in Hamburg, Germany